The biology of the cell cycle

J. M. MITCHISON, Sc.D.

PROFESSOR OF ZOOLOGY, UNIVERSITY OF EDINBURGH

Cambridge at the University Press
1971

Published by the Syndics of the Cambridge University Press
Bentley House, 200 Euston Road, London NW1 2DB
American Branch: 32 East 57th Street, New York, N.Y.10022

© Cambridge University Press 1971

Library of Congress Catalogue Card Number: 72–160100

ISBNS: 0 521 08251 X   clothbound
       0 521 09671 5   paperback

Printed in Great Britain
at the University Printing House, Cambridge
(Brooke Crutchley, University Printer)

Contents

Preface

My aim in the first nine chapters of this book is to provide a reasonably comprehensive survey of work on the cell cycle from bacteria to mammalian cells, with the main emphasis being on patterns of synthesis and their control. The last chapter does not attempt to be comprehensive and includes only a few aspects of division control which, to my personal taste, seem important at present. I have paid more attention than is usual to the lower eukaryotes since they have sometimes suffered from the concentration of interest at the two ends of the cellular scale – mammalian cells and *Escherichia coli*. Various subjects have not been included or have only been mentioned briefly because, although they are interesting, either they are marginal to the field or they are not normal cell cycles. These include meiotic cycles, virus replication, spore germination, and cell and tissue kinetics in normal organs or in those stimulated into mitosis. One outstanding omission is of mitosis and cleavage. These processes are essential parts of the cycle but they are not discussed in this book because they have been so well analysed in the review by Mazia (1961).

I have a particular debt of gratitude to the members of the Department of Zoology of the Berkeley campus of the University of California for their kindness to me when I was a Visiting Professor there in the spring and summer of 1969. If I had not taught the framework of this book as a graduate course in Berkeley, I doubt whether it would have seen the light of day. I am also grateful to a number of my friends and colleagues for reading and commenting on various parts of this book. These include W. D. Donachie, A. H. Maddy, M. Masters, G. J. Mitchison, D. Mazia, L. Rasmussen, R. C. Rustad, R. R. Schmidt, and E. Zeuthen.

J. M. MITCHISON

MAY 1971

To Daniel Mazia and Michael Swann,
with affection

1 Introduction to the cell cycle

The cell cycle of a growing cell is the period between the formation of the cell by the division of its mother cell and the time when the cell itself divides to form two daughters. It is a fundamental unit of time at the cellular level since it defines the life cycle of a cell.

Diagrams of the cell cycle can be made either in a circular form (Fig. 1.1(a)) or, more conveniently for many purposes, in a linear form (Fig. 1.1(b)). In most animal cells, the majority of the cycle is occupied

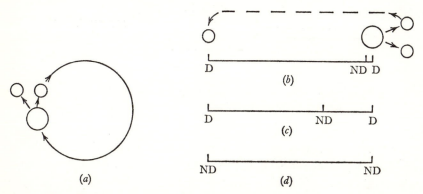

Fig. 1.1. The cell cycle. (a) and (b) Most higher eukaryotes. (c) Some micro-organisms, e.g. *Schizosaccharomyces pombe*. (d) Syncytia, e.g. *Physarum*. D, cell division. ND, nuclear division.

by interphase, the period between cell division and the start of the following mitosis. At the end of mitosis there is nuclear division followed very shortly afterwards by cell division. In plant cells and many micro-organisms where there is a cell wall external to the cell membrane, there may be an appreciable period between nuclear division and cell division (Fig. 1.1(c)). During this period, the daughter cells may become physiologically separated by newly-formed cell membranes before the cell walls have been sufficiently developed to allow the final cleavage. Fig. 1.1(d) shows the special case which applies to multinucleate cells (syncytia) in which there is

1

nuclear division but no cell division. The most important example of this in cell cycle work is the myxomycete *Physarum polycephalum*.

Two separate classes of events occur in a normal cell cycle. One of them, which will not concern us in this book, includes the events of mitosis and cell cleavage which are processes of splitting and spatial separation. The other class includes the synthetic events which in principle double the quantity of all the components of the cell. Another way of putting this is to say that as much new material is added during a cycle as there was old material at the beginning. This prompts us to ask whether the new material is equally distributed between the two daughter cells. We assume in most

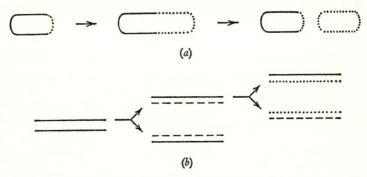

Fig. 1.2. Partitioning of new and old material. (*a*) Growth of new cell wall (dotted) in *S. pombe*. (*b*) Semi-conservative replication of the two strands of a DNA double helix.

cases that the new is mixed up with the old and partitioned equally, though the factual evidence for this is thin. There are, however, some examples where this rule of biological parity is broken and the two new daughter cells are not equivalent in the age of their components. In *Tetrahymena*, the mouth of one daughter cell has been formed in the previous cycle and that of the other daughter has been formed in cycles before that (p. 211). New cell wall is usually laid down at one tip of a growing cell of the fission yeast *Schizosaccharomyces pombe* (Fig. 1.2(*a*)). Division will therefore separate newly made wall from older wall. The semi-conservative mode of replication of DNA ensures that one strand of a double helix has been made in the preceding round of replication and the other has been made either two or more rounds earlier (Fig. 1.2(*b*)). How far this is important is unknown but it is worth bearing in mind that cells of a growing culture may contain molecules and structures that are of very different ages.

Our main concern in this book will be to follow the patterns of growth and synthesis through the cycle. For this we will have to rely on two main bodies of evidence. One of them consists of the work done with single cell techniques which all rely on microscopic measurements. These have the advantage that they can reveal the finer detail of growth patterns and the variation between individual cells, but the disadvantage that what can be measured is strictly limited. The second body of evidence comes from synchronous cultures where the cycles of individual cells have all been brought into phase. The great merit of synchronous cultures is that they provide enough material to allow bulk biochemical measurements to be

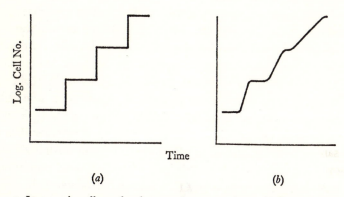

Fig. 1.3. Increase in cell number in a synchronous culture. (*a*) Perfect synchrony, as an ideal situation. (*b*) Imperfect synchrony, as in practice.

made. But they have attendant faults. Most synchronous cultures are imperfect in their degree of synchrony so that cell divisions are spread over a considerable period of time and are not simultaneous (Fig. 1.3). The finer details, therefore, of synthetic patterns are obscured, as are also the variations between individual cells. Synchronous cultures lose their synchrony after a few cycles if they are left to run free because of the natural variation of cycle time from cell to cell.[1] Lastly and most important, the method of making the synchronous culture may affect the patterns of growth – which brings us to a classification of methods of synchronisation (James, 1966).

There are two broad groups of synchronising procedures, *selection synchrony* in which cells at one particular stage of the cycle are selected out of a normal asynchronous culture and grown as a separate culture, and *induction synchrony* in which some treatment is applied to an asynchronous

[1] Burnett-Hall & Waugh (1967); Cook & James (1964); Engelberg (1964*b*).

culture which makes all the cells synchronous.[1] Let us see how far the cycles in these synchronous cultures are normal in the sense that they are identical with those in the cells of a random asynchronous culture growing exponentially in a constant environment. The cycles after selection synchrony do appear to be normal provided the separation procedure does not affect cell metabolism. Here bacteria seem to be more sensitive than eukaryotes. Some kinds of induction synchrony also produce normal cycles, for example synchrony after feeding a starved culture, particularly if the first cycle is ignored. But other induction methods distort the cycles. Multiple heat shocks produce excellent synchrony in *Tetrahymena* (p. 201) but the cells are both larger and more variable in size than in asynchronous cultures, and the cycles are shorter. Temporary use of an inhibitory block (e.g. of DNA synthesis) synchronises division and DNA synthesis, but may not synchronise some aspects of growth (pp. 29, 244). Another method of induction synchrony is to repeat an environmental shift (cold to hot, or light to dark) once per cycle throughout the experiment. The cells may appear normal but the problem is to decide whether a synthetic event in these synchronous cultures is caused by the shift or whether it is under internal controls which operate in the same way in an unstressed culture. For this reason, the results from this kind of cyclical synchronisation have largely been left out of this book. It is important, however, to make the point that although these methods of induction synchrony produce abnormal cycles, they may be very valuable in dissecting the control mechanisms behind the cycle. The most striking example of this is the *Tetrahymena* system which is considered in detail in Chapter 10, and it may be that equally interesting results will emerge from studies of the other methods.

Another criterion that has been used in assessing the normality of a synchronous culture is whether it shows 'balanced growth'. The original definition of this is 'that growth is *balanced* over a time interval if, during that interval, every extensive property of the growing system increases by the same factor' (Campbell, 1957*a*); for a good discussion, see also Anderson *et al.* (1967). This is a criterion which should be used with discretion. As Campbell points out, it does not apply to single cells or to synchronous cultures except when the interval is an integral multiple of the cycle time. Nor does it imply that the factor of increase over one cycle time must be ×2. It should be ×2 in a synchronous culture maintained

---

[1] Abbo & Pardee (1960) described a culture produced by induction as 'synchronised' and one produced by selection as 'synchronous'. This is not a satisfactory classification since the terms are too similar and are also used indiscriminately in the literature.

for several generations in the constant conditions of a chemostat, but these conditions are seldom realised except with a one shock per cycle system. Most synchronous cultures are batch cultures in which the medium is changing during the course of an experiment. Even during the exponential phase of such a culture, when the logarithm of the cell numbers is increasing at a constant rate, the cells may be changing in composition and response (e.g. Prescott, 1957). It would be making too strict a judgement to reject the results from an experiment where the factor of increase was $\times 1.8$, though a factor as low as $\times 1.5$ would raise doubts. This applies even more to single cells where there are very likely to be changes in the factor of increase and the cycle time in succeeding cycles in normal growth. The main exception here would be the DNA and protein of the chromosomes which would presumably have to show a factor of increase of exactly $\times 2$.

There are large variations in the factors of increase of different cell components in the early cell cycles of embryos. During each cycle there is a doubling of the nuclear contents and a smaller increase in the cell membrane, but there is no increase in the total dry mass until the embryo starts to feed. Protein and RNA synthesis does, however, take place and these new macromolecules are necessary for development, either immediately after their formation or later, depending on the organism.

The most important use of synchronous cultures in recent years has been to provide the evidence for a surprising degree of differentiation within the cycle. To appreciate this, we must go back a little in the history of this field of cell biology. Twenty years ago, attention was largely focused on the dramatic changes at mitosis which were easily visible in the light microscope and which had been known for more than fifty years. There was little interest in interphase and even less information about it. If asked what happened during interphase, most biologists would probably have guessed that the cell doubled itself by a process of uniform growth with the relative proportions of the various components remaining constant. There was, after all, no sign of visible change in cells going through interphase. Then, in the early fifties, two important advances were made. One was the development of the first systems of induction synchrony in *Tetrahymena* and *Chlorella*. A major interest here was the mechanism of synchronisation, and this pathway has lead to the theories of the control of division which are set out in Chapter 10. The second advance was the introduction of autoradiography, followed a few years later by tritiated thymidine. This enabled the patterns of DNA synthesis to be determined in asynchronous cultures and tissues, and showed that replication occupied a restricted period of the cycle and did not take place throughout interphase. For some time, however, DNA seemed to be the only cellular component that was synthesised periodically in interphase. Most cells showed continuous synthesis of total protein and RNA, except during mitosis. But total cell protein is made up of many hundreds of component proteins, and, in the last six years, a considerable body of evidence about the synthesis of individual components has come from measurements on cultures which have mainly been synchronised either by selection techniques or by induction methods, such as starvation, which do not markedly distort the cycle. Over the whole range of prokaryotic and eukaryotic cells,[1] the majority of enzymes (Chapter 8) and of other specific proteins such as histones (Chapter 7) appear to be synthesised periodically giving a 'step' pattern if they are stable or a 'peak' pattern if they are unstable (Fig. 8.1). A step pattern for an enzyme is like the DNA cycle but with a period of synthesis at a stage in the cycle which varies from one enzyme to another.

The widespread presence of this periodic enzyme synthesis is in some ways surprising. Our expectation might well be that the simplest way to grow a cell would be to have a collection of enzymes and other components which increased in amount without changing their proportions. Instead, we now find systems in which the enzymes as well as the DNA seem to double suddenly at different points in the cycle. This would appear to be an excellent way of unbalancing the whole process of growth, yet the growth of the cell is a smooth continuous process in terms of bulk measures such as total protein or dry mass. We do not know enough to resolve this apparent paradox, since we are only now beginning to emerge from the phase of straightforward measurement of synthesis patterns – 'molecular bird-watching' to borrow a felicitous phrase from Daniel Mazia. This planned order of periodic synthesis may be the expression of a sequence of structural changes which are too fine for us yet to resolve, or it may be a necessity of regulation which the cell has to put up with and adapt to, or it may be a fundamental timing mechanism.

Whatever the reason for these periodic syntheses, their presence shows that the composition of a growing cell is changing continuously through interphase in an orderly way. This is chemical differentiation, which precedes the morphological differentiation during mitosis. We can now draw a close parallel with embryology. Both the growing cell and the growing

---

[1] This fundamental classification, which will be used frequently, separates bacteria and blue–green algae, which are prokaryotes, from all other cells, which are eukaryotes (Stanier, 1961).

embryo show an ordered pattern of chemical syntheses. Both show morpho-genesis, that of the cell occurring not only in mitosis but also in the growth of organelles. In both, the same pattern of events recurs in the next genera-tion, and, in both, the fundamental problem is gene regulation and expression. To study the cell cycle, therefore, is to study the developmental biology of the cell.

2 Single cell methods

The next two chapters will survey the methods that can be used to study the cell cycle. These techniques are important because they set the practical limits of what can be measured in experiments. Chapter 3 deals with synchronous cultures, whereas this one is concerned with single cell measurements and how they can be related to the cycle.

Measurements on single living cells

There are a limited set of measurements that can be made on single living cells. They are usually time-consuming but have the advantages of avoiding the effects of fixation and in some cases allowing a sequence of measurements to be made on a single growing cell. A sequence of this kind gives a higher degree of precision to growth curves than is possible with any other method. As an example, volume measurements on single growing cells of the fission yeast *Schizosaccharomyces pombe* show that there is a plateau in the growth curve for the last quarter of the cycle before division (Fig. 7.6), but this plateau is scarcely detectable in an average growth curve derived by the frequency distribution method described on p. 19 (J. M. Mitchison, unpublished).

Volume is one of the easiest properties to measure either by direct observation under the microscope with a micrometer scale or from photographs. A proviso is that the cell must be a solid of revolution whose volume can be calculated from the linear dimensions in the plane of the microscope field. Calculation of absolute volume is only possible either with a spherical cell or with one which has a fairly rigid cell wall or cortex as in plants and some micro-organisms. Relative volumes, however, can be derived from linear dimensions of cells with a less regular shape (e.g. bun-shaped) provided the shape does not change during growth. Volume measurements are likely to be more accurate in cells which grow only in one dimension (e.g. bacterial rods or fission yeast) than in cells which grow in three dimensions. A bacterial rod will double in length over a cycle whereas a spherical cell will only increase its diameter by 26 per cent.

Most animal cells have an irregular shape and their volume cannot be calculated from microscopic observation. In some cases, they will round up and form a shape which is roughly spherical. Their volume can then be calculated from the diameter but not with any great accuracy. Recently, however, a new technique has become available which gives a measure of the volume of an irregular cell. This is the electronic particle counter (e.g. the Coulter counter) where the change in electrical resistance as a cell passes through a narrow aperture is proportional to cell size. Cells can be passed through rapidly, so it is possible to measure and record the volume of several thousand cells in a matter of minutes. This method is not without its technical difficulties which are discussed by Harvey (1968) and Kubitschek (1969a). In addition, the volume of an individual growing cell cannot be measured at successive intervals so that growth curves have to be derived either from average volumes in a synchronous culture or by frequency distribution in an asynchronous culture.

Another property that can be measured optically without damage to a living cell is the total dry mass by interference microscopy. The problem here is that it is easy to determine the dry mass per unit area in a particular region of the cell but much more difficult to determine the total dry mass integrated over the whole area of the cell (Davies, 1958; Ross, 1967). Two types of technique are available for measuring the integrated dry mass. One of them uses comparatively simple apparatus but is limited to small thin cells (Davies & Deeley, 1956; Mitchison *et al.*, 1956). The other uses a scanning technique which will work with any size of cell but requires complex instrumentation (Caspersson & Lomakka, 1962). So far, only the first technique has been used to follow the growth of single living cells (Mitchison, 1957; 1958; 1961; Mitchison *et al.*, 1963).

The Cartesian diver, developed at the Carlsberg Laboratory in Copenhagen, is another micro-technique that can be used with single living cells. It can function either as a respirometer (Holter, 1961) or as a balance (Zeuthen, 1961). In the latter case what is measured is the weight of the cell in water or 'reduced weight'. Changes in this are proportional to changes in total dry mass, provided the density of the dry components does not alter. Living cells have been followed through the cycle with a diver by Frydenberg & Zeuthen (1960), Løvtrup & Iverson (1969), Prescott (1955), and Zeuthen (1953; 1955; 1960).

Two other quantitative optical techniques that have been used for cell cycle studies on living cells are polarised light microscopy (e.g. Swann, 1951) and immersion refractometry (Barer & Joseph, 1957; Ross, 1961; 1967). Changes in apparent surface rigidity have also been measured with a 'cell elastimeter' (Mitchison & Swann, 1955; Selman & Waddington, 1955).

Measurements on single fixed cells

Most cytological measurements are made on cells which have been killed and fixed. Fixation affects a number of cell properties. Apart from the possibility of artefacts of precipitation and condensation, the pool of low molecular weight components is removed and there are often substantial changes in volume (e.g. fission yeast cells shrink to 45 per cent of their live volume on fixation; Mitchison & Cummins, 1964). The most serious limitation of fixed cells for cell cycle work is that it is not possible to make sequential observations on single growing cells. Against this can be set the advantage that a much wider range of measurements can be made on fixed cells than on living cells.

A number of the methods described above for living cells can also be used on fixed cells, e.g. volume from linear dimensions, dry mass and re- duced weight – though the two latter measurements will exclude the pool. In addition, there are two major techniques which cannot be applied to living cells, cytochemistry and autoradiography. Quantitative cytochemistry involves spectrophotometry combined with microscopy. The most widely used method in studies of the cell cycle is the measurement of the DNA content of nuclei stained by the Feulgen reaction. This and other tech- niques are well described in review articles and books such as those by Danielli, 1958; 1961; Glick, 1949; 1963; Oster & Pollister, 1956; and Walker & Richards, 1959.

Autoradiography is a well-established and popular technique in cell biology both at the level of the light microscope and at that of the electron microscope (see Rogers, 1967; Feinendegen, 1967; and articles in Prescott, 1964c; 1966a; 1968a). It can be used at a qualitative level to detect and locate the presence of a tracer, and at a quantitative level to estimate the amount of a tracer by grain counting. In most cases, it is used on fixed cells so the only tracer left is that incorporated into macromolecules, but it is also possible to make autoradiographs of cells with an intact pool of unincorporated tracer (e.g. Cummins & Mitchison, 1964; Miller *et al.*, 1964; reviewed by Feinendegen, 1967). The most important use of autoradiography in cell cycle work has been to detect DNA synthesis by the incorporation of tritiated thymidine. This DNA precursor has been a very valuable tool. It is rapidly taken up by most cells from the surrounding medium or from the blood *in vivo* though relatively impermeable cells like Amphibian eggs have to be micro-injected and brain cells are more easily labelled via the cerebro-spinal fluid than via the blood (Watson, 1965). It can be given in short pulses because many cells have small pools of thymidine which are only filled during DNA synthesis. It is also removed rapidly from the circulating blood in whole animals. Most important of all, it is incorporated specifically into DNA in most cells. This is somewhat surprising since it is not on the normal pathway of DNA synthesis which goes directly from deoxyuridine monophosphate to thymidine monophosphate (further details are in Cleaver, 1967). Exogenous thymidine is phosphorylated by thymidine kinase, an enzyme which is widely distributed. This enzyme, however, is not universally present and where it is absent, as in *Neurospora* and yeast (Grivell & Jackson, 1968), tritiated thymidine does not label DNA specifically.

There are a number of difficulties about autoradiography, some of which are at a technical level and are considered in the general references given above, and some at a more general level which applies to all forms of labelling. One of the latter is the conversion of a precursor so that the label appears in an unexpected macromolecule (e.g. tritiated arginine can label DNA: Comings, 1969). But perhaps the most important of them is the problem of pulses and pools, which we should now consider.

Pulses and pools

If we wish to follow the pattern of synthesis of a macromolecule through the cell cycle, one of the most powerful techniques is to give pulses of a labelled precursor which is incorporated into the macromolecule. The precursor is added to the medium and after a short time the pulse is terminated by killing the cells, or by washing off the precursor, or by adding an excess of unlabelled precursor (a 'cold chase'). The amount of incorporated precursor which is left after acid treatment or fixation of the cells is then measured by grain counting in autoradiographs or by bulk counting. With some important provisos discussed below, the measured radioactivity is proportional to the *rate* of synthesis of the macromolecule. The fact that pulses measure rates means that it is possible with successive pulses to define the pattern of a synthesis curve with much greater precision than with successive measurements of macromolecular quantity (Fig. 2.1). Another way of putting this is that many of the features of the curve are better determined from the first derivative.

Although a series of pulses through the cycle can be used to define the pattern of synthesis, this procedure is only valid if certain conditions are met. One way of expressing these conditions is that the specific activity of the pool of the immediate precursor for the macromolecule should be constant through the cycle. This, however, is somewhat loosely defined

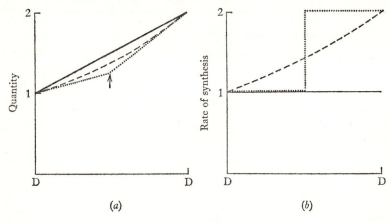

Fig. 2.1. Patterns of synthesis of a macromolecule during a cell cycle. (*a*) Increase in quantity. Continuous line, linear increase. Dashes, exponential increase. Dots, linear pattern with doubling in rate in mid-cycle. Note that the differences between these patterns have been exaggerated. (*b*) Corresponding rates of synthesis. These show much greater differences than the curves in (*a*). D is division.

since the specific activity of this pool may be changing during the course of the pulse. To follow this further, we should consider the model systems in Fig. 2.2. These systems are considerably simplified and make the following assumptions: (1) the mechanisms of uptake through the cell membrane are not saturated; (2) there is no back-exchange of precursor across the membrane or macromolecular turnover; (3) the endogenous supply of precursor to the pool is cut off; (4) all the pools that may be involved in synthesis can be represented as one pool; (5) the time scale is small compared to the length of the cycle – if it were not, the straight lines on the graphs would be exponential curves; (6) the precursor is not depleted from the medium.

Fig. 2.2(*a*) shows the situation when there is a small pool which is unaltered in size when the exogenous precursor is added. The rate of entry of precursor into the pool is equal to the rate of exit into macromolecules. The total activity (pool + incorporated) rises linearly from zero time when the precursor is added. The incorporated activity parallels the total in the later stages of the pulse, but there is a short delay before it reaches this rate, while the specific activity of the pool is rising. If there is a larger pool of fixed size, the delay will be longer (Fig. 2.2(*b*)). If, on the other hand, the

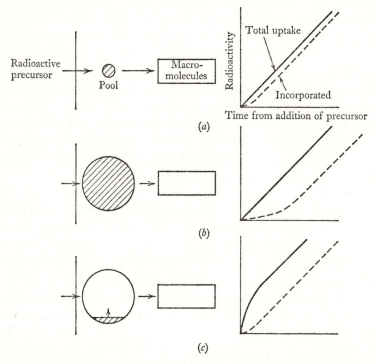

Fig. 2.2. Models of pool situations and corresponding kinetics of uptake and incorporation. (*a*) Small fixed pool. (*b*) Large fixed pool. (*c*) Expandable pool.

pool is expandable in the sense that it increases in size in the presence of exogenous precursor, Fig. 2.2(*c*) shows what will happen. The incorporated activity has a short delay dependent on the initial amount in the pool as in the other graphs, but the total activity rises rapidly in the early stages of the pulse while the pool is filling up. The size of the expanded pool can be dependent on the external concentration of precursor (e.g. amino-acids in the budding yeast *Candida*; Cowie & McClure, 1959) or independent (e.g. adenine in the fission yeast *Schizosaccharomyces pombe*; Cummins & Mitchison, 1967). If in the former case the radioactive precursor is added without effectively changing the external concentration, or in the latter case there has been non-radioactive precursor present before the labelling, then the kinetics of uptake and incorporation will be like Fig. 2.2(*a*) or (*b*).

How far might these situations affect the validity of pulses through the cycle? The most important factor, with a pulse which is short compared to the time taken to reach a constant specific activity in the pool, is the size of the fixed pool, or of the pre-existing pool if it is expandable. Providing these pools increase through the cycle in proportion to the increase in the rates of uptake and synthesis, the amount of precursor incorporated in a short pulse is a valid measure of the rate of synthesis. If the pool sizes change in any other way, the measure is invalid. There are two ways of getting round this difficulty. One is to measure the total uptake in the pulse rather than the amount incorporated. With a fixed pool, this gives a value for the rate of synthesis which is independent of pool size (compare Fig. 2.2(*a*) with 2.2(*b*)) but it will not be so with an expandable pool. The other is to use a long pulse. This will reduce any error due to changing pool size, but it is also a less sensitive measure of rate.

We should also consider briefly what will happen when the initial assumptions for these models are changed. If the uptake mechanism is saturated, there will be no increase in the rate of uptake or incorporation in a growing system until the uptake mechanism itself has grown – for example, by the synthesis of more transport carriers for the membrane. Back-exchange of precursor across the membrane will tend initially to produce decelerating uptake curves similar to that in Fig. 2.2(*c*). Macromolecular turnover back to the pool will also tend to produce decelerating incorporation curves whereas these curves with stable macromolecules are always straight or accelerating. The exact shape of the uptake and incorporation curves under these conditions will depend on the reaction rates in both directions and on the relative size of the pool and macromolecular compartments. If the endogenous supply of precursor is cut off slowly, there will be a delay in incorporation similar to the delay produced by a pre-existing pool. If the endogenous supply continues but maintains a constant ratio to the exogenous supply, this will not affect the validity of pulses. But if the ratio changes through the cycle, this will cause serious errors. The presence of multiple pools will not seriously affect the conclusions that can be drawn about rates of synthesis from kinetic graphs similar to those in Fig. 2.2. If the precursor is depleted from the medium so that the pool fills and then empties, it will produce curves like Fig. 2.3 which are more difficult to interpret than those in Fig. 2.2.

There is no doubt that pool changes *can* lead to erroneous interpretation of incorporation results. In human cells after irradiation (Smets, 1969) and virus infection (Newton *et al.*, 1962), there are changes in the rate of thymidine incorporation which are primarily due to pool changes. The question as to whether this also happens during the cell cycle is an open one since very little work has been done on it. As yet, the most thorough investigation is that done on the incorporation of bases into RNA during the cell

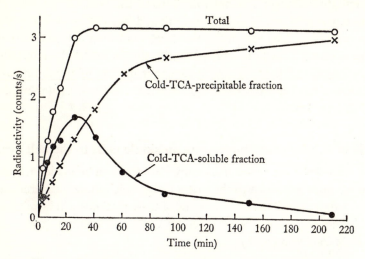

Fig. 2.3. Kinetics of incorporation of tracer quantities of ^{14}C-glutamic acid in the yeast *Candida*. From Cowie & McClure (1959).

cycle of *Schizosaccharomyces pombe* (Mitchison *et al.*, 1969). This showed that pool effects were not likely to have caused serious discrepancies between the rate of incorporation from pulses and the rate of RNA synthesis, but there were no measurements of the endogenous production of precursors. Whether this is a general conclusion remains to be seen, though pool effects might have been responsible for the different increases in the incorporation rate of two RNA precursors through the cycle of mammalian cells (p. 119) and for the differences in the 'uptake' through the bacterial cycle of glycine and leucine according to whether or not they were added *de novo* (p. 144). Precursor pulsing remains a powerful tool but we should regard its interpretation with caution until more is known about the pool position in the cases where it is used. A good deal of the necessary information can be got from the shape of uptake and incorporation curves which can be measured easily in synchronous cultures and with greater difficulty in autoradiographs (Mitchison *et al.*, 1969). It is harder to determine the endogenous contribution but it can be done (e.g. Cowie & Walton, 1956).

A final point on pools is that knowledge about them helps to decide whether a chase of excess unlabelled precursor is an effective way of stopping a pulse. In some cases it may be, and in others not. Where there is an expandable pool (as with thymidine in mammalian cells; Cleaver & Holford, 1965; Adams, 1969*a*) the addition of a small amount of high specific activity precursor will not fully expand the pool. If this tracer is now chased, the pool will expand to its limit and it may take longer to stop the macromolecular labelling from this large pool than it would if there had been no chase. Exactly this happens with thymidine labelling of DNA in *Vicia* (Evans, 1964). Other problems about the use of thymidine as a chase and as a label are discussed by Nachtwey & Cameron (1968) and Feinendegen (1967).

Methods of relating measurements on single cells to the cell cycle

We have considered the types of measurement that can be made on single cells and we must now turn to the methods by which these measurements can be related to the cell cycle. Only some of these methods will be outlined here, and more complete lists are given by Watanabe & Okada (1967) and Nachtwey & Cameron (1968).

Isolated single cells

The most direct method is to follow a single growing cell through one or more cycles making successive measurements of a kind which do not damage the cell (e.g. Prescott, 1955; Mitchison, 1957). This is precise but it can only be done with a restricted range of measurements and it is certainly time-consuming.

If the cell has to be killed, as in most measurements, one way of determining where it is in the cycle at the moment of death is to have followed its previous history since the preceding division by observation or, more usually, from time-lapse films (Showacre, 1968). If, for example, it is killed 10 hours after division and the mean cycle time from other measurements is 20 hours, the cell can be assumed to be half way through the cycle. Examples of this method are in Walker & Yates (1952) and Killander & Zetterberg (1965*a*).

A major source of error in this method is the spread of cycle times between individual cells. This can be reduced by comparing the time from division with the sister cell cycle time rather than the mean cycle time because sister cells have a higher correlation of their cycle times than un-related cells (e.g. Killander & Zetterberg, 1965 *a*). This method can only be used when the sister cell can be separated off from the experimental cell and grown on to the next division, as in *Paramecium* (Kimball *et al.*, 1959).

Cell morphology can also be used to position cells in the cycle. The fission yeast *Schizosaccharomyces pombe* has the useful property of growing

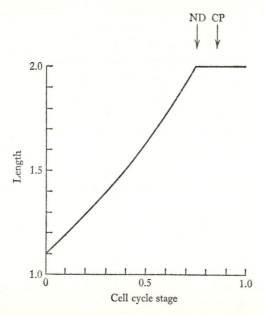

Fig. 2.4. Cell length (or volume) of *Schizosaccharomyces pombe*, in arbitrary units, through the cell cycle. ND, Nuclear division. CP, the first appearance of cell plate. From Mitchison (1970).

only in length, so a length measurement is sufficient to give the cycle stage of a cell, except in the last quarter of the cycle (Fig. 2.4). Nuclear volume has been used in the same way in plant cells (Woodard *et al.*, 1961), although the small changes in volume in the first half of the cycle would make it difficult to position a cell with any accuracy. Two stages of the cycle, mitosis (or nuclear division) and cell division, can be recognised in all cells so it is relatively easy to tell what is happening to events such as RNA synthesis at these stages. It is also possible to recognise cells in the S period by their ability to incorporate tritiated thymidine. The capacity of these cells to perform other syntheses can be followed if another tracer is used (e.g. ^{14}C-uridine for RNA) together with a double-layered auto-radiograph (Baserga, 1962).

Cells from populations

Age distribution and mitotic index. A good deal of information about the cell cycle can be obtained from observations or measurements on a series of single cells from a growing cell population even though this population is asynchronous. As a first example, the proportion of the cell cycle spent in mitosis can be derived from the proportion of cells in the population which are observed in mitosis (the mitotic index). Intuitively, it might seem that these two proportions should be equal, but this is not so. The frequency distribution is skewed and there are always more young cells (early in the cycle) than old cells. Qualitatively, this is because a single old cell becomes two young cells as it passes through the point of division. Quantitatively, the age distribution, which is shown in Fig. 2.5, is defined by:

$$y = 2^{(1-x)}, \tag{1}$$

or

$$Log_e y = (1 - x) \, Log_e 2, \tag{2}$$

where y (varying from 1 to 2) is the relative number of cells at cycle stage x (varying from 0 to 1). The derivation of this equation (in a slightly different form) is given by Cook & James (1964). This and later equations assume that all the cells in the population are growing, that growth is exponential and that there is no synchrony. Another useful relation, that can be derived from Eqn (1) by integration, deals with a terminal event which occupies the last period t_{term} of the whole cell cycle T. If the number of cells showing this event is N_{term} out of the total population N, then:

$$\frac{t_{term}}{T} = \frac{Log_e(N_{term}/N) + 1}{Log_e 2}. \tag{3}$$

In terms of a mitotic index M and a mitotic time t_m, this equation becomes:

$$\frac{t_m}{T} = \frac{Log_e(M + 1)}{Log_e 2}. \tag{4}$$

If M is small, then $Log_e(M + 1) \simeq Log_e M$, and

$$\frac{t_m}{T} = \frac{M}{0.693} = M \times 1.44. \tag{5}$$

As an example, if the mitotic index (M) = 0.1 (10 per cent), then the relative time spent in mitosis (t_m/T) = 0.138 (Eqn 4) and 0.144 (Eqn 5). Eqns (4) and (5) have been given in a number of papers (e.g. Hoffman, 1949; Stanners & Till, 1960; Smith & Dendy, 1962; Painter & Marr, 1968). Eqn (3) is given in this form in Nachtwey & Cameron (1968). They, together with Watanabe & Okada (1967), give similar equations and methods for deriving the length of phases of the DNA cycle after thymidine labelling experiments.

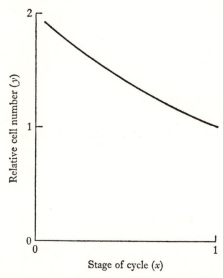

Fig. 2.5. Relative frequency (y) of cells at stage of the cell cycle (x) for cells from an asynchronous exponential phase culture.

A rough estimate of the length of the DNA synthetic period can be found from the proportion of cells which incorporate a pulse of labelled thymidine. But this will not be accurate unless the position of this period in the cycle has been determined, for instance by measuring the length of the G2 phase.[1] This is discussed by Cleaver (1965). A formula is given in Bostock (1970a) for positioning the synthetic period from measurements of the mean value of DNA/cell.

Frequency distributions of cell components. Knowledge of the age distribution makes it possible to predict the frequency distribution of the amount of a cellular component in cells from a growing culture, provided

---

[1] G1, S and G2 are the phases of the DNA cycle and are defined on p. 58.

the pattern of increase of this component through the cycle is also known. If, for example, total cell protein increased linearly through the cycle from one unit to two units in all cells, the frequency distribution of a sample of cells with varying amounts of protein would be identical with that in Fig. 2.4 except that the abscissa would be from one to two units of protein instead of the cycle stage. If, on the other hand, the protein doubled instantaneously in mid-cycle, there would be some cells with two units,

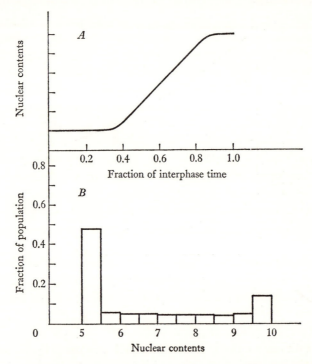

Fig. 2.6. (*A*) Theoretical synthesis curve for nuclear contents in arbitrary units. (*B*) The frequency histogram derived from (*A*). From Walker (1954).

more cells with one unit and no cells with intermediate amounts of protein. A more realistic situation is shown in Fig. 2.6 where the pattern of increase of a component resembles that of DNA in many eukaryotic cells with a synthetic (S) period shorter than the whole cycle. The resulting histogram has peaks at either end and low intermediate values.

Frequency histograms can be derived from patterns of synthesis but in practice the process is reversed. Measurements of quantity (e.g. of DNA from Feulgen staining) are made on a number of cells in a sample from an asynchronous culture in exponential growth, a frequency histogram is then constructed and from this an average curve of synthesis is derived by a method first worked out by Walker (1954). Several assumptions have to be made: that all the cells are growing exponentially and asynchronously; that the component measured does not decrease at any point in the cycle; that the component increases by a known factor (usually × 2) over the cycle; and that the shape of the synthesis curve is the same for all cells. It is not, however, necessary to assume the same total cycle time for all cells.

This method is at its best when used for constructing DNA synthesis curves where there are large rate changes and where there is an exact doubling from the 2C to the 4C value in every cell. It is less accurate when used with parameters such as cell volume or dry mass where it is very doubtful whether it can discriminate between curves as similar as a linear and an exponential pattern of increase. These parameters do not necessarily double exactly and they vary in quantity between cells at the same stage of the cycle – the 'momentary variation' of Scherbaum & Rasch (1957). If there is other evidence about this variation, it can to some extent be allowed for, as in the derivation of the curves in Zetterberg & Killander (1965 a) which are illustrated in Figs. 6.1 and 6.2. Another method which also takes account of this variation has been developed by Collins & Richmond (1962) and used on bacterial length distributions.

Labelled mitoses. One of the most popular methods for determining the phases of the DNA cycle (G1, S, G2) has been the scoring of labelled mitoses in successive samples. This technique was first described in a fully developed form by Quastler & Sherman (1959) though simplified versions were used by earlier workers (e.g. Howard & Pelc, 1953). There are recent descriptions by Sisken (1964), Thrasher (1966) and Van't Hof (1968). In essence, the method consists of giving a pulse label of tritiated thymidine either *in vivo* or *in vitro*, sampling the cells at intervals thereafter, and then counting the proportions of mitoses which are labelled in autoradiographs (Fig. 2.7). The earliest samples will have no labelled mitoses. The proportion of labelled mitoses will then rise to a peak as the cells which were in S at the time of the pulse come through to division. Following this peak, there will be a trough as the cells originally in G1 come to the end of their cycle. The second cycle will show as a similar wave but it will be lower because of the spread of cycle times.

The average S period is taken as the time between the two points in the first wave where 50 per cent of the mitoses are labelled. The time between the start of the experiment and the first of these points is taken as $G_2 + \frac{1}{2}M$. Half the mitotic time is added to G_2 because the mitoses are normally scored in metaphase, which means that the cells have to go through not only G_2 but also prophase before they are scored, and prophase lasts roughly for half the time of mitosis. The total cycle time (T) is the time between two similar points (e.g. the peaks) in the first and second cycle, and $G_1 + \frac{1}{2}M$ is obtained by difference $(= T - S - G_2 - \frac{1}{2}M)$.

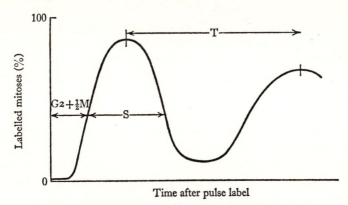

Fig. 2.7. Diagram of labelled mitoses (metaphases) in successive cell samples after a pulse of tritiated thymidine.

This method has the great advantage that it can be used on cell cultures or tissues where there are non-growing cells, though here the labelled mitotic peak will not reach 100 per cent. Other criteria, however, have to be satisfied. The growing cells in the population must have approximately the same cycle times. There must be no synchrony or diurnal peaks of mitoses as occur sometimes in living animals. Care must also be taken to use the lowest possible doses of thymidine since it can cause abnormalities in chromosomes, in mitosis and in the mitotic index (e.g. Bassleer & Chèvremont-Comhaire, 1964; Greulich *et al.*, 1961; McQuade & Friedkin, 1960; Natarajan, 1961).

Nearly all the labelled mitoses experiments have been concerned with DNA, but Bloch *et al.* (1967) have used this method with tritiated lysine and arginine for following histone synthesis and it should be possible to use it with other cell components which are synthesised during restricted period of the cycle.

Double labelling of DNA. An ingenious method of determining the length of the S period by double labelling was developed by Baserga &

Nemeroff (1962), Pilgrim (1964) and Wimber & Quastler (1963) (see also Thrasher, 1966). A growing tissue is first labelled with a pulse of tritiated thymidine and then, Q minutes later ($Q < G_2$, $Q < S$) labelled again with ^{14}C-thymidine, fixed and made into autoradiographs. All the cells which were within the S period at the time of the first pulse will be labelled with ^{3}H. At the time of the second pulse, most of these cells will still be in the S period and will be double labelled with ^{14}C as well as ^{3}H, but a small group will have proceeded into G2 and will not have a ^{14}C label (Fig. 2.8). This group can be distinguished from the double labelled cells on an ordinary autoradiograph or on one made with two emulsion layers. Then:

$$\text{Length of S period} = Q \times \frac{\text{No. of cells labelled with } ^{14}\text{C}}{\text{No. of cells labelled with } ^3\text{H only}}.$$

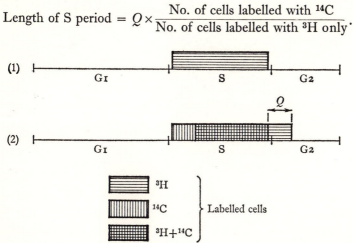

Fig. 2.8. Double labelling of cellular DNA with ^{3}H- and ^{14}C-thymidine. (1) At ^{3}H-thymidine pulse. All cells in S labelled with ^{3}H. (2) At ^{14}C-thymidine pulse, Q mins after ^{3}H pulse. All cells in S labelled with ^{14}C. Most of these will also have ^{3}H label. Some cells in G2 will only have ^{3}H label.

This method could be extended to other phases of the DNA cycle by increasing Q (Wimber & Quastler, 1963).

As a means of determining the length of S in tissues or cell cultures, this method is much easier and quicker than constructing a curve of labelled mitoses, but yields less information. It can also be used on tissues where there is a diurnal mitotic rhythm, since the time for an experiment is relatively short.

Other methods. Puck & Steffen (1963) developed a method of determining the phases of the DNA cycle in cell cultures which is an alternative to the conventional labelled mitoses curve. Tritiated thymidine and colcemid are added simultaneously to an asynchronous exponential culture. The cells are blocked in mitosis by the colcemid and successive counts are made of the accumulating mitoses, labelled and unlabelled. The phases of the cycle are calculated from a plot of the 'collection function' against time. A somewhat similar method is outlined by Okada (1967).

Two ways of measuring the cycle time in plant meristems are described by Van't Hof (1968). A short treatment is given either with colchicine which induces tetraploids in mitotic cells or with caffeine which induces binucleate cells. These marked cells reappear in mitosis a cycle time later. There are only slight differences between the cycle times of normal cells and of binucleates or tetraploids (González-Fernández *et al.*, 1966; Giménez-Martín *et al.*, 1966).

3 Synchronous cultures

Synchronous cultures have been discussed in general terms in the first chapter, and it is only necessary at this point to emphasise that they have the great merit over single cell techniques that they open the cell cycle to a much wider range of analytical techniques, especially those of modern biochemistry. They can be divided into two broad classes, 'induction synchrony' and 'selection synchrony' (James, 1966). In the first case, all the cells of a culture are induced to divide synchronously by some treatment. In the second, the cells in a growing culture which are at a particular stage of the cycle are separated off and then, in most of the techniques, grown up separately as a synchronous culture. Selection synchrony is more recent and inherently has a lower yield than induction synchrony (though see p. 50). But it does produce less distortion of the cycle and for this reason is becoming increasingly popular. Two recent reviews on techniques of synchronisation are by Nias & Fox (1971) for mammalian cells and by Helmstetter (1969c) for micro-organisms.

It is not worthwhile discussing here the analytical methods which have been used on synchronous cultures since they are mostly straightforward biochemical measurements, but we should remember that the section in the previous chapter on pulse labelling applies to whole cultures as well as to single cells.

Induction synchrony

Inhibitor blocks

In these techniques, which have been mostly used on mammalian cell cultures, an inhibitor which acts at one stage of the cycle is added to a growing asynchronous culture. The cells then accumulate at this stage and, after release from the inhibitor, divide synchronously. The block can be either at the time of DNA synthesis or at the time of mitosis.

Inhibitors of DNA synthesis. Xeros (1962) found that high concentrations of thymidine (2 mM) will stop DNA synthesis in mammalian cells. Other deoxynucleosides have the same effect but are more toxic. The mechanism appears to be through a feedback inhibition from an increased thymidine triphosphate pool affecting the deoxycytidine triphosphate pathway (Gentry *et al.*, 1965). The effect of the thymidine can be reversed by washing or by adding deoxycytidine.

The most effective use of thymidine is in a double blockade (Bootsma *et al.*, 1964; Petersen & Anderson, 1964; Puck, 1964). The rationale of this is best explained by summarising the method used by Petersen & Anderson (1964) on CHO cells which have the following phases (p. 58) of the DNA cycle – $G_1/S/G_2/M = 4.5/4.0/3.0/0.5$ hours. A random growing culture is treated with thymidine (25 mM) for 9 hours (a little more than $G_2 + M + G_1$). Cells are blocked in S at whatever point they had reached when the thymidine was added (Puck, 1964). Cells which were in other phases of the cycle are stopped at the G_1/S boundary when they reach it – which will take a maximum of about 8 hours for cells which were just into G_2 when the thymidine was added. The blockade is then lifted by washing and resuspension for 5 hours, rather more than the length of S. All the blocked cells will pass out of the S phase but will be caught at the G_1/S boundary in the next cycle by a second blockade of thymidine put on at the end of the 5 hours and lasting for a further 9 hours. At the end of the second blockade, the culture is well synchronised and shows mitotic index peaks for at least four generations (Fig. 3.1). The final yield can be gram quantities of synchronised cells.

This is clearly an efficient way of getting synchronous DNA synthesis and synchronous division, though there is a problem about what is happening to other cell processes which will be discussed after the end of the next section. Even with the DNA cycle, however, there is some evidence of differences between thymidine-synchronised cells and normal cells. Synchronised human kidney cells showed a shorter G_1 and G_2 in a careful study by Galavazi *et al.* (1966) and Galavazi & Bootsma (1966). HeLa cells also have a shorter $S + G_2$ after synchronisation (Firket & Mahieu, 1966) and a longer metaphase after treatment with 2 mM thymidine (Barr, 1968). Tobey *et al.* (1967*a*) found that CHO cells showed a large variation in G_1 (and cycle times) according to which batch of serum was used in the media, yet the addition of thymidine reduced all the G_1 periods to the minimum time. Another effect of thymidine is to produce mitotic and chromosomal abnormalities (Firket & Mahieu, 1966; Yang *et al.*, 1966). Finally, there are considerable differences between cell lines in their sensitivity to thymidine. An interesting example is that the normal C13 strain of BHK 21 hamster cells is blocked by 10 mM thymidine but the P 183 strain, transformed by polyoma virus, is not (Tobey *et al.*, 1967*b*).

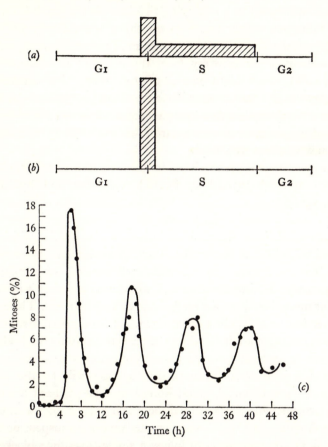

Fig. 3.1. Double thymidine blockade. (*a*) Presumed cell distribution at the end of the first thymidine block. Most cells are at the G1/S boundary but some are spread through S. (*b*) Presumed cell distribution at the end of the second thymidine block. All cells are at the G1/S boundary. (*c*) Mitotic index of synchronised Chinese hamster ovary cells through four successive generations following release from double thymidine blockade (25 mM). From Petersen & Anderson (1964).

Other inhibitors of DNA synthesis have been used with higher eukaryotes to produce synchronous cultures after a single blockade. The first is fluorodeoxyuridine (Rueckert & Mueller, 1960; Littlefield, 1962). According to Till *et al.* (1963) the pattern after a block is like that in Fig. 3.1(*a*), so a double blockade might improve synchrony as it does with thymidine. The second inhibitor is amethopterin (methotrexate) which is a folic acid antagonist (Rueckert & Mueller, 1960; Mueller & Kajiwara, 1966a). The third is 5-aminouracil which works well in higher plants (Smith *et al.*, 1963; Wagenaar, 1966) and, less well, in HeLa cells (Regan & Chu, 1966). All these inhibitors appear to act on the thymidine pathway and can be reversed by adding thymidine to the medium. In budding yeast, however, there is a spontaneous recovery from the effect of fluorodeoxyuridine (Esposito, 1968), which might be used to produce synchrony. There is the same recovery in the fission yeast *Schizosaccharomyces pombe* with fluoro-deoxyuridine, hydroxyurea and deoxyadenosine, and the last of the three can be used to give quite good synchrony after a single 3 hour blockade (Mitchison & Creanor, 1971c).

Partial synchrony has been induced in plant suspension cultures both by aminouracil and by thymidine, fluorodeoxyuridine and hydroxyurea (Eriksson, 1966).

Metaphase inhibitors. Colcemid, a derivative of colchicine which is less toxic to animal cells, will block cells in metaphase but can be reversed by washing. It is not possible to accumulate all the cells of a mammalian culture in metaphase since the colcemid will cause damage if it is left with the cells for more than 2–3 hours (e.g. Kato & Yosida, 1970). The result of a short reversible block will be a synchrony which is only partial, so colcemid treatment is usually combined with another synchrony method. One way of doing this is to treat growing monolayer cultures with colcemid for 3 hours and then harvest the blocked metaphase cells by the wash-off method (p. 47) (Stubblefield, 1968; Romsdahl, 1968; Scharff & Robbins, 1966, with colchicine). The yield will be much greater than with detachment only. Another way, which has been used on cells in suspension culture by Doida & Okada (1967a) is to follow a single thymidine block with a colcemid block. This will give better synchrony than a single thymidine block alone. In *Tetrahymena*, there is a spontaneous recovery from division blocks caused by colcemid and colchicine even when the agents are present. The effect of this is to induce synchrony in a growing culture (Wunderlich & Peyk, 1969).

Another reversible agent which will block in metaphase is nitrous oxide under pressure (Rao, 1968). This has the advantage over colcemid that it appears to cause less damage and HeLa cells can be accumulated for 8 hours. It gives good synchrony when combined with a thymidine block.

Vinblastine is another metaphase blocking agent which has been employed in synchrony work but it is not easily reversible in mammalian cells. Marcus & Robbins (1963) used it to get a large yield of metaphase cells on wash-off, but the cells may well have been damaged since the culture was treated for 20 hours. An inverse way of using vinblastine has been described by Kim & Stambuck (1966). HeLa cells in monolayers are incubated with vinblastine for about a generation time (23–24 hours). The blocked metaphases are then washed off and discarded, leaving a small population of late G2 cells. These are incubated for a further 4 hours which will bring them into G1 of the next cycle, and then given a final wash to remove any further blocked metaphase cells. The yield by this method is small and the degree of synchrony is only fair. The synchrony can be improved by a double vinblastine treatment which leaves a more sharply defined 'window' of cells in progress through the cycle (Pfeiffer & Tolmach, 1967). These last two techniques will only yield normal cells if vinblastine has no effect on cells in interphase. But there is evidence in fact of some changes in the fine structure of interphase cells caused by vinblastine and colchicine (Robbins & Gonatas, 1964).

In *Tetrahymena*, the effect of vinblastine can be reversed by washing off. If a random culture is treated for 4–6 hours, division (but not growth) is inhibited during treatment and at the end the culture divides synchronously after a delay of about 1 hour (Stone, 1968). The mechanism here is uncertain. The vinblastine may act as a blocking agent at one stage of the cycle or it may act through variable 'excess delay' (Chapter 10).

Inhibitor blocks and 'unbalanced growth'

There has been an argument for some time about the protein and RNA synthesis which continues in many situations when DNA synthesis has been blocked with an inhibitor such as excess thymidine. On the one side, it has been pointed out that this represents 'unbalanced growth' in the sense that it is different from the 'balanced growth' (p. 4) of a normal asynchronous culture where the cell components will increase by the same factor over a given interval of time. Mueller (1969) says 'synchronisation at the point of entry of cells into DNA synthesis (i.e. chemical blocks) synchronises only those processes which depend on or are coupled to the process of DNA synthesis'. The other side rejects balanced growth because it is a concept which is clearly invalid for a single cell or a synchronous culture, and emphasises that the accumulation of cells at the G1/S boundary means that cells will have to progress round the cycle for some time until they have all reached this boundary and that this progress implies continued synthesis of all components except DNA (Anderson *et al.*, 1967).

The resolution of this argument needs a quantitative illustration. Let us consider a growing asynchronous culture of mouse L cells with a total cycle time of 20 hours and with these phases of the DNA cycle: $G_1/S/G_2 + M = 9/6.5/4.5$ hours. It can be calculated from Eqn (2) in Chapter 2 that 30 per cent of the cells will be in S at any given moment. If the effect of adding a DNA inhibitor, e.g. thymidine, is to block all progress round the cell cycle of those cells in S, then growth, and with it protein and net RNA synthesis, will decrease in the whole culture by 30 per cent as soon as the inhibitor becomes effective. Thereafter there will be slow and continuous decrease as cells reach the G_1/S boundary, and all growth will cease after 13.5 hours $(G_2 + G_1)$ when the last cell will have reached the boundary. If, on the other hand, the inhibitor only affects DNA synthesis and does not block other progress in the cycle, growth together with protein and RNA synthesis will continue without a check for some time – though there may be a long-term effect of the inhibitor on growth, as in the 'thymidine-less death' of HeLa cells left for more than a generation time in fluoro-deoxyuridine (Rueckert & Mueller, 1960).

The evidence is definitely in favour of the second alternative. Protein and RNA synthesis continue at the same rate as before for 16 hours after the addition of amethopterin to HeLa cells (cycle time of *c.* 24 hours) and for 24 hours after fluorodeoxyuridine (Rueckert & Mueller, 1960). A similar result was found with thymidine by Kim *et al.* (1965), though there was a delay here of about 12 hours before DNA synthesis was stopped. Mouse L5178Y cells show no change in the rate of volume increase for a genera-tion after the addition of fluorodeoxyuridine or hydroxyurea, another DNA synthesis inhibitor (Rosenberg & Gregg, 1969). In an experiment done with HeLa cells synchronised by detachment, Studzinski & Lambert (1969) showed that thymidine added in G_1 did not affect the rate of synthe-sis of protein and RNA or the increase in cell volume, for at least 16 hours. This system of synchronised cells would have revealed an S period block much more clearly than an unsynchronised culture. There was also some DNA synthesis in the presence of thymidine, and the ideal situation in Fig. 3.1(*a*), where all the cells stop at the G_1/S boundary probably does not apply in this case. Another important aspect of the growth after DNA inhibition is that it has been shown in a range of cells to be normal in the sense that periodic enzyme synthesis continues and shows a pattern similar to that in the undisturbed cycle (p. 168).

Continued cell growth in the absence of DNA synthesis and of division should produce abnormally large cells with some proportionate spread of

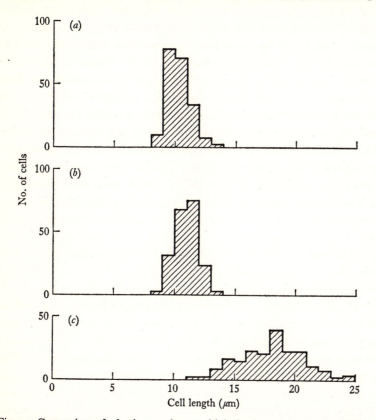

Fig. 3.2. Comparison of selection synchrony with induction synchrony by a DNA synthesis inhibitor in *Schizosaccharomyces pombe*. (*a*) Length distribution of dividing cells (cell with cell plates) in an asynchronous culture. (*b*) Length distribution of dividing cells during the first synchronous division after selection synchronisation by sucrose gradient sedimentation (p. 49). (*c*) Length distribution of dividing cells during the first synchronous division after induction synchronisation (this division occurs $2\frac{1}{2}$ h after a 3 h pulse of 2 mM deoxyadenosine applied to an asynchronous culture). From Mitchison & Creanor (1971*c*).

size as an unsynchronised population. This has been found in mammalian cells by Rosenberg & Gregg (1969; see also Kraemer, 1966). It is also shown in Fig. 3.2 for the fission yeast *Schizosaccharomyces pombe* in which cell length is a measure of cell volume. With selection synchrony, the cell length at division is the same as it is in an unsynchronised culture. With induction synchrony produced by blocking DNA synthesis and then reversing the inhibition, the cell length at the first synchronous division is both much longer and more variable.

DNA synthesis and division are synchronised because they are first blocked and then later unblocked. Growth, on the other hand, is not blocked and so it is not synchronised. This leads to the suggestion, which is expanded in Chapter 10 (p. 244), that there are two dissociable cycles in the growing cell – the 'DNA-division cycle' and the 'growth cycle'. Here lies the particular value of induction synchrony by a DNA inhibitor, since we may be able to use it to distinguish which cellular events belong to which cycle. If they are associated with the growth cycle, they should be synchronised by selection synchrony but not by this type of induction synchrony. If they are associated with the DNA-division cycle, they should be synchronised by both methods. In practice, however, the situation may not be as simple as this since a DNA inhibitor may act on other processes as well. There is, for example, evidence that excess thymidine in mammalian cells may depress RNA synthesis as well as DNA synthesis (Kasten *et al.*, 1965; Painter *et al.*, 1964). It may need further exploration in any one system to find an inhibitor which only affects DNA synthesis.

The situation with metaphase arrest is less clear, and may depend on the strength of the colcemid or colchicine that is used. Since there is no RNA synthesis at metaphase, we would expect blocked cells to show no increase in RNA and only as much protein synthesis as could be coded for by pre-existing messages (though there is also some evidence of repression of translation at metaphase – p. 133). Rosenberg & Gregg (1969) found that cell volume increased at only slightly less than the normal rate for 7.8 hours (80 per cent of the cycle time) after the addition of colchicine to mouse L5178Y cells. This could be explained in terms of pre-existing messages were it not for the fact that actinomycin was found to reduce volume increase immediately. In any case, a problem arises about metaphase arrest if there is any substantial use of stored messages for protein synthesis. Consider a case where metaphase arrest is stopped after 3 hours by washing off the colcemid. The first cells to be accumulated in metaphase will spend 3 hours in using up their messages whereas the last cells to be caught will spend no time in doing this. It is by no means certain that these two groups of cells will follow the same patterns of protein synthesis in their subsequent cycles. Another factor in metaphase arrest is that cycles of chromosome condensation and other cellular events may occur in the long term even though division is blocked. This has been known for some time in sea-urchin embryos (Zeuthen, 1952; Swann & Mitchison, 1953) and Whitmore *et al.* (1961) found that DNA and RNA synthesis continued for 120 hours in mouse L cells treated with colchicine and blocked from division. There is also a slow recovery in the presence of colcemid in Chinese hamster cells (Stubblefield, 1964). There can be little doubt that we need to know more about what happens in metaphase arrested cells before we can safely interpret the results of experiments in which this technique is used to synchronise cells.

There is no particular reason to believe that synchronous cultures made by DNA blocks or metaphase arrest are distorted in their patterns of DNA synthesis or in the events associated with division. But the arguments above suggest that such cultures may not show the true patterns of synthesis of protein and RNA through the cycle since the cultures may be partially or completely asynchronous for these components, quite apart from other effects on synthesis caused by the inhibitors. For this reason, I have marked with an asterisk all the papers in Chapters 6 and 7 which use these techniques. In Chapter 8, on enzyme synthesis, the methods of synchronisation are given in the Tables.

Selective lethal agents

A method which is on the borderline between induction and selection synchrony is to use an agent which kills cells in certain stages of the cycle and leaves a minority of cells to grow up as a synchronous culture. One example is the use of high specific activity tritiated thymidine by Whitmore & Gulyas (1966) on L cells with these phases of the DNA cycle: $G_1/S/G_2/M = 5/8/4/1$ hours. Thymidine (1 μc/ml of ^{3}H-thymidine with specific activity of 6.7 c/mM) was added for 6 hours to a normal asynchronous culture. Cells which were in all phases except G_2 at the start of the labelling period incorporated the thymidine and were so damaged by the radiation that they did not divide and ultimately died. Only the unlabelled cells survived to grow as a synchronous culture.

Hydroxyurea can be used in a similar way since it has a selective lethal action on cells in the S phase (Sinclair, 1967). In addition, it has some of the effects of thymidine since it appears to block cells at the G_1/S boundary. An interesting application of hydroxyurea is to induce synchrony of mouse lymphoma cells *in vivo*, a technique which might be of use in cancer therapy when combined with X-ray treatment (Mauro & Madoc-Jones, 1969; see also Rajewsky, 1970).

Selective killing may be of use in some circumstances, but it has the great disadvantage for biochemical work that the cell cultures contain large numbers of dead or moribund cells.

Starvation and growth

A method of synchronising which has been used widely with micro-organisms and less frequently with higher cells, is to starve a culture or let it run into stationary phase, and then to add fresh medium. Under the right conditions, the culture will grow synchronously for one or more divisions and the yield will be large. The method depends on cells starting into growth at the same time and the same point of the cycle, so we can presume that they align themselves at this point in stationary phase or, more likely, in the final growth period before stationary phase (Nilausen & Green, 1965; Nishi *et al.*, 1967). Not surprisingly, the results vary according to which component of the medium is exhausted, when the culture becomes stationary and, in bacteria, which strain is used (Matney & Suit, 1966).

In many cases, the time between the addition of fresh medium and the first division is abnormally long for a cell cycle. It is unwise only to analyse this first cycle since an event here may be caused by the synchronising procedure rather than by a normal cycle mechanism. If, however, the event recurs in several successive cycles, it is unlikely to be due to the initiation of fresh growth.

The earlier results with this method when applied to mammalian cells were not very satisfactory (Littlefield, 1962; Whitfield & Youdale, 1965; Yoshikura *et al.*, 1967). The degree of synchrony was low, the cell number did not always double at division, and sometimes there was no second cycle. Much better results, however, have been obtained recently in Chinese hamster cells where the key components in the medium which have to be depleted are isoleucine and glutamine (Tobey & Ley, 1970; Ley & Tobey, 1970).

Yeoman & Evans (1967) found that explants of artichoke tubers were partially synchronous for the first two or three divisions after being stimulated into growth by a nutrient medium. In the first experiments, there was a substantial proportion of non-growing cells but it was found that this could be reduced to 10 per cent or less by excluding light (Fraser *et al.*, 1967). This is a valuable system since it is one of the few synchronising procedures, apart from DNA inhibitors, which has been applied to higher plants.

One of the best examples of starvation synchrony among eukaryotes is that developed for budding yeast by Williamson & Scopes (1960, 1961*a*). A stationary phase culture is treated with alternating cycles of cold growth medium and warm starvation medium accompanied by decanting-off of small cells. The resulting population is composed of cells of uniform size which grow as a synchronous culture when inoculated into warm growth

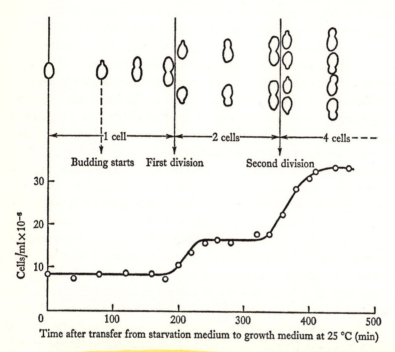

Fig. 3.3. Growth of a synchronised yeast culture (*Saccharomyces cerevisiae*) on inoculation into synthetic medium. From Williamson & Scopes (1961*a*).

medium (Fig. 3.3). The first cell cycle is abnormally long because of the lag phase of about 75 minutes which precedes the start of budding.

There are other examples of starvation synchrony in budding yeast though the degree of synchrony is somewhat less than it is in Fig. 3.2 possibly because small cells were not separated off (Campbell, 1957*b*; Hilz & Eckstein, 1964; Ogur *et al.*, 1953; Sylven *et al.*, 1959). Nosoh & Takamiya (1962) did remove the small cells by repeated centrifugation of yeast-cake and got a fair degree of synchrony in growth medium. Sando (1963) synchronised fission yeast (*Schizosaccharomyces pombe*) by a process that is analogous to starvation. Exponential phase cells are concentrated to a dense suspension of 10^9 cells/ml, incubated in fresh medium for a generation time (after which the medium will be depleted) and then diluted 100 times to give a synchronous culture. The cycles, however, are shorter than normal and the cell number does not double at division. A modified and improved version of this technique has been used on budding yeast by Tauro & Halvorson (1966). They compared this method both with that of Williamson & Scopes and with the gradient selection method (p. 49) and found all three techniques gave the same patterns of periodic enzyme synthesis. Comparisons of this kind between synchronous cultures produced in different ways are valuable and unfortunately rare.

Another lower eukaryote which can be synchronised by starvation is the diatom *Navicula* (Coombs *et al.*, 1967b). Silicon depletion blocks the cell cycle after cell division but before the formation of the new wall between the daughter cells. If silicon is now added, there is synchronous uptake of silicon, wall formation and cell separation.

There are a number of cases in which bacteria will grow for a time as a synchronous culture when stationary phase cells are put into fresh medium (e.g. Bergter, 1965; Houtermans, 1953; Masters *et al.*, 1964). The exact stage of the cells may be important. Cutler & Evans (1966), working with *Escherichia coli* and *Proteus vulgaris*, only got successful synchrony from cells in the very early stationary phase. Cells which were taken before or after this phase of culture growth showed little or no synchrony (Fig. 3.4). This was also the phase in which there were slight synchrony steps in an Hfr strain of *E. coli*. These steps are probably not a universal property of bacterial cultures since they were absent from an E64 strain of *E. coli* (Nishi *et al.*, 1967). This latter strain also gave good synchrony with an inoculum taken from a late stationary phase culture (unlike the Hfr strain in Fig. 3.4) which was consistent with other results which showed that the cells finished growth with most of them blocked at the start of the cycle.

The method used by Sando (1963) for synchronising yeast was originally developed for bacteria by Yanagita & Kaneko (1961). Incubating *E. coli* B at high cell density and then diluting gave a good synchronous culture.

Bacteria can also be synchronised by removal and re-addition of a specific nutrient. If DNA synthesis is temporarily blocked by the removal of the required nutrient in thymine-requiring *E. coli* (Barner & Cohen, 1956) or in thymidine-requiring *Lactobacillus acidophilus* (Burns, 1959), a burst of

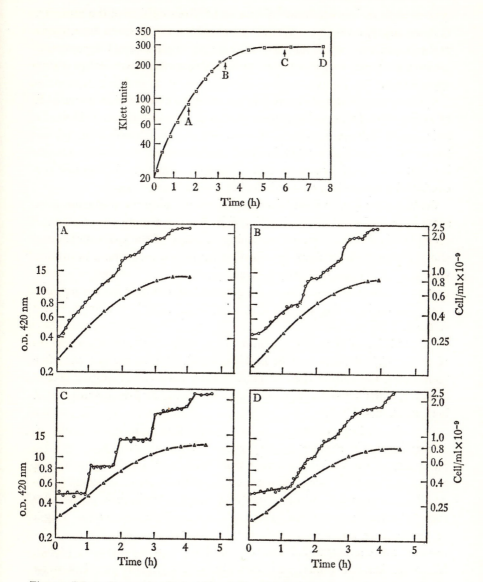

Fig. 3.4. Effect of harvesting *Escherichia coli* K12 Hfr CS 101 met⁻ at different points of the growth curve. Four separate experiments are shown. Cells were first grown in 100 ml of medium in a side-arm flask and harvested at different stages of their growth curves (points A to D on top diagram). Cells were then inoculated into a volume of fresh medium that would give each of the four experiments the same initial optical density. Symbols: □, Klett units; △, optical density at 420 mμ: ○, cells per ml. From Cutler & Evans (1966).

synchronous divisions follows (after a lag) the re-addition of the nutrient. It is not likely, however, that these synchronous cycles are representative of the normal cycle since they are much shorter. In any case, the point that has been made earlier in this chapter about unbalanced growth very probably applies here.

Cycles that appear to be more normal occur after the temporary removal of amino-acids from amino-acid-requiring mutants of *E. coli* (Matney & Suit, 1966; but see Inouye & Pardee, 1970) and *Streptococcus faecalis* (Stonehill & Hutchison, 1966). Even so, the question of unbalanced growth is raised by some of the strains of *E. coli* in which RNA and DNA synthesis continued after protein synthesis had stopped.

Scott & Chu (1958) found that depletion and addition of glucose gave some degree of synchrony with *E. coli* B. This is a more clearly defined example of what is probably happening in many of the cases mentioned above where fresh medium is added and synchrony occurs. The results were similar to those of Cutler & Evans (1966) in that the stage of culture growth at which glucose was added affected the resulting synchrony, but they differed in giving the worst synchrony from early stationary phase cultures.

The addition of fluorophenylalanine to growing *E. coli* causes a reduction of the DNA/cell (Brostrom & Binkley, 1969). If it is washed out after 80 minutes, the culture then becomes synchronous. The resulting cycles, however, differ from the normal cycle in showing step-wise synthesis of both RNA and DNA.

Periodic medium changes can also be used to induce synchrony in cultures grown in chemostats. Cultures of *E. coli* grown in this way with phosphate as the limiting nutrient will divide with partial synchrony if phosphate pulses are added to the medium with a periodicity equal to the doubling time of the culture (Goodwin, 1969a). Division happens after each pulse in a manner analogous to the effect of multiple shifts of light or temperature (cf. Fig. 3.9). Somewhat similar systems have been developed for chemostat cultures of budding yeast by Dawson (1965), Müller & Dawson (1968) and Meyenburg (1969); and for *Astasia* by Morimoto & James (1969). The cell cycle in these chemostat cultures is very likely to be different from the normal cycle, but, if the technique is used with insight, it can provide elegant ways of examining control mechanisms (Goodwin, 1969b).

Single changes of temperature or light

Some of the oldest and best known methods of inducing synchrony consist of a series of temperature 'shocks' or light changes, as we shall see in the next section. But even a single shock will synchronise a culture, though this only works with some cells and usually does not produce a high degree of synchrony. In most cases, there has been no exploration into the reasons for this effect but they may involve the phenomenon of variable 'excess delay' which is discussed in Chapter 10.

If HeLa cells are kept at 4 °C for an hour and then taken up to their normal growing temperature, they will in some situations divide synchronously about a generation time (17–18 hours) later. Newton & Wildy (1959) who found this effect called it 'parasynchrony' because there was usually a proportion of non-dividing cells. Chèvremont-Comhaire & Chèvremont (1956) also found that a longer cold shock (24 hours) with chick embryo cells would produce a burst of synchronous mitoses three hours after transfer to the warm. The ease of giving a cold shock coupled with the almost complete absence of published papers that use this method (an exception is Koch & Stokstad, 1967) suggests that it is difficult to apply to other mammalian cell systems. Rao & Engelberg (1966), who have made a considerable study of the effects of temperature on mammalian cells, could not induce synchrony in HeLa cells by long or short cold shocks and they suggest that nutritional factors were involved in the experiments of Newton & Wildy.

Early fish embryos (middle blastula) can be synchronised by a two hour cold shock at 3 °C followed by transfer back to the normal temperature of 18 °C (Neifakh & Rott, 1958). This causes two mitotic index peaks of 40 per cent separated by a time (25–30 minutes) which is less than the normal cycle time (40–45 minutes).

Among the lower eukaryotes, *Tetrahymena* cultures can be partially synchronised by a 1–2 hour shock at temperatures greater than 31 °C or less than 18° (Zeuthen, 1964, p. 103). The synchrony is much improved by a series of temperature shocks (Chapter 10). The amoeba *Naegleria* will go through a single sharply synchronous division after a 100 minute heat shock at 38–39 °C (Fulton & Guerrini, 1969). The cells do not divide during the heat shock and they reach nearly twice the normal size by the time of the synchronous division on return to 30 °C – a very similar effect to that in *Tetrahymena* synchrony. A period of 24 hours in the dark will produce synchronous division in the diatom *Cylindrotheca* (Coombs *et al.*, 1967*a*).

Single cold shocks have been widely used in the past for synchronising bacterial cultures.[1] They have not however been so popular in recent years for two reasons.[2] One is that the degree of synchrony is not very high in most cases, and the other is the doubts that have been expressed as to whether these cultures are similar to the normal cycle in their synthetic patterns. As an example, the pattern of DNA synthesis in *Salmonella* after a cold shock (Lark & Maaløe, 1956) is not the same as that in the closely related *E. coli* after selection synchrony (Helmstetter, 1967; Clark & Maaløe, 1967).

Multiple changes of temperature or light

Two of the earliest and best known synchronous systems date from the mid-fifties when Zeuthen and Scherbaum showed that *Tetrahymena* could be synchronised by repetitive temperature changes (Scherbaum & Zeuthen, 1954), and Tamiya and his colleagues found that cycles of light and darkness would synchronise *Chlorella* (Tamiya *et al.*, 1953). This method of synchronising *Tetrahymena* will be discussed in detail later in this book (Chapter 10) both because it yields valuable clues to the control of division and because the procedure is radically different from that used with *Chlorella* and with the other systems described in this section. In the latter case, the cycles of light or temperature are roughly the same length as the cell cycle and there is a synchronous cell division after each of them. But with this *Tetrahymena* method, the temperature shifts are shorter than the cycle, and cell division is completely suppressed while they are happening. The synchronous divisions take place after the end of the temperature cycles. Not that this is the only way of synchronising *Tetrahymena*. It can also be done with division taking place between each temperature shift or 'shock'. Zeuthen and Scherbaum showed this both for cold shocks and hot shocks (see Zeuthen, 1964, p. 105), and the cold shock technique has been further developed by a group of American workers (Padilla & Cameron, 1964; Padilla, Cameron & Elrod, 1966; Whitson *et al.*, 1966*a*; 1966*b*). Nevertheless the main body of work on synchronised *Tetrahymena* has been done with the first technique described above.

---

1 For *E. coli*: Doudney (1960); Kogoma & Nishi (1965); Hegarty & Weeks (1940); Scott & Chu (1958). For *Salmonella typhimurium*: Bruce *et al.* (1955); Lark & Maaløe (1956). For *Pneumococcus*: Hotchkiss (1954). For *Bacillus megaterium*: Falcone & Szybalski (1956); Hunter-Szybalska *et al.* (1956). For *Azotobacter*: Lin & Wyss (1965).

2 An exception is the important paper by Smith & Pardee (1970) which is discussed later (p. 228). This shows that a heat shock synchronises cell division but does not synchronise DNA replication.

In the systems where there is a division between each temperature or light cycle, there are difficulties in comparing the results with those from other ways of analysing the cell cycle. These difficulties have been discussed in Chapter 1, and they account for the scant mention of the results elsewhere in the book. But we should consider here the techniques of synchrony that have been used since they are interesting in themselves and may prove valuable in the future.

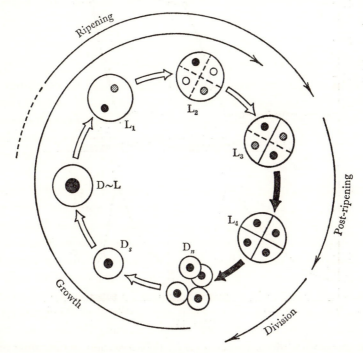

Fig. 3.5. Diagram of the life cycle of *Chlorella ellipsoidea*. The white arrows indicate light-dependent processes, and the black arrows show the transformations which occur independently of light. From Tamiya (1963 *b*).

Pride of place should go to the *Chlorella* cultures developed and exploited by Tamiya and his colleagues, if only because more biochemical work has been done on them than on any other synchronous system (reviewed by Tamiya, 1963 *a, b*; 1964; 1966; Iwamura, 1966). Cell components which have been followed through the cycle include proteins, peptides, amino-acids, RNA, DNA, nucleotides, carbohydrates, lipids, sterols, coenzymes, pigments and vitamins. The basic life cycle of *Chlorella ellipsoidea* is shown in Fig. 3.5. Starting with D_n cells, light-dependent growth takes place up to the 'ripened' L_3 cells which are mostly quadrinucleate. The processes which are independent of light are the production of the 'fully matured' L_4 cells and the release of four new D_n daughter cells from each mother cell. There are a number of ways of producing synchronous cultures, one of which is shown in Fig. 3.6. The culture is started from D_s cells. These are small daughter cells with diameters around 3.0 ($\pm$0.3) μm which are

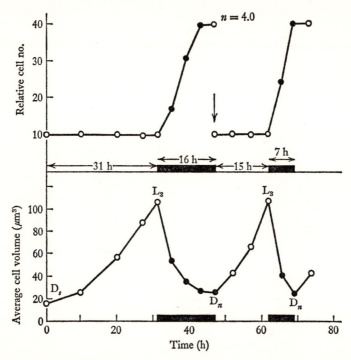

Fig. 3.6. Changes in relative cell numbers and average cell volume in a synchronous culture of *Chlorella ellipsoidea*. The culture is started with D_s cells (see text). The arrow indicates where the culture was diluted to the original cell concentration. From Tamiya (1963 b).

produced by differential centrifugation of a 7–9 day culture (Tamiya & Morimura, 1964). There is, therefore, a degree of selection and of growth after starvation in the start of these *Chlorella* cultures. The D_s cells are then illuminated for 31 hours at 21 °C and reach the L_3 stage. After a further 16 hours in the dark, cell separation has taken place and all the cells are in the D_n stage – almost identical with the D_s cells except that they are a little larger and can grow faster in the light. The sequence is called a D_sLD cycle. It is then followed by a DLD cycle with 15 hours in the light and 7 hours in the dark. The alternation of light and dark is not essential for synchrony since there are other cycles (D_sLD' and DLD') where the culture is illuminated throughout and the synchrony must be due to the regrowth of the selected small daughter cells.

TABLE 3.1. *Nuclear division at different stages of the light period with D_sLD and DLD cycles in* Chlorella ellipsoidea. *Data from Figs. 9, 10 and 14 in Tamiya et al.* (1961).

		D or D_s	$D{\sim}L$	L_1	L_2	L_3
D_sLD cycle	Time (h)	0	15	21	28	35
(35 h light + 17 h dark)	Per cent of light period	0	43	60	80	100
DLD cycle	Time (h)	0	5	10	13	17
(17 h light + 9 h dark)	Per cent of light period	0	29	59	76	100
Percentage of cells with:						
1 nucleus		100	100	60	25	0
2 nuclei		0	0	35	50	5
4 nuclei		0	0	5	25	95

A major problem in comparing these results with other work on the cell cycle is the position of nuclear division and cell division in the *Chlorella* life cycle. Fig. 3.5 indicates that nuclear division starts considerably before the final production of daughter cells and this is borne out in Table 3.1. By the time the cells have reached the L_1 stage (60 per cent of the way through the light period in both D_sLD and DLD cycles) 35 per cent of them have gone through one nuclear division and 5 per cent have gone through two. This means that nuclear division must start about half way through the light period or one third of the way through the whole cycling regime of light + dark periods. Nor are the nuclear divisions well synchronised since it takes about half the light period for them to be completed.

'Cell division' is an ambiguous phrase in *Chlorella* systems. It is used by Tamiya (e.g. in Fig. 3.5) for the period of release of daughter cells when the cell numbers increase. But there is no doubt that the mother cell contains four separate daughter cells, each with a cell membrane round it, for some time before they are released. 'Cell division' therefore is the release of pre-existing cells (see discussion after Tamiya, 1963a). Exactly when true cell division occurs is more obscure. It has already taken place in the L_3 cells shown in the electron micrographs in Tamiya, 1963a, b, and it probably starts in the L_2 cells. If so, there is a gap of 24 hours in a D_sLD cycle (nearly half the complete cycle) between the start of cell division and the completion of release of the daughter cells. There are other cases, for example fission yeast, where there is a time gap between the physiological separation of daughter cells by a cell membrane and the final physical separation when a cell wall has been completed, but in none of them is it so large. The presence of this big gap, the early advent and relatively poor synchrony of nuclear division, and the fact which will be discussed below that all *Chlorella* systems show an increase over the cycle not of a factor of two but of four to sixteen, all make it very difficult to draw useful comparisons between this system and other cell cycles. For this reason, I have not discussed the results from it in any detail in this book, but it is clearly an interesting system for the cell biologist and one where there is a great deal of background knowledge.

Another *Chlorella* system which has considerable merits is that developed by Schmidt and his colleagues and used for a series of biochemical investigations (reviewed by Schmidt, 1966 and 1969). A random culture of a high temperature strain of *Chlorella pyrenoidosa* is brought into synchrony by four alternating cycles of light and dark (11 hours each) at 38·5 °C (Baker & Schmidt, 1963). The culture is then left in continuous light where it shows a series of synchronous cell releases. After each release, the culture is diluted back to its original cell concentration. Recently, however, a method of continuous dilution has been developed which gives a higher yield and eliminates any shock which might occur with periodic dilution (Hare & Schmidt, 1968). From the point of view of comparisons with other cell cycles, this system has two great advantages over Tamiya's. One is that the culture is synchronous for at least three cycles without any changes in the environment. If an event recurs in all these cycles it is unlikely to be due directly to an environmental shift. The second is that nuclear division is later in the cycle and more synchronous (Fig. 3.7).

These are not the only ways of producing synchronous cultures of *Chlorella*. Other systems are described in reviews by Kuhl & Lorenzen (1964), Lorenzen (1964), Pirson & Lorenzen (1966), Senger & Bishop (1969), Sorokin (1963; 1964). Most of these methods involve alternating cycles or light and dark with cell release taking place in the dark.

Chlorella and some other algae raise a problem about the relation of their cell cycles to those of most other cells. The normal cell cycle has a two-fold increase followed by division into two daughters. In a *Chlorella* culture, such as that in Fig. 3.6, there is an eight-fold increase followed by three rapid divisions into eight daughters. There are two ways of regarding the *Chlorella* interphase. One of them is to look at it as equivalent to a single normal cell cycle but with the difference that every synthetic event involves an eight-fold rather than a two-fold increase. The other is to regard it as equivalent to three successive normal cycles with their divisions dissociated and occurring at the end of the third cycle. This distinction is important for discontinuous events like the synthesis of DNA or of a 'step enzyme' (see Chapter 8) which occur once in the normal cycle. Looking at the *Chlorella* cycle in the first way, we would expect there to be a single period of synthesis but with more material being synthesised. Looking at it in the second way, we would expect three separate periods of synthesis each with a two-fold increase. Which of these ways is correct is at present unknown, but if it is the second one it will be important to look at the fine detail of

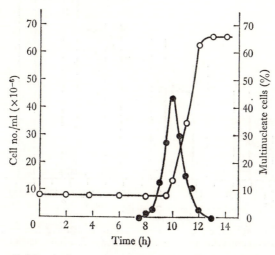

Fig. 3.7. Nuclear division during synchronous growth of *Chlorella pyrenoidosa*. ●—●—● per cent multinucleate cells. ○—○—○ cell number. From Curnutt & Schmidt (1964*b*).

a synthesis curve. Unless the measurements are frequent and accurate, it is easy to ignore three small steps and draw a smooth curve through them.

Another system which has similarities with *Chlorella* is *Chlamydomonas* which can also be synchronised by light–dark cycles (Bernstein, 1968; Chiang & Sueoka, 1967; Kates *et al.*, 1968). It is relatively easy to change the 'burst size' or number of daughter cells by altering the light cycles, and it would be interesting to see if this also altered the number of any synthetic steps.

Other Algae which can be synchronised by light-dark cycles are *Euglena* (Edmunds, 1965; James, 1964; Padilla & Cook, 1964) and *Ulva* (Lövlie, 1964), and a number of others mentioned in reviews by Hoogenhout (1963), Pirson & Lorenzen (1966) and Tamiya (1966). Many of the light–dark cycles have a periodicity of about 24 hours and suggest a connection between the cell cycle and an endogenous circadian clock (e.g. Edmunds, 1966; Edmunds & Funch, 1969*a, b*). A recent striking example is that a persistent and free-running circadian rhythm of synchronous division can be induced in a mutant of *Euglena gracilis* by a single switch from dark to light (Jarrett & Edmunds, 1970).

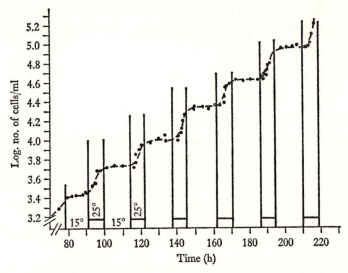

Fig. 3.8. Synchronous culture of *Astasia longa* grown on 1 per cent proteose peptone with a cycle of 16 hr at 15 °C and 8 h at 25 °C. From James (1964).

A number of the lower eukaryotes which are not photosynthetic can be synchronised by warm–cold cycles. *Tetrahymena* has been mentioned earlier. Another good example is the flagellate *Astasia* (James, 1964; Padilla & James, 1960; Padilla & Cook, 1964). It can be synchronised by alternating periods of 16 hours at 15 °C and 8 hours at 25 °C (Fig. 3.8). Division and nucleic acid synthesis take place in the warm, but the cold period is not a simple period of rest since most of the dry mass increase happens then. *Euglena* can be synchronised in a similar way (Terry & Edmunds, 1970), and some degree of synchrony has been obtained by temperature cycles in *Amoeba* (James, 1959) and budding yeast (Louderback *et al.*, 1961).

As we have seen, a single cold shock will synchronise bacteria but will not do it very well. Multiple shocks are much more efficient, and the effect of 8 minutes at 37 °C alternating with 28 minutes at 25 °C on *Salmonella* is shown in Fig. 3.9 (Lark & Maaløe, 1956). This also demonstrates very clearly that the cell cycles with this type of induction synchrony are not the same as the normal cell cycle. Turbidity ($\simeq$ dry mass) and DNA increase are periodic and only occur in the warm periods, whereas they are continuous in the normal cycle in closely related bacterial rods (pp. 100, 144).

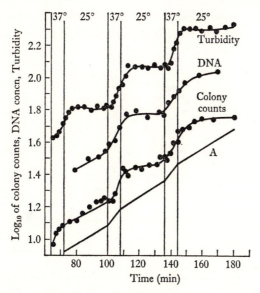

Fig. 3.9. Synchronous culture of *Salmonella typhimurium* produced by temperature cycles of 8 min at 37 °C and 28 min at 25 °C. Curve A is constructed from the exponential growth rates of *S. typhimurium* in broth at 25 °C and 37 °C and represents a theoretical exponential course of growth. From Lark & Maaløe (1956).

Selection synchrony

Wash-off

Cultured mammalian cells which are growing on a solid surface round up in the later stages of mitosis and loosen their attachment to the surface. It is therefore possible to separate off the mitotic cells by gentle washing of a cell culture, and then grow them up as a synchronous culture. This method was first used by Terasima & Tolmach (1961; 1963) on HeLa cells. Slight variations and improvements of the technique were made for HeLa cells by Robbins & Scharff (1966), for Chinese hamster lung cells by Sinclair & Morton (1963), for Chinese hamster ovary cells by Tobey *et al.* (1967c) and for chick embryo cells by Liébecq-Hutter (1965). A good recent description of the method is that by Petersen *et al.* (1968). A somewhat complex apparatus which works on this principle is outlined by Lindahl & Sörenby (1966).

This is undoubtedly the method of choice for producing synchronous cultures from cells growing on solid surfaces. The degree of synchrony is excellent and the cells are selected because they are at a narrowly defined part of the cell cycle. Most other methods of selection rely on characters like cell size which vary between cells at any one stage of the cycle. On the

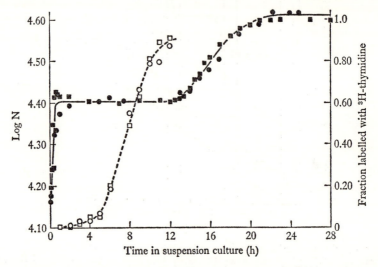

Fig. 3.10. Comparison of the rates of cell division and entry into the DNA-synthetic period in synchronous cultures of Chinese hamster ovary cells prepared by chill-accumulation and continuous harvest after wash-off. The circles represent non-chilled cultures, and the squares represent cultures which were chilled during the collection period. Cell concentrations are represented by solid figures, and the open figures represent the fraction of cells incorporating ^{3}H-thymidine. From Enger & Tobey (1969).

other hand, this selection from a very short part of the cycle carries with it the inherent disadvantage of low yield – only a few per cent of the population is washed off.

Three approaches have been used to compensate for the low yield. One is to increase the size of the original population by using a number of large surface area bottles (Robbins & Scharff, 1966). The second is to increase the proportion of mitotic cells by a preliminary synchronisation with a DNA inhibitor such as thymidine, or by prior accumulation of metaphases with colcemid (see p. 28). Colcemid treatment increases the yield of mitotic cells from 1 per cent to 10 per cent, but it and other chemical inhibitors carry with them the risk of distorting the cycle by unbalanced growth (p. 29). Since the great advantage of selection synchrony is that it should not distort the cycle, it is unwise to sacrifice this advantage by combining it with chemical inhibitors. The third way of increasing the yield is to collect successive samples of washed-off cells (every 10 minutes for several hours), chill them, pool them together and finally grow them up as a synchronous culture (Petersen *et al.*, 1968; Lesser & Brent, 1970). Fig. 3.10 shows that the growth in numbers and the timing of the S period is the same for a synchronous culture from 'chill-accumulated' cells as it is for one from a single sample collected over one 10 minute period. This evidence suggests that the chilling does not affect the resulting synchronous culture, but it would be dangerous to assume without further tests that this applies either to other synthetic patterns in these cells or to other varieties of cell, in view of the effects of cold shock which have been mentioned earlier (p. 39).

Gradient separation

If the cells from a random growing culture are collected and concentrated, they can be layered in a small volume on the top of a sucrose gradient and then centrifuged. After a few minutes of centrifugation, the cells will have spread down through the gradient to form a broad layer with small cells at the top and large cells at the bottom. This layer is a linear representation of the cell cycle since the small top cells are early in the cycle and the large bottom cells are late. Any part of the gradient could be removed and used to start a separate synchronous culture, but, in practice, it is easiest to use the small cells at the top (up to 5 per cent of the total cells). This method of producing synchronous cultures was first developed by Mitchison & Vincent (1965) and tested on fission yeast, budding yeast and *Escherichia coli*. Fig. 3.11 shows a synchronous culture of fission yeast made by this technique. It is important to realise that this is velocity sedimentation which separates cells by size and density, and not equilibrium sedimentation which would separate cells by density alone.

This method has been used on micro-organisms by other workers, for example, on *E. coli* by Donachie & Masters (1966) and on budding yeast by Tauro & Halvorson (1966) who compared it with other techniques of synchronisation. Malinovsky & Mitjushova (1966) tried it on budding yeast with a glycerol gradient. It has also been used on mammalian cells with a plain sucrose gradient by Sinclair & Bishop (1965), and, more successfully, with an isotonic sucrose gradient by Schindler *et al.* (1970) or with a Ficoll gradient by Ayad *et al.* (1969).

A variation on this technique is to exploit the fact that the order of cells in the gradient is the same as the order in the cycle. Samples separated by spatial intervals down the gradient are the same as samples separated by temporal intervals in a synchronous culture. This has been applied to *E. coli* by Kubitschek (1968 *a*, *b*) and by Manor & Haselkorn (1967), and to mammalian cells by Morris *et al.* (1967), and by MacDonald & Miller (1970) who used a protein gradient and gravity sedimentation. The

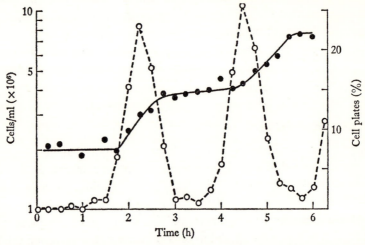

Fig. 3.11. Synchronous culture of the fission yeast *Schizosaccharomyces pombe* prepared by gradient separation. Closed circles are cell numbers and open circles are cell plate index (equivalent to mitotic index). From Mitchison (1970).

yield and, with mammalian cells, the degree of separation can be increased by the use of zonal rotors (Halvorson *et al.*, 1971; Warmsley & Pasternak, 1970). The advantage here over setting up a synchronous culture is that the yield is much higher since all the cells of the original population can be analysed. It also has the advantage, shared by the membrane elution technique, of allowing a culture in normal exponential growth to be treated in some way, for example pulse-labelled, and then afterwards sorted out into cycle stages. This is a way of countering the objection that the process of gradient selection may have an adverse effect on the cells. The disadvantage is that the cells cannot be followed through successive cycles to see whether the same event recurs at the same point in each cycle. It is also difficult to use if there is cell adhesion or clumping.

Separation of different sized cells by centrifugation or sedimentation under gravity is also a part of other techniques of synchrony which have been described earlier, e.g. for *Chlorella* (Tamiya & Morimura, 1964) and for budding yeast (Williamson & Scopes, 1960). Recently, Schmidt and his colleagues have separated cells of *Chlorella* by centrifuging to equilibrium in a Ficoll gradient (Hopkins *et al.*, 1970; Sitz, Kent, Hopkins & Schmidt, 1970; Sitz, Hopkins & Schmidt, 1970; Anderson, 1970).

There are two general points against gradient separation. One is that the selection is less precise than wash-off or membrane elution and therefore the degree of synchrony is less good. This is partly because the size separation is imperfect and partly because cells at the same stage of the cycle vary in size. The other objection is the one mentioned above – that the cells may be affected by their initial concentration and by the material of the gradient. The great merit of the method is that it should be applicable to any cell system that can be grown in suspension, whereas the other methods of selection synchrony are severely restricted in the systems to which they can be applied. It is, for instance, easier to grow large quantities of mammalian cells in suspension than in monolayers, and the gradient technique is the only selection method which can be applied to suspension cultures.

Membrane elution

An ingenious method of synchronising *Escherichia coli* strain B was developed by Helmstetter & Cummins (1963; 1964, see also Cummings, 1965). A growing culture is collected on a membrane filter or ion exchange paper. The filter is then inverted and fresh warm medium is run through slowly and continuously. At first, the excess cells are washed off leaving only those cells which are firmly absorbed onto the surface of the filter. Thereafter there is a continuous flow of small young cells which are released from the filter surface into the circulating medium (hence one of the popular names for this technique – 'the baby factory'). The presumption is that all the absorbing sites are saturated with cells so that when there is a cell division, one of the daughter cells remains attached to the surface but the other is released. Samples of the young cells can be collected and grown up separately as a synchronous culture (Fig. 3.12). The yield can be up to 5×10^8 cells/min though this needs a large membrane filter with a diameter of 293 mm (Cummings, 1965).

Cultures of *E. coli* are certainly sensitive to environmental changes and it could be argued that the changes involved in the initial harvesting for this technique would affect the resulting synchronous cultures. Helmstetter (1967) therefore devised an elegant variation of the method in which a random culture was labelled with a pulse of ^{14}C-thymidine *before* loading and eluting the filter. This should eliminate any distortions to metabolism caused by the synchronisation procedure, and should give an accurate picture of the rate of DNA synthesis through the cycle. The method is outlined in Fig. 3.13. The ideal distribution of cell ages in the random

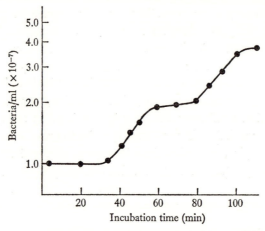

Fig. 3.12. Synchronous culture of *Escherichia coli* B/r prepared by membrane elution. From Cummings (1965).

culture should be as shown in Fig. 3.14(*a*) (cf. Fig. 2.5), and the relative concentration of cells in the eluate after loading and eluting should follow the fluctuating curve in Fig. 3.14(*b*). If DNA is being synthesised at an increasing rate through the cycle, the radioactivity per cell in the eluate should drop steadily and continuously (curves *A* in Fig. 3.15). If, on the other hand, there is a sharp doubling in rate in mid-cycle, the radioactivity should show a step-wise decline (curves *B* in Fig. 3.15). These diagrams assume that there is no dispersion of individual cycle times and that exactly one half of the new-born cells are freed from the filter surface. In practice, neither of these assumptions are fully valid and the experimental results are somewhat different from the ideal case. The method, however, is still a powerful one and gives a fairly clear picture of the synthetic pattern of DNA in different media (Chapter 5). It is not, of course, restricted to DNA synthesis and it has also been used to measure the potential of three

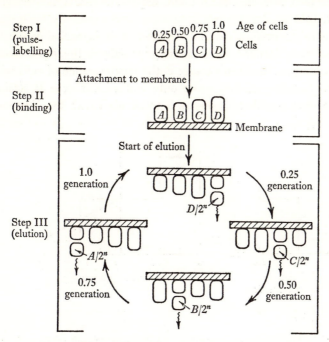

Fig. 3.13. Pulse-labelling before membrane elution with *Escherichia coli*. Outline of the procedure for determining the rate of incorporation of a labelled molecule into cells of different ages in an exponential unsynchronised culture. From Helmstetter (1967).

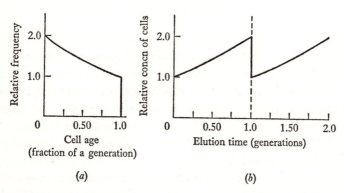

Fig. 3.14. Pulse-labelling before membrane elution with *Escherichia coli*. (*a*) Idealised age distribution in an exponential phase culture containing no dispersion of generation times of individual cells. (*b*) Theoretical concentration of cells in the eluate from a membrane-bound culture with an age distribution as shown in (*a*). From Helmstetter (1967).

inducible enzymes with induction being applied before synchronisation (Helmstetter, 1968).

Provided a fairly small yield is acceptable, this is probably the best selective synchrony method for bacteria, especially if the pre-treatment variation can be used. Its disadvantage is the narrow range of cells which can be used. The original technique only worked with the B strain of *E. coli*, but a new modification allows it to be used with the K12 strain (Cummings, 1970).

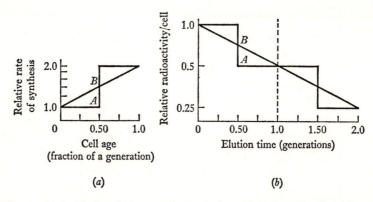

(a)   (b)

Fig. 3.15. Pulse labelling before membrane elution with *Escherichia coli*. (a) Rate of synthesis of two hypothetical macromolecules through the cell cycle. (b) Theoretical radioactivity per cell in the eluate from a membrane-bound culture if it has been pulse-labelled with the radioactive precursors of the macromolecules in (a). From Helmstetter (1967).

Filtration

The earliest successful method of separation synchrony was developed for *E. coli* by Maruyama & Yanagita (1956). Growing cultures are concentrated and filtered on layers of filter papers or on a thick mat of paper pulp. Small cells pass through first into the filtrate and can be used to start a synchronous culture. Alternatively, the top layers of paper can be eluted to give large cells at the end of the cycle, or the middle layers to give mid-cycle cells (Lark & Lark, 1960).

Although this technique does not give such good synchrony as membrane elution and is open to the objection that harvesting may cause metabolic disturbance, it is attractively simple and has been used for several studies of the cycle (e.g. Abbo & Pardee, 1960; Maruyama, 1956; Maruyama & Lark, 1962; Nishi & Kogoma, 1965; Rudner *et al.*, 1965). It works on a wider range of bacteria than membrane elution (e.g. *E. coli* K12 as well as B; *Alcaligenes faecalis*) but does not work on fission yeast (J. M. Mitchison & K. G. Lark, unpublished).

A preliminary report that cultures of *E. coli* could be selectively filtered through Millipore membranes with 1.2 μm pore size has not been confirmed (Anderson & Pettijohn, 1960). In theory, it should be possible to separate cells by size with filters or grids with appropriate pores, but in practice it seems difficult to avoid clogging.

Electronic sorting

An instrument has been devised by Fulwyler (1965) which will sort individual cells by volume. Individual cells are distributed in small droplets which are electrically charged and pass between deflector plates. The deflection for each cell is proportional to its volume which has been measured a short time before when it passed through the aperture of a Coulter counter or its equivalent. Each droplet is collected in one of a series of containers according to the amount of deflection. Up to 1000 cells per second can be sorted this way and they remain viable. This is an ingenious instrument but it is complex and still rather slow for collecting sufficient cells for biochemical assay.

Natural synchrony

Quite apart from synchronous cultures produced by induction or selection, there are many cases of naturally synchronous groups of cells throughout the animal and plant kingdoms. These have been reviewed by Agrell (1964) for animals, and by Erickson (1964) for plants. The synchronous tissue most widely exploited in animals is the early embryo, although it is a special case in having cell cycles without overall growth. The number of synchronous cycles which occur after fertilisation varies in different species. In the holothurian *Synapta digitata*, there are nine synchronous cycles up to the stage with 512 cells. There are, however, only three synchronous divisions in the best known of all early embryos, that of the sea urchin (much of the earlier work on sea urchin embryos is summarised in the admirable book by Harvey, 1956).

A number of plant tissues, for instance endosperm, show synchronous divisions but they have not been much used for biochemical work. An exception is the slow synchronous meiosis in the anthers of liliaceous genera, *Trillium* and *Lilium*, which has been worked on by several groups, including in particular Stern and his colleagues (e.g. Hotta *et al.*, 1966).

The myxomycete or acellular slime mould *Physarum polycephalum* has been extensively exploited in recent years for cell cycle work, principally by Rusch and his colleagues. Large single plasmodia with thousands of nuclei can be grown on sterile medium and will show a high degree of synchronous mitosis. One difference between the *Physarum* cell cycle and other cell cycles is there is only nuclear division and no cell division in the growing plasmodium. Cell division does occur in other stages of the life cycle (sporulation and sclerotium formation) and these stages may prove useful material for the study of differentiation. There is a recent general review by Rusch (1969) and earlier descriptions of culture methods by Daniels & Baldwin (1964) and Guttes & Guttes (1964). A suggestion made later (p. 248) should be borne in mind when interpreting results from cultures of *Physarum*.

Another naturally synchronous system occurs in spore germination in fungi and bacteria (Sussman & Halvorson, 1966). As with some of the regeneration systems in animals, e.g. mammalian liver, it is uncertain how far these processes are homologous with the normal cell cycle.

Synchrony indices

At intervals in this chapter and elsewhere, I have used the phrase 'degree of synchrony' to describe the efficiency of synchronisation by various methods. The qualitative concept here is that this degree is lowest in a random culture with no synchrony and highest in the perfectly synchronised culture in which all the cells divide at exactly the same time. I have not quantified the phrase largely because very few of the papers on the production of synchrony do so, but this not for want of a suitable index of synchrony. If anything, there are too many rival indices (e.g. Scherbaum, 1963 a; Engelberg, 1964 a; Engelberg & Hirsch, 1966; review and tests by Burnett-Hall & Waugh, 1967). The difficulty here is the separation between theory and practice. Most of those who develop methods of synchrony appear to feel that a graph of cell number increase is an adequate demonstration of the efficiency of their method and do not bother to calculate a synchrony index. Equally, most of the papers on indices do not apply them to the various synchronous systems (an exception is Scherbaum, 1963 a).

At the present time, a comparison can be made on a rough and ready basis between synchrony methods by inspection of the cell number curves, and this is why a number of such curves have been reproduced in this book.

One point about them, however, should be borne in mind. In a synchronous culture, the degree of synchrony falls with time because of the variation in individual cycle times (for the theory of this, see Engelberg, 1964*b*; Burnett-Hall & Waugh, 1967). If division happens very shortly after the culture is set up, there will be an impressively sharp doubling in numbers (e.g. the wash-off synchrony in Fig. 3.10). If, on the other hand, the culture is started with young cells at the beginning of the culture, as in gradient separation, the first doubling will be much less sharp and will take an appreciable amount of time even though the initial degree of synchrony might have been the same in both methods. Another point about comparing methods is that caution should be used in interpreting peaks in mitotic index or division index. A peak value of 50 per cent of cells in mitosis indicates much better synchrony if mitosis only lasts for 1 per cent of the cycle time than it does if mitosis lasts for 10 per cent.

4   DNA synthesis in eukaryotic cells

General patterns of DNA synthesis

The pattern of DNA synthesis in eukaryotic cells is the most deeply explored region of our territory. The importance of DNA as the genetic material of the cell justifies much of the effort that has been spent here, and the work has been made easier by two biological features, that the doubling of DNA only occupies a part of the cycle and that it is exact; and by a technical feature, that newly made DNA can be easily and specifically labelled by tritiated thymidine. Mammalian cells, both *in vitro* and *in vivo*, have been the most favoured material and the number of papers in this field runs into hundreds. I shall only try to discuss some of the general patterns that emerge from this work so that they can be compared with the much less complete picture in the lower eukaryotes and in bacteria. More information can be found in the general reviews mentioned on p. 201, in Lark (1963), J. H. Taylor (1963), John & Lewis (1969), Prescott (1970) and, in particular, in the excellent monograph by Cleaver (1967) which covers a wider field than its title indicates.

Higher animals

In the early fifties, it became apparent from microspectrophotometry (e.g. Swift, 1950; 1953; Walker & Yates, 1952) and autoradiography (Howard & Pelc, 1953) that the DNA of higher cells was doubled during a restricted period of interphase. This period of synthesis (the S phase) was preceded by a gap (the first gap – G1) in which there was no DNA synthesis, and followed by another one (the second gap – G2) in which there was also no synthesis. The G2 phase ran on into mitosis (M) and the end of the cycle. This nomenclature of G1, S, G2, M, introduced by Howard & Pelc (1953), is convenient and fully accepted but needs a few comments. The boundary between G2 and M is often shadowy since the point where prophase starts is difficult to define in some cells, and it may be a matter of taste for the observer. Mitosis is a morphological event and falls into a different category from the synthetic event of the S period. For this reason, several of the methods used to define G1, S and G2 do not give the mitotic period M and it either has to be determined separately or it is included in G1 and G2 (commonly, half of M is included in G2 and the other half in the next G1). The beginning and end of this DNA-nuclear cycle should be taken as nuclear division rather than cell cleavage. This makes little difference in most cases, but it is important in cells like sea urchin eggs or fission yeast where DNA synthesis takes place between nuclear division and cell division.

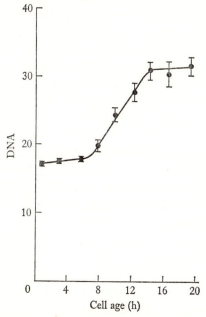

Fig. 4.1. Feulgen DNA content plotted against interphase time in mouse L-929 cells. The points are means of duplicated measurements on 39 to 129 cells. 95 per cent confidence intervals of the mean are shown as vertical bars. From Killander & Zetterberg (1965a).

The DNA cycle of mouse L cells can be taken as a typical and well worked out example of the pattern in mammalian cells. Killander & Zetterberg (1965a) measured the relative amount of DNA in cells at different stages of the cycle using microspectrophotometry and the Feulgen stain. The cycle stage of each cell was found by filming the slide culture and following the history of the cell from its previous division until the moment of fixation. The results in Fig. 4.1 give a G1 of about 8 hours, an S of 6 hours and a G2 of 5 hours. In a second paper, Zetterberg & Killander

(1965 a) used the same method of DNA measurement but another way of finding the length of the cycle stages – by frequency distributions (p. 19). This gave G1/S/G2 times of 9/6/4 hours. Stanners & Till (1960) used the quite different technique of following labelled mitoses in autoradiographs (p. 21) and found an S period of 6–7 hours and a G2 of 3–4 hours. By difference, the G1 was about 10 hours, giving G1/S/G2 of 10/6–7/3–4 hours. Finally, Mak (1965) used two methods on L cells. One was the labelled mitoses technique which gave periods of 8.0/6.2/3.8 hours, and the other was a single slide method involving both autoradiography and Feulgen spectrophotometry which gave similar periods of 8.2/6.4/3.4 hours. These four papers, using a variety of methods, are in substantial agreement in giving for these cells a pattern which is representative for mammalian cells in culture – G1 of 8–10 hours, S of 6–7 hours and G2 of 3–5 hours. The total cycle time is 18–20 hours, and mitosis takes about an hour, as in most mammalian cells. But this is by no means the universal pattern even with one cell line, and two other workers have found much longer S phases with L cells, both in hours and as a proportion of the cycle. Cleaver (1967; see also Dendy & Cleaver, 1964) found periods of 5.7/12.2/4.4 hours from labelled mitoses and 4.0/13.5/4.8 hours from the single slide method mentioned above. Fujiwara (1967) gives similar figures of 5.5/10.5–11/6 hours for L cells synchronised by wash-off (p. 47). These differences may be due to variations in the sub-lines of the L cells or to differences in the media and culture conditions.

A wider picture of the variation in the cycle phases of mammalian cells in culture is shown in the twenty-three cases in Fig. 4.2(a). The majority of S phases last for 6 to 8 hours, but there are cases where synthesis takes twice as long. The distribution is markedly skewed, suggesting that 6–8 hours is not only the most frequent time but also near the minimum time. G2 is less variable and there are few cases where it is less than 2 hours or greater than 6 hours. G1, however, is the most variable of the phases. It can be completely absent, as in the Chinese hamster lung cell strain of Robbins & Scharff (1967), or as large as 16 hours. As a general rule (though not a rigid one) cells which grow fast have short G1 periods. The cells of Robbins & Scharff have a short cycle time of 9–10 hours with phases 0/6–7/1.5–2 hours.

This tendency for cells with different cycle times to have a varying G1 and relatively constant S + G2 has been apparent for some time (e.g. Aoki & Moore, 1970; Cleaver, 1967; Defendi & Manson, 1963) but most of the information has come from surveys of different types of mammalian cells.

What is particularly interesting is that this also occurs when the cycle time of the same cell strain is altered by changing the growth medium, but keeping the temperature constant. Tobey *et al.* (1967 *a*) found that the cycle time of Chinese hamster cells varied from 13 to 24 hours according to the batch of calf serum used in the medium, and this variation was almost entirely in G_1. $S + G_2 + M$ was constant at about 8 hours. Sisken & Kinosita (1961) have show that changing the pH of the medium affects the cycle time of human amnion and kitten lung cells mainly by changing the G_1.

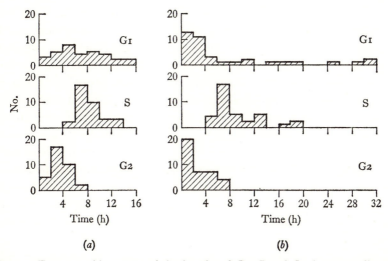

Fig. 4.2. Frequency histograms of the lengths of G_1, S and G_2 in mammalian cells. (*a*) *In vitro*. Results from 15 cases cited by Cleaver (1967, Table 4.2) plus the following papers: Fujiwara (1967), Gaffney & Nardone (1968), Johnson *et al.* (1967), Maekawa & Tsuchiya (1968), Rao & Engelberg (1966), Robbins & Scharff (1967), Watanabe & Okada (1967), Wheeler *et al.* (1967). (*b*) *In vivo*. Results from 38 cases cited by Cleaver (1967, Table 4.3), excluding spermatogonia. There is an uncertainty of about half a mitotic time (*c.* 0.5 h) in the lengths of G_1 and G_2 since in some cases this has been included and in others excluded.

Deliberate control of the phases of the DNA cycle has not yet been exploited in higher cells in the way that it has in bacteria, but it may well be a good way to approach the problem of pin-pointing those events in G_1 which are directly connected with the initiation of DNA synthesis.

The other way of changing cycle times is by altering the growth temperature (Sisken *et al.*, 1965; Rao & Engelberg, 1966; Watanabe & Okada, 1967). The detailed responses vary to some extent between the cell types used, but the important general point is that all the phases of the DNA cycle change and the variation is not restricted to G_1.

Many of the mammalian cell lines that are commonly used originated in tumours but there is no systematic difference in the DNA cycle between tumour cells and normal cells either *in vitro* or *in vivo* (Baserga, 1965; Cleaver, 1967). Nor do tumour cells necessarily have shorter cycle times.

As might be expected, the DNA cycle of mammalian cells *in vivo* has many similarities with the situation in cultures. Considerable emphasis has been put on the constancy of S or S + G2. There is a constant S + G2 of about 7.5 hours in twenty-seven types of mouse and rat tissue cells (Pilgrim & Maurer, 1965). S and G2 are roughly constant (at 5–6 hours and 1.2–1.9 hours) in nine foetal rat tissues, but G1 varies from 5.6 to 31.6 hours (Wegener *et al.*, 1964). Cameron & Greulich (1963) and Cameron (1964) found S periods of 6–8 hours in various mammalian tissues with widely different cycle times, and slightly shorter S periods of 5–6 hours in birds where there is a higher body temperature (40.5 °C rather than 37 °C). There is a striking case in mouse endothelia where the S period is 5–8 hours and G2 is 2.8–4.6 hours, but the cycle time varies from 380 to 8000 hours (Blenkinsopp, 1969). But although 6–8 hours may be the commonest time for the mammalian S period, it is by no means universal and some tissues have much longer S periods. One example is mouse ear skin with an S of 18 hours (Pilgrim *et al.*, 1966) or 34.5 hours (Blenkinsopp, 1968). Part of the reason for this elongated S period may be the lower temperature of the ear, but it is unlikely to be the sole cause since the slowly proliferating epithelium of the mouse forestomach also has a long S period of 13.5 hours (Wolfsberg, 1964). Nor is G2 a constant in all tissues, since Pflueger & Yunis (1966) have found it varying from 0.5–1 hour in lymph nodes to 5 hours in regenerating liver. An overall picture of the position in mammalian tissues is shown in Fig. 4.2(*b*), which is taken from the useful tables in Cleaver (1967). G1 is by far the most variable part of the cycle. In some tissues it may be absent, e.g. dog normoblasts (Alpen & Johnston, 1967), and in others it may extend into days. G2 is the least variable and seldom lasts longer than 6 hours. S has a skewed distribution similar to that in cultured cells but rather broader, with 6–8 hours being the commonest and also nearly the minimum time. The concept of a minimum time for DNA synthesis has been suggested several times. For example, Bresciani (1965) found that injection of ovarian hormones shortened the S period of mouse mammary gland from 20 to 10 hours but massive doses did not reduce it any further. A minimum time, however, can only apply to adult tissues. In late embryos, the S period is somewhat shorter than in adults (Wegener *et al.*, 1964) and increases during development (Kauffman, 1968). In many early embryos, as we shall see, the S period is very much less.

There is much less information about Vertebrates other than mammals. The DNA cycle in bird cells *in vitro* is similar to that in mammals (e.g. Firket & Verley, 1957; Liebecq-Hutter, 1965; Bassleer, 1968) and bird tissues have a slightly shorter S period than mammals (Cameron, 1964, see above). Amphibia, at lower temperatures, have much longer cycles and S periods. Frog kidney cells *in vitro* have $G_1/S/G_2$ phases of $16/22.3/7.7$ hours at 25 °C (Malamud, 1967), and adult newt regenerating lens has phases of $40/19/2$ hours at 22 °C (Zalik & Yamada, 1967). Newt meiosis has a very long S phase of 9–10 days at 16 °C (Callan & Taylor, 1968).

The Invertebrates (excluding Protozoa) are largely untouched apart from a little information about insects – for example, grasshopper neuroblasts (Gaulden, 1956) and epidermal cells in *Oncopeltus* (Lawrence, 1968), both of which lack a G_2 but have a long prophase.

Early embryos

The early photometric work of Lison & Pasteels (1951) suggested that the S period in the early cleavage stages of sea urchin eggs occupied a short time near telophase. This has been confirmed by Hinegardner *et al.* (1964) in *Strongylocentrotus* using more satisfactory labelling techniques (Fig. 4.3). DNA synthesis starts a few minutes after the beginning of telophase and lasts for 13 minutes (at 15 °C). There is then a G_2 of 20 minutes followed by 40 minutes of mitosis, so the whole cell cycle lasts 70 minutes. Most of the S period precedes the completion of the furrow which marks the end of the cell cycle. This emphasises the point made earlier, that the DNA cycle should start and end with nuclear division rather than cell division. If the phases of the DNA cycle were referred to the cell cycle, the sea urchin egg would have a long G_1 and no G_2. But this is obviously misleading since the nucleus spends most of the cycle with a doubled load of DNA The situation in the interval between fertilisation and first cleavage is different from the pattern described above. There is a period of DNA synthesis at about 30 minutes after fertilization (the time of fusion of the sperm and egg nuclei) in preparation for the first nuclear division, and there are G_1 and G_2 phases before and after. The first 'cycle' is not a typical one since it is more realistic to regard it as starting with fertilisation rather than a preceding division. It is 70 per cent longer than the subsequent cycles and contains the unique event of nuclear fusion.

Very similar results have been found in the sand dollar *Echinarachnius* (Young *et al.*, 1969), and the snail *Limnaea* also has a G2, a short S and little or no G1 (Camey & Geilenkirchen, 1970).

Turning to the Vertebrates, there are similar patterns in the early embryos of the Amphibian *Xenopus* (Graham, 1966*a*). DNA synthesis starts at about 15 minutes after fertilisation and finishes at 40 minutes after fertilisation or 20 minutes before first cleavage. In the first typical cycle, the S period starts in telophase and continues through most of the 25 minutes until the second mitosis. This is similar to the sea urchin in showing no G1 but differs in that there is also no apparent G2. A short G2 may

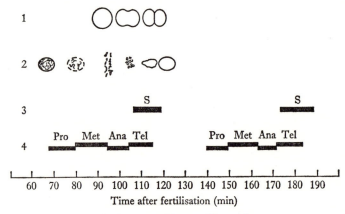

Fig. 4.3. DNA synthesis in the sea urchin egg (*Strongylocentrotus purpuratus*)
(1) Appearance of the egg at the time of DNA synthesis.
(2) Appearance of the nucleus and chromosomes at the beginning of each mitotic phase.
(3) Time and duration of DNA synthesis.
(4) Time and duration of each mitotic phase. From Hinegardner *et al.* (1964).

in fact exist but it would have to last considerably less than 10 minutes and in that case would not be detectable by the labelling methods which were used. The cycle time, however, is shorter in *Xenopus* (*c.* 25 minutes at 20 °C) than in the sea urchin (70 minutes at 15 °C).

Graham & Morgan (1966) have also followed the DNA cycle in the later development of *Xenopus* (Fig. 4.4). At stage 7 (several hundred cells), the cell cycle is even shorter – about 15 minutes – than it is in the first few cleavages. There is no G1, the S period is of the order of 10 minutes and there may or may not be a short G2. Thereafter, all the stages of the cycle increase except that after stage 9 (blastula), the time of mitosis remains constant at about half an hour. The proportionate increase of G1, S and

G2 is the same after gastrulation though the greatest absolute increase is in G2 which becomes unusually long for higher cells. A G1, which is relatively long, also appears in sea urchin embryos by the late blastula stages (Villiger *et al.*, 1970).

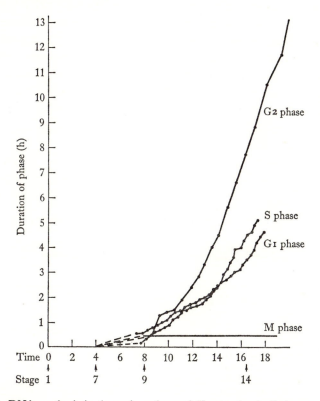

Fig. 4.4. DNA synthesis in the early embryo of *Xenopus laevis*. Estimates of the length of each phase of the DNA cycle in a limited region of the endoderm. From Graham & Morgan (1966).

The cells of early embryos are exceptional in showing division without growth, so it is not surprising that their DNA cycles differ from those of normal cells. Freed as they are from what Daniel Mazia has called 'the friction of growth', most of them divide even faster than bacteria. They demonstrate that neither a G1 nor a G2 phase is a necessary part of the DNA cycle, though the absence of one or other (but not both) also occurs in some adult cells. But the most striking aspect of their rapid divisions is the speed at which they double their DNA. If the adult cells of *Xenopus* have an S period of 20 hours, as in the Amphibian cells mentioned on

p. 63, then the embryonic cells can go through the same process in a time which is less by two orders of magnitude.

Mammals provide a notable exception to this general rule of short cell cycles in early embryos. The cycle time in mice varies between strains but is about 10–20 hours from the second to the seventh cleavage, which is not far short of the times in adult somatic tissues (data from Graham, 1971). The earliest cycles probably lack a G1, as in the other cases discussed above, but it is uncertain at what stage a G1 appears. Gamow & Prescott (1970) argue that it does not occur from the 2- to the 16-cell stage and is probably absent until the blastocyst. Samoshkina (1968) derives a G1 phase of 4–5 hours after the 8-cell stage (but see a criticism in Graham, 1971). The relative lengths of S and G2 are also uncertain, though Gamow & Prescott suggest that G2 is probably short.

Higher plants

Frequency histograms of the DNA cycle in angiosperms are shown in Fig. 4.5, where most of the measurements were made on the proliferating cells in the root tip meristem. There is less information about higher plants than about mammals and these results cover a wider systematic range with a much greater variation in cellular DNA content. The experiments were also done at lower and more variable temperatures. Even so, the most conspicuous point is the similarity between these histograms and those for mammalian cells in Fig. 4.2. The ranges and mean values for the cycle phases do not differ widely from the animal cells despite the fact that the growth temperatures were about 15 °C lower. The main differences are that there is not the predominant 6–8 hour S period in plants, that the G2 is a little more variable, and that the mitotic time is considerably longer than the hour of mammalian cells.

A diagram like Fig. 4.5 may conceal significant relations between the cycle phases. Van't Hof

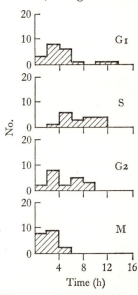

Fig. 4.5. Frequency histograms of the lengths of G1, S, G2 and M in higher plant cells at temperatures between 19 and 25 °C. Results from twenty cases from the following papers: Clowes (1965), Dewey & Howard (1963), Evans & Scott (1964), Howard & Dewey (1960), Howard & Pelc (1953), Prasad & Godward (1965), Van't Hof (1963; 1965a; 1967), Wimber (1960; 1966), Wimber & Quastler (1963).

(1965*b*) has measured the S periods in the root meristems of seven plant species which differ in cycle time and in DNA content. The cycle times increase with DNA content, and this increase is due almost entirely to longer S periods. The time for the other phases of the cycle remains constant at 5.25–6.5 hours. It would be an elegant simplification if this relation was due to a constant rate of DNA synthesis in g/hour. But this is not so, since the rate of synthesis also increases with the DNA content, though not proportionately. The constancy of $G_1 + G_2 + M$ in these experiments does not seem to hold for most of the results in Fig. 4.2 although some of the variation might be due to the different growth conditions and techniques used in other experiments. The work of Clowes (1965), however, shows that within one species (*Zea mays*) the variation in cycle time in different regions of the root is due primarily to changes in G_1. The fast growing cap initials have no G_1 ($G_1/S/G_2/M$ = 0/8/5/2 hours) whereas the slow growing cells in the quiescent centre have a G_1 of 151 hours ($G_1/S/G_2/M$ = 151/9/11/3 hours). This variable G_1 is like the mammalian situation and is in contrast with the results of Van't Hof, though the experiments are not strictly comparable since in one case the variation is intraspecific and on the other it is interspecific with different DNA contents.

Wimber (1966) has examined the effect of temperature changes on the DNA cycle in *Tradescantia* root tips. On raising the temperature from 21 °C to 30 °C, the cycle shortens with the reduction being almost entirely restricted to G_1. On lowering the temperature to 13 °C, all the phases of the cycle are elongated, as in mammalian cells.

Lower eukaryotes

Ciliates. The best known DNA cycle in lower eukaryotic cells is that of the macronucleus of *Tetrahymena pyriformis* in normal unsynchronised cultures.[1] There is general agreement that there is a G_1, an S and a G_2, apart from the early experiments of Walker & Mitchison (1957) who used a relatively insensitive technique, but there are considerable variations in the absolute and relative times spent in these phases. Much of this variation is undoubtedly due to differences in strains, in medium and in temperature, but it should be remembered that there is evidence (p. 215) in the Ciliates of a looser relation than usual between macronuclear replica-

---

[1] Calkins & Gunn (1967), Cameron & Nachtwey (1967), Cameron & Stone (1964), Charret (1969), Cleffman (1967), Mackenzie *et al.* (1966), McDonald (1958; 1962), Prescott (1960), Stone & Prescott (1964), Walker & Mitchison (1957).

tion and the cell cycle. Cameron & Stone (1964) found that strain HSM has cycle phases of $G_1/S/G_2 = 29/59/42$ minutes whereas strain EU 6002 with the same medium and temperature has a G2 which was 50 per cent longer (24/58/68 minutes). As the medium is made poorer, the S phase starts later in the cycle and lasts proportionately longer (Cameron & Nachtwey, 1967). Changes in temperature alter all the phases but experiments differ in the exact details of these alterations. Mackenzie *et al.* (1966) and Cleffman (1967) found that the relative proportion of the cycle spent in each phase remained constant with changing temperature while Cameron & Nachtwey (1967) found that it varied. An interesting result from Cameron & Nachtwey (1967) is that cells grown with different temperatures and media can have the *same* cycle time but *different* DNA phases. HSM cells have the same generation time of 225 minutes in a rich medium at 23 °C and in a poor medium at 29 °C, yet their cycle phases are $G_1/S/G_2/M = 51/39/105/30$ minutes and 100/60/40/25 minutes. As mentioned earlier, these methods of altering the time of initiation of DNA synthesis may provide useful ways of investigating control mechanisms.

The micronucleus starts DNA synthesis at a different time from the macronucleus. McDonald (1962), working with a micronucleate strain, found that there was no G1 in the micronucleus which went into the S period immediately after nuclear division. The macronucleus, on the other hand, showed a definite though variable G1.

A series of careful experiments using different techniques have shown that the macronucleus of *Paramecium aurelia* has a G1 phase which occupies the first half of the cycle and a G2 which is either short or absent (Kimball & Barka, 1959; Kimball *et al.*, 1960; Kimball & Perdue, 1962; Woodard, Gelber & Swift, 1961). The micronuclear S also starts about half way through the cycle but it lasts a shorter time and there is an appreciable G2 (Woodard, Gelber & Swift, 1961). Similar DNA cycles have been found in *Paramecium caudatum* for the macronucleus (Rao & Prescott, 1967; Walker & Mitchison, 1957) and for the micronucleus (Rao & Prescott, 1967). The measurements of Walker & Mitchison, however, suggested that micronuclear S started rather later than macronuclear S.

In *Euplotes eurystomus*, DNA synthesis occurs at two visible 'reorganisation bands' which move inwards from each end of the elongated macronucleus. This S period occupies most of the cell cycle (10 out of 14 hours) but there is a definite G1 of 3 hours (Prescott *et al.*, 1962; Prescott, 1966*b*; Fig. 7.14). Macronuclear division is complete about an hour after the end of the S period. As in other Ciliate macronuclei, there can be no sharp separation of G2 and M because there is no mitosis. G2 is either absent in *Euplotes* or it is part of the preliminaries of the amitotic splitting. The DNA cycle in the micronucleus is quite different. The S period starts immediately after mitosis and is complete before the beginning of the macronuclear division, so the cycle here is no G1, a short S and a long G2.

DNA synthesis in the macronucleus of *Stentor coeruleus* occupies most of the cell cycle of 1–2 days (Guttes & Guttes, 1960; De Terra, 1967) but there is a short variable G1 ranging from 2 to 12 hours. The division of the complex macronucleus lasts 8 hours and the S period runs into this, so there is no apparent G2, as in *Euplotes*.

Blepharisma (*americanum* and *intermedium*) has a relatively longer G1 with cycle phases G1/S/G2 = 17/7/1–2 hours (Minutoli & Hirshfield, 1968). *Urostyla weissei* has a macronuclear cycle G1/S/G2/Div. = 5–6/4/ 111.5 hours and a micronuclear cycle with no G2, and S taking place during the late S + G2 phase of the macronucleus (Jerka-Dziadosz & Frankel, 1970). *Stylonichia notophora* has somewhat similar macronuclear cycle with S lasting from 0.20 to 0.95 of the cycle, and with the micronuclear S happening in the middle of the macronuclear S (Sapra 1968, cited by Jerka-Dziadosz & Frankel, 1970). *Stylonichia mytilus* has a micronuclear cycle G1/S/G2 = 8–9/2/0 hours similar to *Urostyla* but different from *Euplotes* (Ammerman, 1970).

In summary for the Ciliates, the DNA cycle of the macronucleus shows an S period which varies from 30 per cent to 90 per cent of the cycle. There is always a G1 but a G2 is short or absent in the larger slow-growing species. The DNA cycle of the mitotic micronucleus is always different from that of the macronucleus. The S period is usually shorter, and in two cases (*Tetrahymena* and *Euplotes*) there is no G1.

Amoeba. *Amoeba proteus* has a DNA cycle without a G1. Prescott & Goldstein (1967) found an S period of 3–6 hours followed by a G2 of 30 hours or more. Ord (1968) investigated this in greater detail and found two waves of synthesis. 75 per cent of the DNA was synthesised in the first 6 hours after division with a peak between 0.5 and 4 hours, and another 15 per cent of the DNA was synthesised between 9 and 13 hours. Thus

90 per cent of the DNA was synthesised within the first quarter of the cycle (total length 48–54 hours) and the remaining 10 per cent later. Ron & Prescott (1969) found a substantially similar result except that delayed synthesis only took place in some of their cells.

Physarum. *Physarum polycephalum* is another lower eukaryote where there is no G1 (Nygaard *et al.*, 1960). The maximum rate of DNA synthesis is reached in more than 99 per cent of the nuclei of a plasmodium within 5 minutes of the uncoiling of the telophase chromosomes (Braun *et al.*, 1965). This rate is maintained for 1.5 hours and then slowly decreases to the low premitotic level at 4 hours after mitosis. During this period of decreasing rate there are differences in the rate of synthesis between individual nuclei, as judged by levels of labelling after tritiated thymidine pulses. After the end of this 4 hour S period, there is a G2 of 4–6 hours.

Yeast and other cells. Williamson (1966) has shown by autoradiography and chemical measurements in synchronous cultures that the DNA cycle in the budding yeast *Saccharomyces cerevisiae* has G1, S and G2 periods of approximately equal length in the 2 hour cycle time. Care should be taken here in discussing cycle periods since the nuclear cycle is almost exactly out of phase with respect to the cell cycle. The final separation of mother and daughter cell takes place nearly half a cycle after nuclear division.

The fission yeast *Schizosaccharomyces pombe* has a different pattern (Bostock *et al.*, 1966; Bostock 1970*a*). There is no G1, a very short S of the order of 10 minutes and a G2 of 2.5 hours.[1] Since there is a gap of about a quarter of the cycle between nuclear division and cell division, DNA synthesis happens between these two processes, as in sea urchin eggs. This DNA cycle can be altered by growing the cells in the presence of phenylethanol (Bostock, 1970*b*) or high glucose concentrations (Duffus & Mitchell, 1970). A G1 period is then inserted and the S period takes place at a different position in the longer cell cycle.

Kessel & Rosenberger (1968) have examined the DNA cycle in the multinucleate hyphae emerging from the germinating conidiospores of *Aspergillus nidulans*. In fast growth (doubling time of 87 minutes), the nuclei have synchronous S periods lasting about 20 minutes. At slower growth rates, the S periods remain the same length but are asynchronous between nuclei in the same hypha. It appears that these nuclei have variable G1 and G2 periods even though they are in the same cytoplasm.

[1] Other strains of *S. pombe* appear to have a slightly later S period and a short G1 (Mitchison & Creanor, 1971*b*).

Early results in the synchronous cultures of *Chlorella pyrenoidosa* used by Schmidt and his colleagues (p. 44) showed continuous synthesis of DNA through the cycle. Recent results, however, with improved methods show a pattern of periodic synthesis more like that in other eukaryotes (Hopkins *et al.*, 1970). The S period is short and takes place towards the end of the cycle. It appears like a cycle with a long G1 and a short G2, but this may not be so because of uncertainty about the exact timing of the three successive nuclear divisions that precede the release of each of eight daughter cells.

Cytoplasmic DNA

The DNA that has been considered up till now is that in the nucleus. In recent years, however, it has been clearly established that a small proportion of the cellular DNA is in the cytoplasm, in mitochondria and chloroplasts. This raises the question of whether this DNA is under the same controls as the nuclear DNA and replicates at the same time. Although the evidence is thin, it all points the same way – that the pattern of replication is different. Mitochondrial DNA synthesis is continuous throughout the cycle in *Physarum* (Guttes *et al.*, 1967; Braun & Evans, 1969; Holt & Gurney, 1969). It is also continuous in *Tetrahymena* though the rate is higher in the nuclear S periods (Cameron, 1966; Parsons & Rustad, 1968). The maximum incorporation of thymidine into mitochondrial DNA in mammalian cells is in G2 (Koch & Stokstad, 1967). The evidence in *Saccharomyces* is conflicting. Williamson & Moustacchi (1971) showed by two techniques that mitochondrial DNA synthesis is continuous, whereas the results of Smith *et al.* (1968) and Cottrell & Avers (1970) suggest that it is periodic, though at a different point in the cycle from the nuclear S period. In all three cases, however, the pattern of synthesis of mitochondrial DNA is *not* the same as that of nuclear DNA. The same is true for chloroplast DNA in *Chlamydomonas* after induction synchrony when the two periods of synthesis are almost exactly out of phase (Chiang & Sueoka, 1967). There are two peaks of incorporation into chloroplast DNA in *Euglena* after induction synchrony, one at the time of nuclear synthesis but the other, sharper one at another time in the cycle (Cook, 1966).

These cytoplasmic organelles have similarities with prokaryotes in a number of features, including their DNA (Roodyn & Wilkie, 1968; Nass, 1969) so it is not altogether surprising to find their periods of synthesis differing from the eukaryotic nuclei. But there is an obvious interesting gap in the present evidence. One explanation of continuous synthesis of total mitochondrial DNA is that there is continuous synthesis in each mito-chondrion, as in fast-growing prokaryotes. The other explanation is that each mitochondrion has a restricted period of synthesis but these periods are asynchronous over the whole mitochondrial population. If mitochondria divide in two, these divisions must normally have the same frequency as the cell cycle but they do not need to be in phase either with each other or with the cell cycle. This problem is discussed in a later chapter (p. 182).

Conclusions

An important generalisation is that DNA synthesis in eukaryotes is periodic and does not occupy all the cell cycle. The only exception may be in the very rapidly dividing cells of early Amphibian embryos.

In mammalian tissues and cell cultures, there is a tendency for the S period to last 6–8 hours and also for the G2 period to be rela-tively constant. As a result, the main difference between fast and slow growing cells is in the length of the G1 period, which can vary from zero to many hours. But although this is the predominant pattern, there are many exceptions. The S period is seldom less than 6 hours but there are a number of cases in which it is much greater (e.g. in skin).

The ranges of G1, S and G2 are similar in higher plants (root meristems) though there is not the same tendency for a 6–8 hour S and there is not sufficient evidence to say whether or not G1 is the most variable phase. This similarity reflects the fact that the total cycle time in these plant cells is of the same order (10–30 hours) as that of mammalian cells even though their growth temperature is about 15 °C lower.

There are very much shorter cycle times in the cells of most early animal embryos, which divide rapidly but do not grow. The S periods are correspondingly shorter, and the DNA of Amphibian nuclei is synthesised about a hundred times faster in early embryos than in adults. There is no G1 in these early cycles.

Data on the lower eukaryotes is scanty, but it seems that the DNA cycle in the micronucleus of Ciliates is always different from that of the macronucleus, and that there is little or no macronuclear G2 in the larger species. Cycles without a G1 are quite common and are not restricted to fast growing cells. They occur in *Amoeba*, *Physarum*, *Schizosaccharomyces*, and in the micronuclei of *Euplotes* and *Tetrahymena*.

The patterns of synthesis of cytoplasmic DNA in mitochondria and chloroplasts are different from those of the nuclear DNA.

DNA synthesis at the chromosomal level

Since the synthesis of DNA occupies a restricted part of the cell cycle, its specific rate of synthesis will in general be higher than that of other bulk components of the cell such as total dry mass, total protein or total RNA, all of which tend to be made continuously through the cycle and without large changes in rate (though there are exceptions). Can we make any other generalisation about the rate of synthesis? Any answer to this question should be deferred until later because it must take into account the evidence that we have about the fine structure of replication at the chromosomal and molecular level. The importance of this can be seen in a simple but misleading model. In *Escherichia coli* in glucose medium, a replicating fork takes 40 minutes to encompass the chromosome and complete a round of DNA synthesis. If there were one replicating fork traversing each human chromosome at the same rate, the S period would last 13 hours since, on average, a human chromosome has about 20 times as much DNA as that of *E. coli*. This S period would be within the normal range, and we might be tempted to take this model seriously, but in fact the mode of replication is quite wrong. As we shall see, there are many replicating forks in each chromosome, they start at different times in the S period, and they move more slowly.

Tritiated thymidine has been as powerful a tool in following DNA synthesis in individual chromosomes as it has been with the nucleus as a whole, and it has been correspondingly exploited with a wide variety of animal and plant cells.[1] In essence, the experimental plan is to label growing cells with a pulse of tritiated thymidine, fix them at varying times

[1] Some references (but by no means a complete list) are: Cave (1967), Evans (1964), Evans & Rees (1966), Froland (1967), Gavosto *et al.* (1968), German (1964), Gilbert *et al.* (1965), Graves (1967), Grumbach *et al.* (1963), Hsu (1964), Lima-de-Faria (1959), Lima-de-Faria *et al.* (1964), Martin (1966), Morishima *et al.* (1962), Peacock (1963), Petersen (1964), Plaut (1963), Plaut *et al.* (1966), Schmid (1963), Schneider & Rieke (1967), Stubblefield (1965), Stubblefield & Mueller (1962), Taylor (1960*b*), Wimber (1961). There are reviews by Cleaver (1967), Hsu *et al.* (1964), Huberman (1967), Lima-de-Faria (1969), and J. H. Taylor (1963).

thereafter, and then examine autoradiographs of labelled metaphase chromosomes (which may have been accumulated by colchicine). If the time interval is a little more than the length of G2, the labelled regions of the chromosomes will be those where DNA was being synthesised in late S. If the time interval is equal to G2 plus most of S, the labelled regions will have been synthesised in early S. With the model outlined above, each chromosome would have one small region of labelling irrespective of the

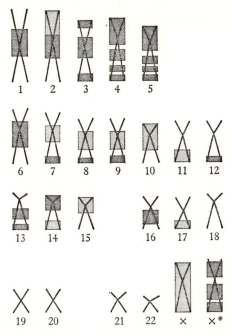

Fig. 4.6. Late replicating segments of human chromosomes. Big dots indicate very high synthetic activity, and small dots indicate high synthetic activity. From Gavosto *et al.* (1968).

time interval. What happens in practice is much more complicated. The labelling in many cases covers substantial regions of the chromosome, indicating a number of replicating sites. There are also differences in the labelling pattern during the S period. Whole chromosomes as well as regions of chromosomes are 'late-replicating' and show DNA synthesis only in the later parts of the S period. This asynchrony in the initiation and completion of synthesis is not a random process which varies from cycle to cycle. It has a remarkably consistent pattern within one cell type which can even be used to identify particular chromosomes (Fig. 4.6).

There is also evidence in some but not all cases, that it is the same between different somatic cell types in one species (Gavosto *et al.*, 1968; Martin, 1966).

Two recent analyses (Gilbert *et al.*, 1965; Gavosto *et al.*, 1968) show that there is not a sharp time transition between early and late-replicating regions in the sense that one stops before the other begins. Rather, the late-replicating regions start later and show higher rates of synthesis than the other regions in late S, but finish at about the same time. The higher rates of synthesis are per unit length of chromosome and they do not of course imply that the molecular rate is any higher – there may simply be more replicating forks in action in those regions. The detailed relations, however, of the initiation and completion times certainly vary between different organisms. In *Drosophila* salivary chromosomes, for example, the different regions appear to start at the same time and finish at different ones (Plaut *et al.*, 1966), and Lima-de-Faria (1969) list a number of cases where late-replicating sex chromosomes finish later than autosomes.

The heterochromatic regions of the chromosomes, which remain condensed and more easily stainable during interphase, are always late-replicating. This relation, which was first discovered by Lima-de-Faria (1959) in grasshopper spermatocytes and rye leaves, holds good in many kinds of cell (reviewed by Lima-de-Faria, 1969) and is particularly important because of the evidence from genetics and RNA labelling that heterochromatin is comparatively inert in gene expression (Brown, 1966). A good example is that one of the two X chromosomes in a number of female mammals is late-replicating, heterochromatic (stainable as the 'Barr body') and is thought to be genetically inert in somatic cells. The other X chromosome is euchromatic and replicates earlier. Although heterochromatin is always late-replicating, the reverse relation does not hold since there are many examples of late-replicating regions which are not visibly heterochromatic. On the other hand, heterochromatin, which is not a precisely defined entity, would be difficult to detect in small amounts.

There has been a good deal of interest in recent years about a fraction of the nuclear DNA of higher organisms which has highly repetitive sequences and which is often separable on density gradients as a 'satellite' with a different density from the main band of DNA (reviewed by Walker, 1971). The function of this satellite DNA is at present unknown, but there is increasing evidence that it is a component of heterochromatin and is located, at any rate in part, near the centromeres in mitotic chromosomes and in the nucleoli in interphase cells (Jones, 1970; Pardue & Gall, 1970;

Schildkraut & Maio, 1968; Yasmineh & Yunis, 1970). In most cases it resembles the rest of the heterochromatin in being late-replicating, though there is some dispute about when its synthesis stops in the S period (Bostock & Prescott, 1971*a*; Flamm *et al.*, 1971; Tobia *et al.*, 1970; but see also Comings & Mattoccia, 1970). However, in the first cycle after infection of mouse kidney cell with polyoma virus, the satellite DNA is synthesised before the main band DNA (Smith, 1970).

The late replication of heterochromatin suggests that the DNA synthesised in late S might be less important for cell survival than that synthesised in early S. There is evidence for this from the experiments of Kajiwara & Mueller (1964) who found that short pulses of BUDR (5-bromouracil deoxyriboside) incorporated into the DNA of HeLa cells caused a greater decrease in cloning efficiency when they were given in early S than in late S. On the other hand, nucleolar DNA, which probably includes the active genes for ribosomal RNA, is late replicating in some cases (e.g. Charret, 1969).

The work in this field has firmly established the existence of multiple replication sites and asynchrony of replication and has also raised several important questions. Perhaps the most interesting is how far conversion to heterochromatin is used to control gene expression and whether this can be recognised in late-labelling patterns. There are, for instance, differences between the amount of heterochromatin in adult somatic cells and that in early eukaryotic and germ-line cells (Kinsey, 1967; Utakoji & Hsu, 1965). Unfortunately, present techniques are severely limited by the resolving power of tritium in autoradiography which prohibits any fine mapping of late-labelling on metaphase chromosomes. The longer the chromosome the higher the resolution, so the precision of the methods would be greatly increased if it were possible to apply them to prophase chromosomes.

DNA synthesis at the molecular level

The patterns of chromosome labelling show that there are multiple replication points but they do not give any estimate of the number of points or of the mode of replication. For this we have to turn to methods which have their origin in the molecular biology of bacteria.

Painter *et al.* (1966) labelled the DNA of HeLa cells with a short pulse of a radioactive label (tritiated thymidine) followed by a longer density label of BUDR. The DNA was then extracted, sheared into fragments small compared to the chromosome, and banded on a density gradient. If there was one or a small number of replicating units per chromosome, there would be little association between radioactive fragments and heavy fragments. But there would be a high degree of end-to-end association of these fragments if there were many units of the same order of size as the fragments. There was in fact a significant degree of association, from which it can be estimated that the number of replicating units per cell at *any one time* in the S period is between 10^3 and 10^4. Later experiments compared the situation in slow-growing rabbit cells with an S period of about 25 hours and in four fast-growing mammalian cell lines with S periods of about 6 hours (Painter & Schaefer, 1969). The speed of replication of the DNA molecules was the same in all the cell strains (in the range 0.5–1.8 μm/min). The slow-growing cells must therefore have a smaller number of replicating units ($c.\ 4 \times 10^3$) in action at one time compared to those in the fast-growing cells (more than 10^4). These experiments, however, do not distinguish between two alternatives: that the replicating units are longer in the slow-growing cells, or that they are the same length but fewer of them are initiated at any one part of the S period. A similar replicating speed of 1–2 μm/min was found in Chinese hamster cells by Taylor (1968) using a density labelling technique.

Another method is that of autoradiography of individual DNA molecules which was developed in the now classical work of Cairns (1963) on the bacterial chromosome. Cairns himself (1966) found that long DNA fibres from labelled HeLa cells showed a series of replicating regions with a speed of replication of 0.5 μm/min or less. But the most detailed study is in an important paper by Huberman & Riggs (1968) on Chinese hamster and HeLa cells, synchronised by treatment with fluorodeoxyuridine and thymidine release (p. 27). The DNA was labelled with tritiated thymidine in pulse and pulse-chase experiments, and then extracted for autoradiography. The autoradiographs showed long fibres which were very probably single DNA double helices (or twin after replication) with many replication sections joined in series (Fig. 4.7). The replication was at fork-like growing points, visible in some of the autoradiographs though not in Fig. 4.7. In a pulse–chase experiment, the activity of the DNA synthesised in the early part of the chase decreased gradually, due presumably to a slow reduction in the specific activity of the precursor pool. This produced a gradient of grain density which enabled the direction of replication to be determined (Fig. 4.7). Surprisingly, each replicating section appeared to have two growing points moving in opposite directions from a common origin. The

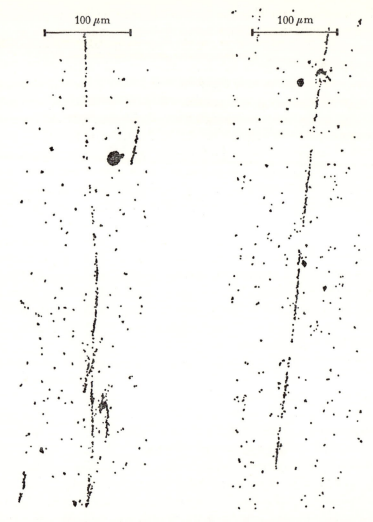

Fig. 4.7. Autoradiographs of replicating sections in series. DNA from Chinese hamster cells after pulse label with ³H-thymidine and chase. From Huberman & Riggs (1968).

commonest length of section was about 30 μm but many of them were shorter and there were some that were so short that they were difficult to resolve. Neighbouring sections could begin replication at different times, and, at any one time, the regions of synthesis were not distributed equidistantly along the fibres. The best estimates of the speed of replication were 0.5–1.2 μm/min, but there was probably a range of speeds. There was not however any great change in the speed during the S period. This speed, which is in agreement with that found by Painter & Schaefer, is much slower than the speed in *E. coli* (30 μm/min). Finally, the fibres were resistant to pronase which suggests that there are no protein linkers joining the sections (cf. Macgregor & Callan, 1962).

These results lead to a bidirectional model for DNA replication which is shown in Fig. 4.8. It includes the assumption that the termini are points where replication stops rather than points where converging growing points meet.

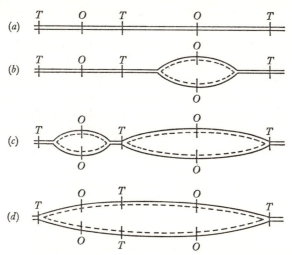

Fig. 4.8. Bidirectional model for DNA replication. Each pair of horizontal lines represent a section of a double helical DNA molecule containing two polynucleotide chains (————, parental chain; - - - - - -, newly synthesised chain). The short vertical lines represent positions of origins (*O*) and termini (*T*). The diagrams represent different stages in the replication of two adjacent replication units; (*a*) prior to replication; (*b*) replication started in right-hand unit; (*c*) replication started in left-hand unit and completed at termini of right-hand unit; (*d*) replication completed in both units; sister double helices separated at the common terminus. From Huberman & Riggs (1968).

One further question arises from this work – what is the total number of replicating units? Any estimate must be very tentative because of the varying lengths of the sections, and the authors do not commit themselves in the paper. But it may be worth doing a piece of rough arithmetic. If the DNA content of the Chinese hamster cell is taken as 1000 times that of the *E. coli* genome, the total length of DNA will be 10^6 μm. There would then be 33 000 units if each were 30 μm long, but since many of them are smaller, a figure of very approximately 50 000 would be a better estimate.

The same technique has recently been applied to Amphibian cells by Callan *et al.* (1971). In the somatic cells of the newt *Triturus*, which have an S period of about 24 hours at 25 °C, the replication speed is 18–24 μm/h and the replicating units are upwards of 100 μm in length, and bidirectional as in mammalian cells. In the same species, the speed of replication is 12 μm/h at 18 °C in the meiotic S period which takes 8–9 days, nearly ten times as long as in the somatic cells. Bearing in mind the temperature difference, these two speeds of replication are very similar, as in the slow- and fast-growing mammalian cells of Painter & Schaefer. In the newt meiotic cells, however, it is clear that the slow rate of overall DNA synthesis is mainly due to the replicating units being much *longer* than those in somatic cells.

Chromosome labelling has shown that there is a sequential and heritable pattern of DNA synthesis in different regions. This can also be shown at the molecular level. Braun & Wili (1969) labelled a synchronous plasmodium of *Physarum* for a small fraction (about 20 per cent) of an S period with tritiated thymidine and then for the same fraction of the next S period with a density label. After extraction, the radioactive fraction of the DNA coincided on a density gradient with the heavy fraction. The experiment was repeated at different positions within the S period and gave the same result. Thus that part of the DNA replicated at one time during the S period is replicated at the *same* time in the next cycle. A similar result was found with HeLa cells by Mueller & Kajiwara (1966*a*).

The view of chromosome replication from the work outlined in the last two sections is of long DNA molecules containing a very large number of replicating units. The units are necessarily short, so each of them completes its replication in a fraction of the S period – even though the rate of replication (in um/min) is an order of magnitude smallerthan in bacteria. Throughout the S period there must be a sequence of some units starting replication while others stop, and this sequence, at the chromosomal level, appears to be genetically determined. The bidirectional mode of replication is different from that in bacteria.

Rate of DNA synthesis

We can detect some uniformities in the time period during which cells synthesise DNA, for instance the fact that it is almost always shorter than the cell cycle, and it is reasonable to ask whether there are similar uniformities in the rate of synthesis during this period. Unfortunately the picture is not very clear. One reason for this is the inadequacy of much of the evidence. Absolute measurements of DNA quantity are insensitive and rate measurements from thymidine pulses are open to objections of changing pool size. A crucial experiment would be a series of thymidine pulses closely spaced through the S period and accompanied by parallel measurements of pool kinetics of the kind made by Cleaver & Holford (1965) but at different times in S.

Within the limitations of the existing evidence, the commonest pattern is one of a constant rate of synthesis during most of the S period.[1] But there is no doubt that other patterns can occur. The rate may rise during the S period (Dendy & Cleaver, 1964; Evans, 1964) or it may fall during the later stages (Braun *et al.*, 1965; Odell *et al.*, 1968; Wimber, 1961). The fastest rate can be in mid-S (Terasima & Tolmach, 1963*a*) but there are also cases where the reverse appears to happen and there is a sharp fall in rate or even a gap in synthesis during the S period.[2] In these cells that have a split S period, it would be interesting to know how much of the genome replicated in the second phase was heterochromatic or functionally inert. Not surprisingly, the method of synchronisation can affect the rate of synthesis in synchronous cultures. Mouse L cells have a slow initial rate at the start of the S period after synchronisation by treatment with excess thymidine, whereas the same cells show a constant rate through the whole S period after synchronisation with aminopterin or when unsynchronised (Adams, 1969*b*).

It would be reasonable to predict a constant rate of synthesis in the simplest of bacterial systems with only one replicating fork moving round a long chromosome. What is surprising is to find it as the common pattern in eukaryotes where there are variations within one nucleus in the size of the replicating units, in the speed at which they replicate and in the time at which they initiate replication. The mechanisms which regulate these variations are unknown but it would seem that a profitable approach might be to extend the molecular autoradiography of Huberman & Riggs (1968) and Callan *et al.* (1971) into early development where there are large changes in the length of the S period. There are interesting implications in a situa-

---

[1] In mammalian cells: Alpen & Johnston (1967), Edward *et al.* (1960), Schwarzacher & Schnedl (1965), Stanners & Till (1960), Zetterberg & Killander (1965*a*). In Ciliates: Kimball & Barka (1959), Prescott (1966*b*), Woodard, Gelber & Swift (1961).

[2] Cohn (1968), Hamilton (1969), Howard & Dewey (1961), Kasten & Strasser (1966), Klevecz (1969*b*), Ord (1968).

tion like the Amphibian embryo where the total DNA is synthesised 100 times faster than it is in adult tissues. This could be due to an increase in the speed of the replicating forks. The polymerase could be different, the intimate structure of the chromosome might change, or there could be a faster supply of precursors from the large embryonic pools. Another way of increasing the speed of overall synthesis would be to eliminate the asynchrony of initiation and to have all the replicating units starting together. If so, the controls which determine the sequence of initiation in adult cells would have to be overridden. Taken together, these two possibilities could give very fast synthesis. If the replicating units in mammalian cells started simultaneously and the forks moved at bacterial speed, the S period would take about one minute. In view, however, of the results of Callan *et al.* (p. 80), the most likely possibility is that the size of the replicating unit is reduced, though if this alone were responsible for the short S periods in early embryos, the units would be only a few microns long. It would, of course, imply that the origins and termini of the units could be changed in an orderly way during development.

One further point about the situation in early embryos is that the initial stimulus for rapid DNA synthesis can come from the cytoplasm. Graham *et al.* (1966) have shown that if gastrula nuclei are transplanted into Amphibian eggs, 50 minutes residence there induces them to complete their cell cycle in 14 minutes where it previously took 12 hours.

Another question that can be asked about the rate of DNA synthesis is whether the S period of a cell is changed if the DNA content is altered. The results of Van't Hof (1965*b*), mentioned earlier (p. 67), do show changing S periods but the comparisons were made between different plant species. A more stringent test is to compare cells of the same species but different numbers of chromosomes. This has been done in plants (Friedburg & Davidson, 1970; Troy & Wimber, 1968), in Amphibian embryos (Graham, 1966*b*), in ascites tumour cells (Oehlert *et al.*, 1962) and in *Tetrahymena* (Cameron & Stone, 1964) with the same result – the S period remains the same with different degrees of ploidy. The absolute rate of DNA synthesis doubles between a diploid and a tetraploid but the specific rate of synthesis remains the same. This is not a very surprising result. One set of chromosomes is the same as the other in the patterns of initiation and it would be reasonable to expect the same time of synthesis provided there was double the supply of precursors. An increased supply of precursors should be available since polyploid cells are usually larger than diploid ones. On the other hand, cycle times are not necessarily the same. Haploid Amphibian embryos, for example, divide faster than diploids in the period from 4 to 12 hours after fertilisation (Graham, 1966*b*). One might therefore expect some deviations from the pattern of a constant S period in polyploids, depending on cycle time and cell size, and these in fact occur. Tetraploid *Phalaris* root cells have nearly twice the S period of the diploids, as well as a longer cycle time (Prasad & Godward, 1965). Alfert & Das (1969) also found that the S period in liver and plant cells increased by about 50 per cent when the DNA content was double (in tetraploids and in binucleate diploid cells).

It is usually assumed that the rate of nuclear DNA synthesis is zero outside the S period. Certainly it is true that most of the DNA is made during this period but it is also possible that a small amount of synthesis goes on at other times in the cycle and that the precise limits of S are difficult to define. In a number of labelling experiments (e.g. Ord, 1968) there is a low level of incorporation outside the main S period, but it is difficult to exclude factors such as cytoplasmic labelling, adsorption and conversion of label. Some of the synthesis may also be repair rather than net increase. Recent papers, however, by Holt & Gurney (1969) and Braun & Evans (1969) describe a fraction (about 1 per cent) of the nuclear DNA in *Physarum* which is synthesised throughout G2, is separable from the rest of the nuclear DNA on a density gradient, and may be associated with the nucleolus (Guttes & Guttes, 1969). It would be interesting to see if there is a similar fraction in other cells which is divorced from the normal controls of nuclear DNA synthesis.

Initiation of DNA synthesis

Relation of nucleus and cytoplasm

The synthesis of nuclear DNA is a discontinuous process in eukaryotes and one of the key questions in cell biology is how it is initiated at a particular point in the cell cycle. We do not know the full answer to this question in either eukaryotes or prokaryotes but several lines of evidence in recent years provide important clues.

One problem in this field is whether the signal for initiation comes from the cytoplasm (reviewed by De Terra, 1969). In situations where cells are stimulated to grow and divide by an external inducer, for instance hormone action on an animal tissue, the problem is trivial since the signal must come from or through the cytoplasm. In the cell cycle, however, the problem is a more important one since, ideally at any rate, the cell is a closed system from the point of view of cycle stimuli.

Multinucleate cells occur frequently among the lower eukaryotes and less frequently among the higher ones. If the nuclei in these cells are similar, they almost always divide synchronously though there are some exceptions (e.g. Evans, 1959; Hegner & Wu, 1921). A similar generalisation applies to the synchrony of the S periods although the evidence is less. Binucleate cells show synchronous DNA synthesis in mouse embryo cultures and endosperm (Church, 1967), in bean root tips (Howard & Dewey, 1960) and in double animals of *Euplotes* (Kimball & Prescott, 1962). Large multinucleate cells behave in the same way in *Physarum* (Braun *et al.*, 1965) and in ascites tumours (Oehlert *et al.*, 1962). But there are exceptions, quite apart from the two kinds of nuclei in Ciliates which are formally excluded from this generalisation because they are dissimilar. Asynchronous S periods have been found in the vitellarium nuclei of rotifers (Birky *et al.*, 1967), in *Aspergillus* (Sandberg *et al.*, 1966), and in some binucleate tumour cells.

This tendency for nuclei in the same cytoplasm to have the same S period suggests that there is a common initiating signal originating in the cytoplasm. It is, however, a rather weak argument because the signal might be one which initiated mitosis, in which case the synchronous S periods would be simply the result of synchronous mitoses.

Much stronger arguments come from experiments on nuclear transplantation and cell fusion. Gurdon and his colleagues (Graham *et al.*, 1966; reviewed by Gurdon & Woodland, 1968) have shown that nuclei from adult liver, brain and blood cells swell very considerably when injected into frog egg cytoplasm and then start DNA synthesis. Since few, if any, of these cells normally synthesise DNA, there must be a stimulus from the egg cytoplasm. This stimulus is not even Class specific, let alone species specific, since more than 90 per cent of injected *mouse* liver nuclei synthesise DNA in *frog* eggs. Similar results, though with a longer time scale, have been obtained by Harris and his colleagues with virus-induced cell fusion (Harris, 1967; reviews, 1966; 1970). Nuclei which do not normally synthesise DNA, such as those of macrophages or erythrocytes, are induced to synthesise DNA and then divide when fused with actively growing cells. Again, the stimulus is not Class specific, since hen erythrocyte nuclei are stimulated after fusion with human HeLa cells. The stimulus can even operate *in vitro* since Thompson & McCarthy (1968) have found that isolated hen erythrocyte nuclei will synthesise DNA if they are treated with extracts of the cytoplasm of mouse L cells.

The effects of cytoplasm on nuclear synthesis have been resolved in greater detail with respect to the cycle in experiments using nuclear transplants, grafting or fusion. Both in *Physarum* (Guttes & Guttes, 1968) and in *Amoeba* (Ord, 1969 – though different results have been reported by Prescott & Goldstein, 1967), it appears that G2 nuclei do not synthesise DNA when transferred to S period cytoplasm, and that S nuclei continue to synthesise DNA when in G2 cytoplasm. The effect on the S nuclei shows that there is not an inhibitor present in the G2 cytoplasm, and that once the nuclei have started synthesis they will continue in a different cytoplasm. This capacity of S nuclei to continue synthesis has also been shown *in vitro* with nuclei isolated from *Physarum* (Brewer & Rusch, 1965) and from mammalian cells (Friedman & Mueller, 1968). The effect on the G2 nuclei suggests either that the initiating stimulus is short-lived and is only present, if at all, right at the start of S (which is at telophase in *Physarum* and *Amoeba*), or that the G2 nuclei which have already completed replication cannot be stimulated again before division. The second alternative seems the most likely in view of the results of Rao & Johnson (1970) with virus-induced fusion of human HeLa cells, which have a G1. When G1 cells are fused with S cells, the G1 nuclei start synthesis earlier than normal, and the more the proportion of S nuclei, the earlier is the initiation in the G1 nuclei. With G2 and S cells, the results are similar to those with *Physarum* and *Amoeba*. So with all three types of cell we can postulate a substance present in S cytoplasm which will initiate DNA synthesis in G1 nuclei but not in G2 nuclei. Once synthesis has been initiated it will continue in the absence of the substance.

The experiments of Rao & Johnson (1970) also throw some light on the control of mitosis and division. Fused cells with the composition G2/S and G2/G1 divided earlier than S/S or G1/G1. The presence of G2 nuclei and cytoplasm accelerated the progress of the S and G1 nuclei and, the more the number of G2 nuclei, the greater was the effect. Conversely, in S/G2 cells, the G2 nuclei were delayed in comparison with those in G2/G2 cells, and, the more the number of S nuclei, the greater was the effect. This is all in accord with the concept of mitotic inducers which have to reach a critical level to be effective, or of the 'division proteins' which are discussed in Chapter 10.

The grafting experiments of De Terra (1967) with the big Ciliate *Stentor* show a pattern with G1 nuclei which is similar to that in HeLa cells. There are, however, some differences with S and G2 nuclei. S nuclei stop synthesis when transferred to G1 cytoplasm, which suggests that in this cell a stimulus is required throughout the S period. Dividing nuclei, which are effectively in G2, are stimulated into further DNA synthesis by transfer into S cytoplasm. This may be because the ciliates seem to have a looser link than usual between the cell cycle and macronuclear replication, and there are other cases where there is more than one S period before division (p. 215).

Relation of protein synthesis and DNA synthesis

There is ample evidence in higher eukaryotes that there is a connection between protein synthesis and DNA synthesis. If cells are treated with inhibitors of protein synthesis such as puromycin or cycloheximide, DNA synthesis either stops or is substantially reduced.[1] The effect is more marked at the beginning of the S period than it is later in S (Kim *et al.*, 1968; Mueller *et al.*, 1962; Taylor, 1965) and it has been suggested that protein synthesis is more important for the initiation than for the continuation of DNA synthesis. It could be that a protein initiator is needed to start a DNA molecule replicating, but, once started, replication could continue without protein synthesis – as happens in prokaryotes. Although protein inhibitors do reduce total DNA synthesis after the S period has started, they could be acting only by stopping new initiations which we know are occurring during the S period at the molecular and chromosomal levels. An interesting point is that cycloheximide stops the synthesis of nuclear DNA in the yeast *Saccharomyces cerevisiae* but it does not stop the synthesis of mitochondrial DNA (Grossman *et al.*, 1969). This underlines the independence of initiation in these two organelles (p. 71). It is conceivable that a mitochondrial initiator is made by mitochondrial ribosomes insensitive to cycloheximide.

Terasima & Yasukawa (1966) found that pulses of puromycin and cycloheximide applied to L cells at different times in G1 caused delays in the start of the S period which were proportional (though not equal) to the length of the pulses but independent of their position in G1. This effect could be due to a delay in building up the concentration of a protein initiator to an effective level but it could also be due to the interruption

[1] Bloch *et al.* (1967), Caspersson *et al.* (1965), Hodge, Borun, Robbins & Scharff (1969), Kim *et al.* (1968), Littlefield & Jacobs (1965), Mueller *et al.* (1962), Powell (1962), Taylor (1965), Wanka & Moors (1970), Young (1966), Young *et al.* (1969).

in a sequence of protein syntheses which have to be passed through before the S period can start.

An interesting series of experiments has been carried out on *Physarum* by Cummins & Rusch (1966). They showed that a limited 'round' of DNA replication (20–30 per cent of the total DNA) could take place in the presence of cycloheximide added during the S period or just before it. The suggestion here is that the S period is divided into several rounds or segments which have to be completed in a temporal sequence, and that protein synthesis is needed for the initiation of a new round but not for its continuation. It might be very informative to follow the relations of these rounds to the patterns of molecular and chromosomal initiation.

One case where DNA synthesis can start in the absence of protein synthesis is in frog egg cytoplasm. Transplanted nuclei will incorporate thymidine in the presence of cycloheximide in quantity sufficient to inhibit nearly all protein synthesis (Gurdon & Woodland, 1968). This is in sharp contradiction to most other results but could be explained by assuming that eggs, as well as having large stores of many other materials, also have a store of protein initiator. Another case, which is more difficult to explain, is the effect of amino-acid deprivation in *Tetrahymena* (Stone & Prescott, 1964). DNA synthesis can not only continue but also be initiated when leucine incorporation has been reduced to a level undetectable by autoradiography. This situation merits further exploration and it would be preferable to use inhibitors which worked faster than the amino-acid starvation.

An indirect method of inhibiting protein synthesis is to use actinomycin D to stop the production of the messenger RNA required for protein synthesis. This method is sometimes more interesting than direct inhibition of protein synthesis, but it is more difficult to interpret. The doses needed to inhibit RNA synthesis differ widely between different cell types, probably because of differences in speed in penetration, and there is also a selective and dose-dependent effect on different kind of RNA. In some cells at any rate, there is a preferential inhibition of ribosomal RNA (Penman *et al.*, 1968). As a result, the total RNA synthesis may be very much reduced at intermediate doses without any marked effect on messenger production. A second problem is that there may be a considerable time lag between stopping messenger RNA synthesis and stopping protein synthesis if the message is relatively stable. If, therefore, protein synthesis is needed for DNA synthesis, we could make the prediction that actinomycin would stop DNA synthesis but high doses might be necessary and there might be a long time lag before it was effective. This prediction is, in general, borne out by the experimental evidence. In all the systems in which it has been tested, actinomycin does inhibit DNA synthesis partially or completely[1] but high doses are sometimes needed which may be larger than those which inhibit cell division (Cleffman, 1965). There is a time lag apparent in some of the experiments where the main or sole effect is in the early S period (Fujiwara, 1967; Mueller & Kajiwara, 1966b; Taylor, 1965). An interpretation of this is that the messages for the necessary proteins are only made in early S but are available throughout S, but it might also reflect the fact, mentioned above, that DNA synthesis in late S is less affected by protein inhibition. The messages may be made even earlier, before the S period. Baserga *et al.* (1965) have found that low doses of actinomycin will only inhibit DNA synthesis if they are applied at least two hours before the S period in ascites cells, and a similar early transition point has also been found by Frankfurt (1968) in mouse stomach epithelium.

Critical cell mass

In the course of their cytochemical studies on mouse L cells, Killander & Zetterberg (1965a) found that there was a significantly smaller variation of the dry mass of individual cells at the start of the S period than there was with cells at the start of the cycle. The variation in cell age at the start of S (i.e. the length of G1) was also much larger than the mass variation. This suggests that there may be a critical mass at which cells initiate DNA synthesis.

These results were analysed in more detail in a second paper (Killander & Zetterberg, 1965b). Cells in populations with a low initial mass at the start of the cycle spent a relatively long time in G1 before they attained the critical mass and then went through a short S + G2. Cells with the same total cycle time but with a high initial mass had a short G1 but a long S + G2. A similar inverse relation between G1 and S + G2 was found by Sisken & Morasca (1965) in human amnion cells.

An interesting aspect of this suggestion of a critical mass for DNA initiation is that the same phenomenon appears to happen in prokaryotes (Donachie, 1968; see p. 108). It seems unlikely that mass *per se* can act as a signal, but it is not hard to construct models in which an initiator, perhaps a protein, reaches a critical concentration at a particular cell mass. Why smaller cells should have a shorter S + G2 is more obscure.

---

[1] In mammalian cells: Baserga *et al.* (1965), Caspersson *et al.* (1965), Fujiwara (1967), Kim *et al.* (1968), Mueller & Kajiwara (1966b), Taylor (1965). In animal tissues: Frankfurt (1968), Prudhomme *et al.* (1967). In *Tetrahymena*: Cleffman (1965). In induced synchrony in *Chlamydomonas*: Jones *et al.* (1968).

These observations have been made on only one type of mammalian cell and we do not know whether they apply to other systems. What is perhaps a trivial point is that critical mass cannot be an initiating factor in early embryos where cell mass is continually diminishing. But the concept of critical mass, or better critical concentration, is one which is very suitable for an exploration at the experimental level with deliberate alterations of the DNA cycle.

The chromosome cycle

Chromosomes first become visible in the cycle when they condense in prophase; and they then disappear again in telophase. This part of the chromosome cycle occupies only the short mitotic period in the cell cycle and we do not know what changes take place in the chromosomes during interphase. An attractive suggestion, made by Mazia (1963), is that there may be continuous changes in chromosome structure throughout interphase. Decondensation of the chromosomes might continue through G1 and the completion of this process could make the DNA accessible to polymerase and so initiate replication. The reverse process of condensation could start after replication and continue until the chromosomes became visible in prophase. The advantage of this mechanism for starting DNA synthesis over the initiator mentioned in the preceding section is that local differences in the speed or degree of decondensation could explain why there are variations in the times of initiation of different replicating units in the euchromatin, of heterochromatin and euchromatin, and of different nuclei such as the macro- and micronuclei in Ciliates.

Evidence for a chromosome change in interphase does exist, though it is not very strong. In some cells, at any rate, anaphase chromosomes can be seen to be double (two 'half-chromatids'; see Wolff, 1969a). Chromosome aberrations produced by X-rays suggest that the target for radiation is single in early G1 but double in late G1 (p. 243). This change from two units to one unit to two units all happens between the end of one cycle and the beginning of DNA synthesis in the next one. There are also indications of chromosome changes in G2. Nuclei from late G2 cells of *Vicia* produce, after trypsin treatment, chromosomes which appear similar to those in prophase which are formed later in the cycle (Wolff, 1969b). Changes in enzyme potential (p. 167) and in radiation responses in *Schizosaccharomyces pombe* may be due to chromosome changes in G2 (Mitchison & Creanor, 1969). Finally there is a clear association between DNA synthesis and structural changes in the nucleus in the reorganisation bands of *Euplotes* (p. 68).

The case for a continuous chromosome cycle will have to be based on firmer evidence than this before it carries much conviction, but this may emerge from the active work that is being done at present on that important problem in cell biology, the structure of the eukaryotic chromosome.

Conclusions

At least three factors are involved in the initiation of DNA synthesis. One is a stimulus, presumably chemical, which is present in some circumstances in the cytoplasm and which can induce synthesis both in transplanted nuclei which are out of the cycle and in those which are in G$_1$. It is not clear whether this stimulus is important in the normal cycle though the widespread, but not universal, occurrence of synchronous initiation in multinucleate cells suggests that it is. Since the stimulus operates across a Class barrier in Vertebrates, it may be mediated through small molecules. It is possible that such molecules have to reach a critical concentration before synthesis is initiated.

A second factor is protein synthesis – though it is not known what species of protein are needed. They might be the enzymes of DNA synthesis, or histones, or proteins which alter the molecular structure of chromosomes and make initiation points accessible. Several kinds of protein may be important and those which are made in late G$_1$ may be different from those made in S.

A third factor is chromosome structure. Since DNA synthesis can be initiated at different times in a common cytoplasm, at the molecular level, at the chromosomal level and at the nuclear level (macro- and micronuclei of Ciliates), it is almost certain that the structural arrangement of the nucleo-protein must affect initiation. The wide gaps in our present knowledge of the architecture of chromosomes make it impossible to say whether there is a cycle of structural changes in interphase which is a continuation of those that happen in mitosis. But, as others have done, we could make a guess that there is.

How these factors interact and what other factors are involved, we do not know.

5 DNA synthesis in prokaryotic cells

One of the aims of this book is to compare the cell cycles of eukaryotes and prokaryotes and to see how far they show similar patterns. This is hardest to do when considering DNA synthesis and chromosome replication since it is in these processes and in nuclear structure that there lie the major differences between these two great groups of cells. Eukaryotes have a nuclear membrane and nucleolus: prokaryotes lack these structures and the chromosome seems to be attached directly to the cell membrane. Eukaryotes have many linear chromosomes which (with a few exceptions) contain histone as well as DNA: prokaryotes have one chromosome which is probably a single DNA molecule without any associated histone. In *Escherichia coli* the chromosome is circular and about 1100 μm in length when unfolded. It may also be circular in *Bacillus subtilis*, though the evidence is not quite so strong (Sueoka & Quinn, 1968; Ramareddy & Reiter, 1970; Yoshikawa, 1970). Eukaryotic chromosomes usually contain much more DNA, but this is not universal since a single chromosome of a budding yeast contains on average about a third of the DNA of a chromosome of *E. coli*. DNA synthesis is almost always periodic during the cell cycle of eukaryotes, while it is continuous in fast-growing *E. coli*. Finally, the eukaryotic chromosome has many thousands of replication forks whereas the prokaryotic chromosome usually has only one.

The attention in this field has been largely focused on the gram-negative rod, *E. coli*, and, unless another organism is mentioned in this chapter, the reader should assume that the work was done on this bacterium. The only other prokaryote which has been used at all extensively is *B. subtilis*.

Some recent reviews on DNA synthesis, chromosome replication and relations with the cell cycle are by Bonhoeffer & Messer (1969), Helmstetter (1969a, b), Kjeldgaard (1967), Kuempel (1970), Lark (1966a, b; 1969), Maaløe & Kjeldgaard (1966), Pardee (1968) and Sueoka (1966). Volume 33 (1968) of the Cold Spring Harbor Symposia on Quantitative Biology contains a series of relevant articles, many of which will be referred to later.

Sequential replication

One of the earlier questions to be asked was whether DNA replication is sequential in the sense that all the DNA is replicated once before any of it is replicated for the second time. This may seem a superfluous question since, with our present knowledge of DNA as the stable genetic material, it is difficult to imagine any answer except a positive one. But a decade or more ago, it was by no means certain that DNA was the genetic molecule or that it was stable.

The experiments of Meselson & Stahl (1958) on density labelling of *E. coli* DNA showed not only that replication is semi-conservative but also that it was sequential. Cultures were grown either in a 'light' medium containing ^{14}N as the nitrogen source or in a 'heavy' medium containing ^{15}N. The DNA was then extracted and centrifuged to equilibrium in a density gradient. The position on the gradient indicated whether the DNA was heavy (all ^{15}N), light (all ^{14}N) or hybrid (both ^{14}N and ^{15}N). After transfer from heavy to light medium, all the newly synthesised DNA was hybrid in density. This went on for a generation time until all the DNA became hybrid and there was no heavy DNA left. Only then did light DNA (twice replicated) begin to appear with further growth of the cells. This means that replication must be sequential.

A similar conclusion can be drawn from the experiments of Lark *et al.* (1963) in which *E. coli* cells were pulsed with a radioactive label for DNA (tritiated thymine) and then transferred to a medium containing a density label for DNA, 5-bromouracil (BU) which replaces thymine. The DNA was then extracted, centrifuged on a density gradient and examined for radioactivity. Doubly replicated DNA (both radioactive and hybrid in density) did not appear until about half the chromosome had replicated, and then only in small amounts. An extension of this experiment with double radioactive labelling of two parts of the chromosome of *Salmonella typhimurium* shows not only that replication of the labelled parts occurs at intervals of a generation time but also that the two parts are replicated in the same sequence for at least two cycles (Chan & Lark, 1969). This result is similar to those with eukaryotic cells (*Physarum* and HeLa, p. 80) which show that DNA replicated at one time in the S period is replicated again at the same time in the next S period. Another experiment with radioactive plus density labelling of *E. coli* shows replication of a part of the DNA at the same time in the cycle for up to six generations (Nagata & Meselson, 1968).

Although these experiments tell us about sequence and order in replication, they do not throw any light on the number of replicating forks present on the growing chromosomes. For that we have to turn to the classic autoradiographs of tritium-labelled *E. coli* DNA made by Cairns (1963; see also Bleecken *et al.*, 1966). Various labelling regimes were used but one of the

$100 \, \mu m$

Fig. 5.1. Autoradiographs of *E. coli* DNA after incorporation of ³H-thymidine for 1 h. A break is postulated between the points marked X. From Cairns (1963).

most informative was to label for about two generations. The autoradiographs (Fig. 5.1) then showed the chromosome as a DNA double helix in the form of a circle (often broken in preparation). There is a fork at one point which separates the chromosome into two daughter chromosomes and they are reunited at another point which is interpreted as the origin of replication. One of the daughter chromosomes has twice the grain density of the other daughter chromosome and of the mother chromosome. This is what would be expected if the mother chromosome is half-labelled (one radioactive and one non-radioactive strand) and so produces, on the second replication in a labelled medium, one fully labelled daughter (two radioactive strands) and one half-labelled daughter. Apart from the important conclusion that there was only one replicating fork on each chromosome under these growth conditions, this work confirmed the earlier genetic evidence for a circular chromosome (Jacob & Wollman, 1961) and also gave the speed of movement of the replicating fork as 20–30 μm/min.

Another important experiment done at about the same time as the molecular autoradiography of Cairns was the radioactive and density labelling of Bonhoeffer & Gierer (1963). This also showed that there was one, possibly two, replicating forks per chromosome.

Sequential replication has also been established in *B. subtilis* through the work of Sueoka and his colleagues (summarised in Sueoka, 1966). One ingenious method exploits the fact that an exponential phase culture contains twice as many young cells at the start of the cell cycle as old cells at the end (Fig. 2.5). If, therefore, there is a single replicating fork which moves during the cycle from one end of the chromosome to the other, the culture should contain twice as many of the first genes on the chromosome as the last ones. Intermediate genes will have ratios between one and two depending on their position on the chromosome. Gene quantity can be measured in *B. subtilis* by transformation, but the efficiency of this process varies from gene to gene. The values, therefore, from the exponential culture have to be normalised with respect to those from stationary phase cultures where it is assumed that replication is complete and all genes are present in equal quantities. When this has been done, the values can be used to map the chromosome. Other experiments both on exponential cultures and on synchronous cultures after spore germination have confirmed the correctness of this map and of the assumption of the sequential movement of a replicating fork along the chromosome. It has also been shown that there is more than one fork during rapid growth in rich media, so that a second round of replicating starts before the first round is complete. This is illustrated in Fig. 5.5 for *E. coli*, but multiple forks were first shown in *B. subtilis* (Yoshikawa *et al.*, 1964; Oishi *et al.*, 1964). This indicates that the rate of DNA synthesis of individual replication forks does not keep pace with the increase in cell growth rate in rich media – a point which will be returned to later.

The chromosome of the very small prokaryote *Mycoplasma hominis* is a ring of DNA about 260 μm in circumference, with one replicating fork (Bode & Morowitz, 1967).

Origin of replication

The evidence in the preceding section indicates that there is usually one replicating fork which moves round the circular chromosome of *E. coli* doubling the chromosome as it goes. It starts at an origin and finishes with the end of a round of replication at a terminus which is presumably very near the origin. Questions then arise as to whether the origin is a fixed point on the chromosome which is invariant in successive cycles, where it is located on the genetic map and what is the direction of replication. Answering these questions has involved a great deal of ingenuity and effort in recent years.

The story starts with work on the triple auxotroph of *E. coli* 15 TAU⁻ which requires thymine, uracil and arginine and can therefore be stopped in the synthesis of DNA, RNA and protein. If protein (or RNA) synthesis is stopped by deprivation of the appropriate precursor, DNA synthesis does not stop at once but continues for a period until the total DNA has increased by about 40 per cent (Maaløe & Hanawalt, 1961). The suggestion was made that this represents the completion of a round of replication in all the chromosomes and that protein (and RNA) synthesis is required for the initiation of a round, but is not required to sustain replication once it has been initiated. When, therefore, protein synthesis is inhibited, replication continues until the fork reaches the terminus. In a random culture, the time required for termination of DNA synthesis will vary from cell to cell – as was found by Hanawalt *et al.* (1961). The final result will be an arrested culture where cell size will vary but the chromosomal replication will all be aligned at the terminus.

If the result of amino-acid starvation were such an alignment, the effect of restarting protein synthesis by subsequent addition of amino-acids would be to start DNA replication synchronously at the origin in most of the cells. What is more, the effect of a second cycle of starvation followed by feeding would again be synchronous initiation at the same point, provided the origin is a fixed point on the chromosome which does not vary from cycle to cycle. This was shown to happen in an ingenious but complex experiment by Pritchard & Lark (1964). A culture of the *E. coli* 15 auxotroph was deprived of amino-acids for a time sufficient to align replication. Amino-acids were then added together with radioactive thymine. The thymine was removed after 10 per cent of the DNA had been labelled (near the origin), and the culture then grown for several generations. This will randomise the position of the replicating forks because of the variation in individual cell cycles (Lark *et al.*, 1963). The procedure was then repeated with a DNA density label (BU) instead of a radioactive label. After the addition of the BU, samples were removed at intervals and their DNA was centrifuged on a density gradient and analysed

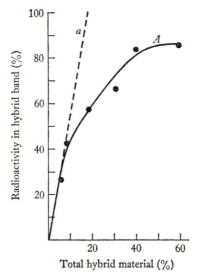

Fig. 5.2. Amount of radioactivity in hybrid (BU-containing) DNA from *E. coli* 15 quadruple auxotroph. Experimental procedure is described in the text. The points are from samples taken at increasing times after the addition of BU and therefore show an increasing proportion of hybrid material (shown on abscissa). The dashed line represents the initial slope expected if no heterogeneity in the rate of DNA synthesis exists in the population and if 100 per cent of the BU is incorporated from the origin of replication. From Pritchard & Lark (1964).

for the amount of radioactivity in the hybrid material which had incorporated the BU. The result, shown in Fig. 5.2, is that the bulk of the radioactive label appears in hybrid material when only a small fraction of the DNA has replicated after the second realignment. This indicates that replication starts from the same origin after each of the two starvation regimes and strongly supports the concept of a fixed origin. There is a divergence from the expected line in Fig. 5.2 which may be due to heterogeneity in the rate of DNA synthesis after amino-acid starvation.

These experiments do not locate the origin on the genetic map nor do they show the direction of replication. There is also the question as to whether the origin after amino-acid starvation is the same as the origin in normal growth. Answers here have only come in the last year or two but the papers are numerous and only some of them will be mentioned.

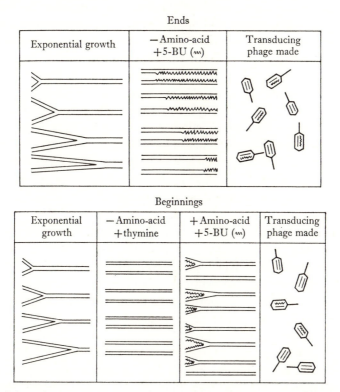

Fig. 5.3. Schematic representation of beginnings and ends experiment. DNA molecules are shown diagrammatically as they exist in cells before and after amino-acid starvation, BU labelling, and incorporation in phage coats to form transducing particles. From Wolf *et al.* (1968*a*).

The generalised transducing phage P1 has proved a useful tool for locating the origin. The method used by Wolf *et al.* (1968*a, b*) started with amino-acid starvation (Fig. 5.3). The chromosome termini ('ends') were density labelled with BU by feeding with BU instead of thymine during the early period of amino-acid starvation when the chromosomes were completing replication. Alternatively, the origins ('beginnings') were labelled by feeding BU together with the required amino-acid after starvation. P1 phage was then prepared from the labelled bacteria, and the heavy phages (containing BU-labelled chromosome segments) separated off by density–gradient centrifugation. Finally, the genetic markers present in the heavy phages were detected by transduction. Heavy phages from 'end' labelling should predominantly contain markers near the terminus; and from 'beginnings' labelling, markers near the origin. This will identify the location of origin and terminus and the direction of replication. The results showed that the terminus and origin were in the same region. For three strains of K12 (two F⁻ and Hfr P7201) the terminus and origin was in the region between *lys* and *xyl*. On the conventional *E. coli* map with *thr* as 12 on the clock face (Sober, 1968), this region is from 7.30 to 9.30 o'clock. The direction of replication was probably clockwise. Another strain, however (Hfr DG 163), had the origin between 12.30 and 2, and anticlockwise replication. Similar experiments with density-labelled P1 phage and amino-acid starvation have been done by Abe & Tomizawa (1967) and Caro & Berg (1969). They also found clockwise replication with a number of F⁺, F⁻ and Hfr strains of K12 and an origin in the lower left quadrant of the chromosome map, though it was rather earlier (5.30 to 7.30) with Abe & Tomizawa than with Caro & Berg (8 to 9.30).

Two objections can be made to these experiments – that there is evidence of toxic effects and alterations of replication with growth in media containing BU (Maaløe & Kjeldgaard, 1966; Abe & Tomizawa, 1967), and that the origin after amino-acid starvation may be different from that in normal growth. These difficulties have been avoided, though in the B/r strain of *E. coli*, by the recent experiments of Masters (1970). The method assumes, and confirms, that there is more than one replicating fork in fast growing cultures (p. 103). Gene frequencies are compared in P1 transducing lysates from cultures grown fast (20 minute doubling time) and slowly (40 minute doubling time). There should then be a greater difference between the fast and slow cultures in the frequency of early genes near the origin than in late genes near the terminus. In the fast cultures, there should be four copies of early genes for every one copy of late genes because of the presence of three replicating forks in each chromosome (Fig. 5.5). In the slow cultures, there should only be two copies of early genes because there is only one fork. The expected difference was in fact found, though the behaviour of genes very near the origin was somewhat anomalous. The origin mapped, as before, in the lower left quadrant between 5.30 and 7.30, and replication was clockwise.

Another method of examining the origin and direction of replication is to use the mutagen nitrosoguanidine, which causes a high frequency of mutations at the replicating fork (Cerdá-Olmedo *et al.*, 1968; Cerdá-Olmedo & Hanawalt, 1968). Mutation frequencies were examined in cultures of 15 T⁻, and K12 Hfr and F⁻, after alignment by amino-acid starvation. The origin was in all cases at about 7 o'clock and replication was clockwise. This method has also been applied to B/r synchronised by membrane elution (Wolf *et al.*, 1968*b*). Here again the origin for three strains was about 7.30 with clockwise replication.

A third method assumes that the inducibility of an enzyme in a synchronous culture doubles when the appropriate structural gene doubles (p. 164). The sequence and position of the inducibility increases for a series of enzymes will then give the origin and direction of replication. For B/r synchronised by membrane elution or sucrose gradient sedimentation the origin is between 7 and 8 o'clock with clockwise replication (Helmstetter, 1968; Pato & Glaser, 1968; Wolf *et al.*, 1968*b*; Donachie & Masters, 1969).

Although there is not universal agreement (e.g. Vielmetter *et al.*, 1968), the overwhelming weight of the evidence is that *E. coli* B/r and most strains of K12 have an origin and terminus lying in the lower left quadrant of the chromosome map (6 to 9 o'clock) and clockwise replication. There are, however, some Hfr strains of K12 which appear to have a different origin and anti-clockwise replication. Earlier suggestions of random origins, alternate directions of replication, or alternate replication of chromosome pairs are not supported by the present evidence. The exact location of the origin in the lower left quadrant is still uncertain. There may be strain differences but it is equally likely that some of the techniques cause small errors in location (reviewed in Caro & Berg, 1968). So far, the evidence is that the origin after amino-acid starvation is the same as the origin in normal growth.

In *B. subtilis*, the work of Sueoka and his colleagues (e.g. O'Sullivan & Sueoka, 1967) has shown that replication starts at the left-hand end of the conventional map (near *ade* 16) and proceeds to the right along the chromosome.

Chromosome attachments

Some years ago, it was suggested that the circular bacterial chromosome was attached to the inside of the cell membrane, with the point of attachment being the replicating fork (Jacob *et al.*, 1963). More precisely, what was attached was the enzyme complex that made up the replicating apparatus through which moved the chromosome. This model has a number of attractions, not least of which is that once the attachment sites have themselves been duplicated, they could move apart by growth of the membrane between them. There would therefore be a mechanism for nuclear separation, a process about which little else can be said in prokaryotes except that it is clearly not the same as mitosis in higher cells. We should not, however, forget that this scheme implies membrane growth in the middle of the cell (there is some evidence against this, p. 188) and that even though the attachment sites are separated there is still the problem of separating the two great circles of DNA which are a thousand times longer than the cell itself. It is difficult to imagine this separation taking place if the chromosomes are tangled up in a random fashion, and it seems much more likely that they condense in some ordered way, such as helices, pleats or folds which would allow easy separation in a few microns. In addition, the condensation might be periodic, as in eukaryotes, with temporary concentration near the attachment site.

The suggestion that the chromosome is attached to the membrane has gained a good deal of support in recent years. Evidence from morphological studies with the electron microscope shows that nuclei appear to be attached to the membrane by mesosomes, which are conspicuous in gram-positive *B. subtilis* but much harder to pick out in gram-negatives like *E. coli* (e.g. Ryter, 1968; Ryter *et al.*, 1968; Pontefract *et al.*, 1969). Labelling experiments have also shown that newly synthesised DNA is attached to cellular material which may be membrane components since they can be solubilised with detergents or lipase (for *E. coli* see Smith & Hanawalt, 1967: for *B. subtilis* see Ganesan & Lederberg, 1965). These labelling experiments indicate that the replication point is attached to the membrane, but there is also evidence that another part of the chromosome, the origin, is attached to the membrane, at any rate in *B. subtilis*. Genetic markers near the origin (such as *ade* 16) are to be found attached to a membrane fraction, as is the first DNA to be replicated and labelled in germinating spores (Sueoka & Quinn, 1968). This has led to the diagrams of chromosome replication in *B. subtilis* which are shown in Fig. 5.4. Whether or not there is this double attachment to the membrane in *E. coli* remains to be seen.

Replication and the cell cycle

The evidence that we have considered so far tells us about replication and attachment of the chromosome but does not relate it to the bacterial cell cycle. Some of the earliest work on the cycle with induced synchrony

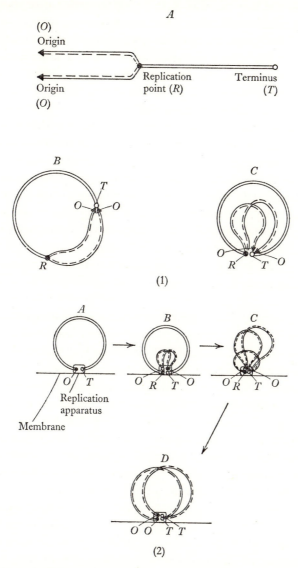

Fig. 5.4. (1) Alternative chromosome arrangements in *B. subtilis*. *A*: simplest linear arrangement. *B*: Juxtaposition of origin and terminus. This is the simplest interpretation of the position in *E. coli*. *C*: Suggested juxtaposition of origin, terminus and replication point in an area of the cell membrane. (2) Diagram of the replication process in *B. subtilis* assuming model *C* in (1) above. The process *A* to *D* completes one cycle of replication for one chromosome. From Sueoka & Quinn (1968).

suggested that replication was periodic and occupied a restricted part of the cycle, as in eukaryotic cells (e.g. Lark & Maaløe, 1956). But it became apparent about a decade ago that this was an artefact of the synchronisation technique and that DNA synthesis was a continuous process in normal growth, at any rate in fast-growing cultures. Pulse-labelling of *E. coli* and *S. typhimurium* with tritiated thymidine showed incorporation over more than 80 per cent of the cycle in cultures with generation times between 28 and 64 minutes (Schaechter *et al.*, 1959). In one of the earlier experiments with selection synchrony by filtration, DNA synthesis was continuous through the cycle of *E. coli* B (Abbo & Pardee, 1960).

Faced with the evidence of continuous DNA synthesis, a cell biologist might be tempted to make a model in which replication started at the chromosome origin at the beginning of the cycle and finished at the terminus at the end of the cycle, to be followed immediately by nuclear and cell division. But there is no good reason why reality should follow this model since the origin of replication and the start of the cycle do not need to be coincident in time. There is in fact good circumstantial evidence against this simple model in the fact which has been known for many years that there is variation in the DNA content and number of stainable nuclear bodies, as well as in cell size, when bacteria are grown at different rates at the same temperature. The implication here is that the pattern of replication may vary according to the growth rate.

This suggestion is borne out in what is now widely accepted to be the best model for the relation of replication to the cell cycle in *E. coli* B/r (Cooper & Helmstetter, 1968; see also Helmstetter *et al.*, 1968 and Helmstetter, 1969 a). This model, which is illustrated in Fig. 5.5, is based on the assumption that there are two constants which are invariant in cultures with doubling times of less than 60 minutes. One constant C = 40 minutes is the time for a replication fork to travel from the origin to the terminus of the chromosome. The other constant D = 20 minutes is the time between the end of such a round of replication and the succeeding cell division. With a doubling time of 60 minutes, the replication round starts at the beginning of the cycle and is completed after 40 minutes. There is then a gap of 20 minutes with no DNA synthesis until the end of the cycle. There is also a gap without DNA synthesis with the slighter faster doubling time of 50 minutes, but rounds of replication start at 10 minutes before the end of the cycle. With a doubling time of 40 minutes, there is no gap in DNA synthesis, and replication rounds start in mid-cycle; so the rate of synthesis doubles in mid-cycle. With doubling times of less than 40 minutes, new rounds of replication begin before the old round is completed so that three replication forks per chromosome appear for a proportion of each cycle – this proportion increasing as the doubling time decreases.

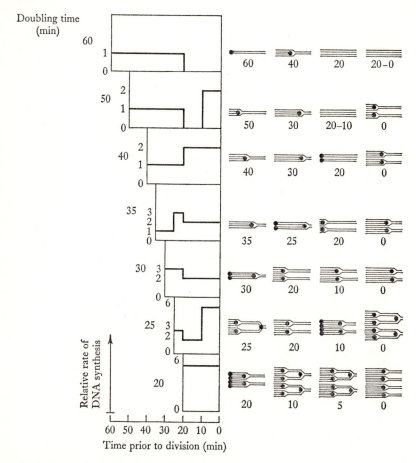

Fig. 5.5. Cooper–Helmstetter model for chromosome replication in *E. coli* B/r. The rate of synthesis of DNA in cells growing with the doubling times shown on the left, and assuming that C = 40 min and D = 20 min (see text). The chromosome configurations are shown on the right. The black dot indicates a replication point, and the numbers indicate the time in minutes prior to cell division at which the chromosome configuration is present in the cell. From Cooper & Helmstetter (1968).

The results are complex and varying patterns in the rates of DNA synthesis. Note that the number of chromosomes at any stage of the cycle alters with the growth rate, but the basic rule of the cycle is kept in that there are always twice as many chromosomes at the end of the cycle as there are at the start and their state of replication is the same. Another way of illustrating the model is in Fig. 5.6(*b*) where the cycles at different growth rates are drawn as the same length. This emphasises two points, that the origin of a replication round occurs at different points of the cycle at different growth rates, and that a round can span three cycles at fast growth rates.

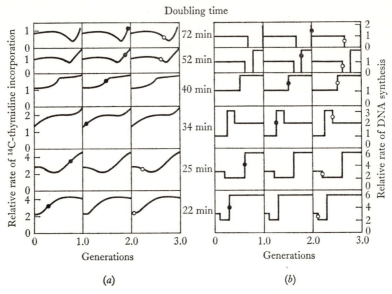

(a)   *(b)*

Fig. 5.6(*a*). The rate of ^{14}C-thymidine incorporation during the cell cycle of *E. coli* B/r growing at various rates. The cells were pulse labelled and then eluted from a membrane. The curves are the best smooth lines through the experimental values and each curve has been repeated three times to give the rate of incorporation during three successive division cycles. The filled and open circles are estimates of the start of a round and the end of this round, respectively. (*b*). The theoretical rate of DNA synthesis according to the Cooper–Helmstetter model during the cell cycles of cultures growing at the indicated rates. The filled and open circles are as in (*a*). From Helmstetter *et al.* (1968).

There are several lines of evidence which support this model. One of the most important comes from measurements of the rates of DNA synthesis in cultures of *E. coli* B/r grown in different media (Helmstetter, 1967; Helmstetter & Cooper, 1968). The cultures were pulse-labelled with radioactive thymidine while in asynchronous exponential growth, and then collected for the membrane elution technique. The radioactivity was followed in the small cells eluted successively from the filter (p. 52) and this was converted into the figures for rates of incorporation during the cycle which are shown in Fig. 5.6(*a*). Accepting the degree of asynchrony which smoothes out the curves in this technique of cycle measurement (and most others), the results are in good agreement with the predictions of the model in Fig. 5.6(*b*). Similar results were also found by Clark & Maaløe (1967) with the same strain of *E. coli*.

The model can be used to predict the average amount of DNA per cell in random cultures and this agrees well with the amounts measured by Cooper & Helmstetter (1968). The results are shown in Fig. 5.7 together with measurements made by Schaechter *et al.* (1958) on *S. typhimurium*. This latter line is higher than that for *E. coli* though Cooper & Helmstetter

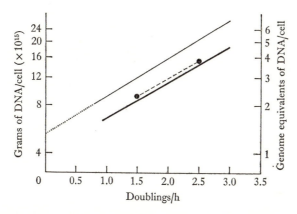

Fig. 5.7. Experimental and theoretical DNA content of cells growing at various rates. The dotted line connects two DNA measurements made on *E. coli* B/r by Cooper & Helmstetter (1958). The thin line shows the results of Schaechter *et al.* (1958) on *Salmonella typhimurium*. The thick line is calculated from the Cooper–Helmstetter model. From Cooper & Helmstetter (1968).

suggest technical reasons which may account for some of the difference. In any case, the fact that the lines are parallel suggests that the parameters of the model may apply to *S. typhimurium*.

The value assumed for C (40 – or more accurately 41 minutes) gives a rate of replication of 27 μm/min with a chromosome 1100 μm long. This is in good agreement with the figure of 20–30 μm/min from the autoradiographs of Cairns (1963 – see p. 93).

The number of termini per cell in the model agrees approximately with the number of visible nuclear bodies. For example, the number of these bodies increases from one to two in mid-cycle in glucose-grown cells with a doubling time of 40 minutes. There is, however, a discrepancy with faster-growing cells which can contain up to four nuclear bodies. It may be that chromosomal material which has already been replicated can fold into two distinct bodies before the end of a round (Fig. 5.8).

Further support for the model comes from shift-up experiments in which cells are changed from poor media with slow growth rates to rich media with fast growth rates. This change was first studied in detail by Kjeldgaard *et al.* (1958) and, among other things, there is a delay of about an hour before the rate of cell division increases to the new rate. This is to be expected from the model since the extra replication points in the rich medium will not express themselves in faster cell division until an interval of $C + D$ ($= 60$ minutes) after the shift-up. There are, however, some

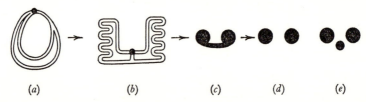

Fig. 5.8. Production of two 'observable' nuclei from one 'true' nucleus. The replicating genome (*a*) may fold in dense masses shown schematically in (*b*) prior to the completion of a round of replication. In the absence of artefacts, a dumbbell shaped nucleus would be visible (*c*). But the strand might at intervals be obscured, resulting in two visible nuclei (*d*) or even three (*e*). From Cooper & Helmstetter (1968).

minor discrepancies in this pattern which have led Helmstetter *et al.* (1968) to suggest that there may be small variations in C and D during the shift. But they have also examined the rates of DNA synthesis after the shift in cells of different ages eluted from membranes and these rates fit well to the predictions of the model.

The constancy of C and D does not hold for cultures with doubling time greater than an hour (Helmstetter *et al.*, 1968). In these slow-growing cultures both C and D increase, keeping the relationship that $C = 2/3$ of the doubling time and $D = 1/3$ of the doubling time. The origin remains at the start of the cycle but the time for a round increases, as does the gap at the end of the cycle. The experiments of C. Lark (1966) and Kubitschek *et al.* (1967) also show a gap in DNA synthesis in slow-growing cultures but at the beginning of the cycle rather than at the end. This might be due to the use of different strains of *E. coli*. It is inappropriate to use the terms G1 or G2 since these refer to the nuclear cycle rather than the cell cycle, and there is no clearly defined nuclear division in prokaryotes.

It is not certain how far the model applies to other strains of *E. coli*. Bird & Lark (1968), from density labelling experiments with *E. coli* 15, also find evidence for a replication round of 40 minutes and for multiple chromosome forks, and prefer at any rate some aspects of this model to one suggested earlier by K. G. Lark (1966b). But they also find discrepancies, especially in slower growing cultures, and suggest that some details of the model are not the same in strains 15 and B/r. It is even less certain whether it is applicable to other species. There is a little evidence (e.g. Fig. 5.7) that a similar model may apply to *S. typhimurium*, but it would be very useful to test its validity for *B. subtilis* which is the only other bacterium where much is known about chromosome replication. Qualitatively, it does seem to apply to *B. subtilis* since multiple forks occur in fast growth (p. 94) and there are gaps in DNA synthesis in slow-growing cultures (Donachie, 1965; Eberle & Lark, 1967) but the quantitative parameters are unknown.

Initiation of replication

We have seen in an earlier chapter (p. 86) that protein synthesis is needed for the continued synthesis of DNA in eukaryotic cells, and there is some evidence that it is the initiation of replication which is the process that is sensitive to inhibitors of protein synthesis. The analysis of initiation has gone further in *E. coli* and it is now clear that the synthesis of protein is required for the initiation of replication in *E. coli*, and it is very probable that this protein has to reach a critical 'level'. But it is not known at present how many proteins are involved, what they are, or how they act.

Some of the earliest evidence came from the *E. coli* 15 auxotroph that has already been described (p. 95). Amino-acid starvation stops the initiation of new rounds of replication but does not prevent the completion of rounds that have already been started. The same thing happens with amino-acid requiring mutants of other strains of *E. coli*, and with *B. subtilis* after inhibition of protein synthesis by chloramphenicol (Yoshikawa, 1965).

The experiments of Maaløe and his collaborators (reviewed by Maaløe & Kjeldgaard, 1966) suggested that the synthesis of initiator protein is a continuous process and that it has to reach a critical level before replication starts. Because of this, an important factor in initiation may be the attainment of a certain ratio of cell mass to DNA content. They also suggest that if DNA synthesis were inhibited while protein synthesis continued, the initiator would accumulate and, when DNA synthesis is resumed, it should do so for a time at a faster rate than normal and with extra replication points. This has in fact been found to occur with a number of different techniques in *E. coli* (Pritchard & Lark, 1964; Hewitt & Billen, 1965; Hardy & Binkley, 1967; Boyle *et al.*, 1967; Donachie *et al.*, 1968; Donachie, 1969) and in *B. subtilis* (Kallenbach & Ma, 1968). In passing, it is worth noting that this temporary inhibition of DNA synthesis does not produce persistent division synchrony as it does in mammalian cells (p. 26).

The principle of a critical cell mass for initiation is supported by Helmstetter *et al.* (1968). They calculate the theoretical initiator content of cells and find that it parallels cell mass and volume over a wide range of growth rates. But the most elegant numerical exposition is in a short theoretical paper by Donachie, 1968 (see also Donachie & Masters, 1969). Cells of *E. coli* are assumed to grow exponentially in mass over the cycle, and to vary their cell mass at division with different growth rates in the same way as the cells of *S. typhimurium* described by Schaechter *et al.* (1958). The rule that there must be an interval of an hour between initiation and division (C + D) is taken from the Cooper–Helmstetter model described earlier. This enables the mass at initiation to be found in the way shown in Fig. 5.9. It can be seen that the cell mass at initiation (M_i) is a constant for doubling times between 30 and 60 minutes, and exactly twice this constant for doubling times between 20 and 30 minutes. The Cooper–Helmstetter model also provides that the number of chromosome origins at the time of initiation (N_i) is two for doubling times between 30 and 60 minutes and four for doubling times between 20 and 30 minutes (Fig. 5.5). As a result, M_i/N_i is a constant – at any rate between doubling times of 20 and 60 minutes. This interesting constancy between the cell mass at initiation and the number of origins could be explained in a number of ways, but one simple one is that initiator protein is synthesised over the whole cycle at a rate proportional to the overall growth rate and reaches a critical level at the time of initiation. This level is not simply a matter of concentration since the number of origins is involved as well as cell size. It could be that the initiator is the material of a structure which would have to be completed at each origin in order that replication could start. Once replication had started, this structure and its components would not be available for further initiations, which would requre *de novo* synthesis of more initiator protein. There are obvious similarities between this concept and that of 'division proteins' which is discussed in Chapter 10, but as we shall see, the work of Smith & Pardee (1970) suggests that division proteins and initiator proteins may be different in *E. coli*.

Pritchard *et al.* (1969) have suggested that replication could be triggered by the dilution of an inhibitor rather than the accumulation of an initiator.

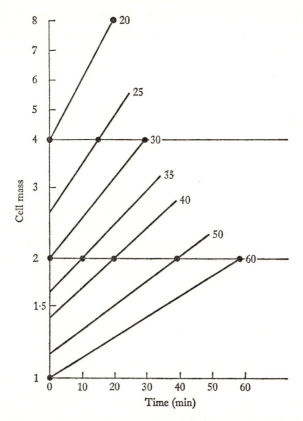

Fig. 5.9. Course of increase in mass of individual bacterial cells with different rates of growth. The cells are assumed to grow exponentially over a single cell cycle which starts at the end of the previous division at time o min. The next division at the end of the cycle takes place after the initial mass has doubled. The doubling takes from 20 to 60 min, as marked. The initial mass at time o is taken to be proportional to the average mass of a population of cells growing at the same growth rate. Each line therefore shows the course of mass increase of individual cells. Because there is a constant time of 60 min from the Cooper–Helmstetter model between the initiation of a round of replication and cell division, it is possible to calculate the time of initiation relative to cell division in cells growing at different rates. These times are shown as solid circles on the corresponding lines of mass increase. It can be seen that the masses at which initiation takes place are the same, or multiples of the same cell mass, for cells growing at all growth rates. From Donachie (1968).

There is little formal difference between the ideas, but it is more difficult to incorporate inhibitor dilution into the unitary initiation structure which explains the relation between cell mass and number of origins.

The process of initiation may require the synthesis of more than one initiator protein. The work of Lark and his collaborators with inhibitors such as phenyl ethanol and chloramphenicol suggest that there are two proteins involved, with different inhibitor sensitivity (C. Lark & K. G. Lark, 1964; K. G. Lark & C. Lark, 1966; K. G. Lark, 1966 *a*). Recent evidence suggests that these proteins are synthesised at different points in the replication cycle and also that there is a third process needed for initiation which does not require amino-acids (Lark & Renger, 1969). This work, and parallel work on phage replication and on the synthesis of bacterial walls and membranes, is reviewed by Lark (1969).

Replication and cell division

The Cooper–Helmstetter model provides for a delay (D) of 20 minutes between the end of a round of replication and the ensuing cell division. DNA synthesis, however, continues during this period in fast-growing cultures, so we can ask whether this continued synthesis is necessary for division. The answer in some strains of *E. coli* seems to be no, but that the completion of the preceding round *is* a prerequisite. Helmstetter & Pierucci (1968) showed that cultures in which DNA synthesis is inhibited by u.v. light, mitomycin C or nalidixic acid, continue to divide for about 20 minutes and then stop. The cells which do divide in the absence of DNA synthesis are those which have completed a round of replication prior to the treatments. Essentially similar results were obtained by Clark (1968 *a*, *b*) with nalidixic acid and with a temperature sensitive mutant for DNA synthesis. The completion of a replication round is therefore a necessary and sufficient condition for division – as far as DNA synthesis is concerned. The end of a round is a 'transition point' for division, again with respect to DNA synthesis, similar to those which will be discussed in Chapter 10. This result, which was obtained from work on strains B/r and K12, does not, however, seem to apply universally. Donachie (1969) stopped DNA synthesis by deprivation of thymine in a 15T⁻ strain and found that division stopped at once and not after a delay. It may be, however, that the difference is a trivial one due to the fact that cells may not separate during thymine starvation even though they do divide. This work also showed what happens to division when the block to DNA synthesis is removed by the addition of thymine. There is a delay before division recommences which increases with a lengthening of the preceding period of inhibition of DNA synthesis. Donachie has explained this in terms of a model which assumes that initiator continues to be made during DNA inhibition. This results in an increased rate of DNA synthesis for a period after the end of the inhibition, and also a block to division which persists until this extra replication is completed.

Another interesting point that is stressed by Clark (1968*b*) is that there may be physiological division of the cell by a weak cell membrane some time before the cell is cleaved in two – a situation analogous to that in some eukaryotes, e.g. *Schizosaccharomyces pombe*. The evidence here comes from experiments with phage infection and sonication, and there is also support from the electron microscope (Cota-Robles, 1963; Steed & Murray, 1966). Physiological division occurs 15 minutes before final cell division, so the daughter cells may become independent units 5 minutes after the end of a replication round.

The constant delay (D) of 20 minutes in the Cooper–Helmstetter model does not happen in all situations. The effect of a short heat shock on a growing culture of *E. coli* B/r (and 15T⁻) is to synchronise division but not the initiation of replication. As a result, D can vary between 12 and 30 minutes (Smith & Pardee, 1970). These experiments are discussed in more detail in Chapter 10 (p. 228).

Other prokaryotes

We know little about patterns of DNA synthesis during the cycle in prokaryotes other than *E. coli*. Leaving aside the evidence from cultures where synchrony has been induced by shock treatments that appear to distort the normal pattern, there is continuous synthesis of DNA in *Proteus vulgaris* (Cutler & Evans, 1966) and in *Rhodopseudomonas spheroides* (Ferretti & Gray, 1968), but it is difficult to tell whether or not there are the changes in rate of synthesis shown in *E. coli*. In the synchronous divisions of *Bacillus cereus* that follow spore germination, there is continuous synthesis for two cycles (Young & Fitz-James, 1959) after an initial lag (Steinberg & Halvorson, 1968). Although *Streptococcus faecalis* shows a different pattern of growth in dry mass and volume from bacterial rods (p. 144), DNA synthesis seems to be continuous in cultures with a doubling time of an hour (Stonehill & Hutchison, 1966). There is also a polarity in the duplication of five markers, which implies sequential replication.

There is periodic synthesis, however, in slow-growing and presumably synchronous cultures of *B. subtilis* (Donachie, 1965) and in *Alcaligenes faecalis* (Maruyama & Lark, 1962). It also occurs in the slow-growing cultures of *Myxococcus xanthus* (Zusman & Rosenberg, 1970), where DNA synthesis lasts for 80 per cent of the cycle and takes place at a constant rate.

Comparison with eukaryotes

The introduction to this chapter outlined the broad differences in this field between prokaryotes and eukaryotes, and we are now in a position to carry this comparison somewhat further. The unique features of the bacterial chromsome – its singularity, circularity and lack of histones – are mirrored in its mode of replication at the molecular level. The single fork takes 40 minutes to complete its circuit of the chromosome. In contrast, the fork in eukaryotes moves much slower but has a much shorter distance to go since the chromosome is divided into many replicating units (p. 76). There are also two forks in each replicating unit moving in opposite directions from a common origin. A similar mode of replication has been suggested for some situations in bacteria (Caro & Berg, 1968) but there is not much support for it. There can be little doubt that the multiplicity of replicating units in eukaryotes is a reflection of their much larger content of DNA, even though a single fork per chromosome moving at a bacterial speed could complete the S period of a mammalian cell in a matter of days rather than weeks. It would be interesting here to know the position in yeast which is a simple eukaryote with smaller chromosomes than those of *E. coli*.

There is more convergence between eukaryotes and prokaryotes in the matter of initiation. In both groups, protein synthesis seems to be required, though the eukaryotic evidence (p. 86) is less precise. There is also evidence in both groups that there is a critical cell mass for initiation (p. 88), and that initiation takes place near a membrane – in eukaryotes, the nuclear membrane (Comings, 1968; Comings & Kakefuda, 1968). In neither group is it clear what exactly happens at initiation, apart from the act itself, or why new proteins are needed. Apart from this mystery, which clearly needs to be resolved, there is a particular problem in the eukaryote which does not arise in *E. coli*. There is usually a long period in the cycle before DNA starts. When it does start, why does it not do so simultaneously in all the replicating units? The answer here may come in terms of a critical quantity of initiator protein which is maintained through the S period but not at a level sufficient to start all the units at the same time. The raising of this level might then account for the dramatic shortening of the S period in early development (p. 63).   On the other hand, the correspondence between heterochromatin and late replication shows that late initiation during the S period must be related to some aspect of chromosome structure and cannot be a matter of chance response to initiator concentration.

It is not easy to judge the importance of the fact that DNA synthesis is periodic in nearly all eukaryotes and continuous in fast-growing pro-karyotes. One could argue that this is a trivial difference dependent only on growth rates which are usually faster in bacteria than in higher cells. DNA synthesis does become periodic when bacteria are grown slowly, and conversely it occupies a greater proportion of the cycle with faster growth rates in mammalian cells. But it is not as simple as this. Some lower eukaryotes have short cycles but even shorter S periods, for example the yeast *Schizosaccharomyces pombe* with a cycle time of 2.5 hours and a S period of about 10 minutes (p. 70). The sea urchin embryo is even more striking with a cycle time of 70 minutes and an S period of 13 minutes at 15 °C (p. 63). We cannot take the argument very far since we do not know why synthesis is periodic in the eukaryotic interphase, or for that matter in slow-growing bacteria. There is, however, a point where the nuclear structure of the two groups has an obvious effect on the synthetic patterns. Whatever the proportions of the cycle spent in G1, S and G2, there is no case where DNA synthesis continues during the mitotic meta-phase when the chromosomes of eukaryotic cells are highly condensed. By contrast, the chromosomes of bacteria can be replicating while they are separating. This, together with the easily variable initiation times and replication patterns in bacteria, gives them DNA cycles which do differ considerably from eukaryotes.

A point that has been made by Helmstetter *et al.* (1968) is that there is considerable resemblance between the constant C + D of their model for the *E. coli* cycle and the constant S + G2 in mammalian cells. Put in other words, there is fixed time for a complete round of replication and the preparations for the succeeding division. This is an interesting similarity, though subject to two provisos, that the constancy of S + G2 is a tendency rather than a rule, and that a new C can start during D whereas a new S does not start during G2.

Conclusions

The genome of *Escherichia coli* is contained in a single circular chromosome which appears to be one long molecule of DNA with-out histone. Except in rapid growth, there is a single replicating fork which moves round the chromosome starting at an origin. In rapid growth, two further forks start at the origin before the first one has completed a round of replication.

In most strains there is fixed origin lying in the lower left quadrant of the conventional chromosome map, and replication proceeds clockwise. In a few strains, however, the origin may be elsewhere, and replication may be anti-clockwise.

The chromosome is attached to the cell membrane in the region of the replicating fork. There may also be a second point of attachment at the origin, but the evidence here comes from *Bacillus subtilis*. Growth of the membrane between these attachments may be the main mechanism of chromosome separation.

In slow-growing cultures (doubling time more than 40 minutes), DNA synthesis is periodic. In fast growing cultures, there is a continuous synthesis. The relations between replication and the cell cycle can best be explained in terms of a model where the number of chromosomes, the number of replicating forks and the time of initiation vary with the growth rate, but where there are two invariant factors – the time for a complete round of replication and the time from the end of a round to division.

Protein synthesis is necessary for initiating a new round of replication but is not necessary for the completion of a round that has already started. The growing cell has to reach a critical mass at the time of initiation and this may be related to the building up of an initiating structure from one or more proteins.

Most of the work has been done on *E. coli*, though there has been some on *B. subtilis*. The information about other prokaryotes is so scanty that satisfactory conclusions cannot be drawn at present.

6 RNA synthesis

A large part (about 80 per cent) of the total RNA in a growing cell is ribosomal RNA, and most of the remainder is transfer RNA. These are stable molecules and we know where they are made, where most of them are located and the process with which they are associated – protein synthesis on the ribosomes. The difficulty comes with the other minor components of the total RNA (for recent reviews, see Darnell, 1968, and Loening, 1968). In the early sixties, there was thought to be only one minor component, messenger RNA. This was rapidly labelled in the nucleus and broken down in the cytoplasm after directing the synthesis of proteins on polyribosomes. Apart from being unstable, it was of varying molecular weight and had a base composition which differed from the ribosomal RNA and approached that of the DNA. It is now clear, however, that the majority of the rapidly labelled nuclear RNA in higher eukaryotic cells breaks down within the nucleus and does not get out into the cytoplasm to act as the conventional messenger for the polyribosome system. The function of this RNA is a major mystery at present, though many cell biologists would hazard the guess that it is concerned with gene control. This leaves us with the problem of the labelling kinetics of the true cytoplasmic messenger. Presumably it is a fraction of the rapidly labelled nuclear RNA, but we cannot be certain of this because, in most cases, it cannot be identified unequivocally. To add to the difficulty of identifying minor RNA components, there is also ribosomal RNA (and probably messenger RNA) in mitochondria and chloroplasts.

This confusing picture will no doubt become clearer, but at present it makes it hard to interpret detailed studies of the rate of RNA synthesis during the cycle. Pulse-labelling with precursors is a sensitive method of following rate changes, but it has to be used with caution in the case of RNA where there is not only the general difficulty of varying pool sizes (p. 11) but also the particular problem that different species of RNA will be labelled with different lengths of pulse. Short pulses are mainly incorporated in the rapidly labelled nuclear RNA and long pulses in the ribosomal RNA, but careful analysis is needed, and seldom made, of exactly what RNA is being labelled with a given length of pulse.

Higher eukaryotes

If a growing culture of mammalian cells is pulse-labelled with uridine and then examined in an autoradiograph, all the cells will be found to have incorporated the label into RNA except those in mitosis (e.g. Prescott & Bender, 1962). Such an experiment brings out the two most important things that we know about RNA synthesis in the cycle of higher eukaryotic cells. One is that mitotic cells do not synthesise RNA, and the other is that the synthesis of RNA, unlike that of DNA, is a continuous process through the rest of the cycle.

There is overwhelming evidence that RNA synthesis is very much reduced in mitotic cells.[1] Does it completely stop? The evidence here is not absolutely clear-cut since some of the work shows a small amount of label incorporated into cells in mitosis. On balance, however, we can judge that RNA is not synthesised on metaphase chromosomes and that the apparent exceptions are caused by factors such as adsorption or conversion of label, terminal end-labelling of transfer RNA, or labelling cells which were not in division at the start of the pulse. The gap in synthesis appears to start in late prophase and continue into telophase. In terms of time, the start is one hour after the end of S and about 10 minutes before metaphase in mouse cells with a cycle time of 9–10 hours, and the finish is at the end of cell cleavage (Doida & Okada, 1967b). The simplest explanation of this gap is that DNA cannot be transcribed when the chromosomes are in their highly condensed state at metaphase.

A minor qualification is that there may be a small amount of synthesis in the cytoplasm of dividing cells (Newsome, 1966; Harris & La Cour, 1963, but see a criticism by Das *et al.*, 1965). This is likely to be mitochondrial or chloroplast RNA synthesised on DNA which does not go through a normal chromosomal cycle in these organelles. Recently, there

---

[1] In mammalian cells in culture: Baserga (1962), Doida & Okada (1967b), Feinendegen *et al.* (1960), Feinendegen & Bond (1963), King & Barnhisel (1967), Konrad (1963), Prescott & Bender (1962), Taylor (1960a), and Terasima & Tolmach (1963a). In mammalian tissue: Linnartz-Niklas *et al.* (1964). In grasshopper neuroblasts: Schiff (1965). In plants: Das (1963), Das *et al.* (1965), Davidson (1964), Kusanagi (1964), Van't Hof (1963).

has been direct evidence that mitochondrial RNA synthesis does continue during mitosis (Fan & Penman,[*1] 1970a).

The spectrophotometric measurements of Zetterberg and Killander on mouse L cells provide good evidence that there is a continuous increase in the total cellular RNA throughout interphase. They used two methods to construct cell cycle plots of total RNA. With one of them, cell histories from films, there were no obvious fluctuations in rate through the cycle (Killander & Zetterberg, 1965a). With the other, frequency distributions,

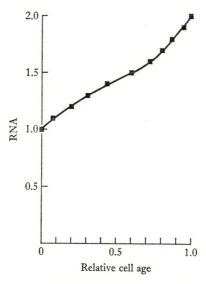

Fig. 6.1. RNA increase through the cycle of mouse L-929 cells. From Zetterberg & Killander (1965a).

there were signs that the rate remained roughly constant during the first half of the cycle and then doubled during the second half at the time when DNA was being synthesised (Zetterberg & Killander, 1965a). The curve derived by the second method is shown in Fig. 6.1. It is clear that ribosomal RNA must be synthesised continuously through interphase since most of the total RNA is ribosomal. The rate change is less clear, but intriguing because it suggests a gene dosage effect. If the rate of production of ribosomal RNA is proportional to the number of RNA genes present in the cell, the rate will double during the S period. This very simple method

---

[1] This asterisk, here and elsewhere in this chapter, marks experiments with cultures synchronised by DNA inhibitors or colcemid arrest. The reasons for this are explained on p. 33.

of rate control has parallels in enzyme synthesis which are discussed in Chapter 8.

If there is a gene dosage effect, it should be easier to detect with pulse labelling than with measurements of total RNA. There are in fact several sets of experiments which show this, of which the most thorough are those of Pfeiffer & Tolmach (1968) on HeLa cells synchronised by mitotic selection. The amount of uridine incorporated in 10–30 minute pulses rises slowly during G1, then doubles sharply in the first half of S, and

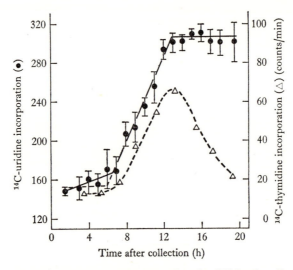

Fig. 6.2. Rate of synthesis of RNA through the cell cycle of HeLa S3 cells. Solid line and circles are from pulses (10–30 min) of labelled uridine. The vertical bars are 95 per cent confidence limits. The dashed line and triangles are the rate of DNA synthesis from pulses of labelled thymidine. From Pfeiffer & Tolmach (1968).

finally stays constant for the second half of S and G2 (Fig. 6.2). This suggests a gene dosage effect with the ribosomal RNA genes doubling early in the S period. Other experiments with inhibitors show that the extent of the acceleration in synthesis depends on the amount of the DNA which has been replicated, and that some of the possibilities of pool variation can be eliminated. There are two problems about these results. One is that the rate of synthesis does rise in G1 even though more slowly than it does later. This would not be expected on a simple gene dosage model, and it would be interesting to see whether there is a similar but longer period of rising rate in a cell strain with a more extended G1. The second problem is the nature of the RNA labelled in the pulses. The rate changes in the total RNA in Fig. 6.1 are ribosomal RNA, whereas the RNA labelled in 10–30 minute pulses will probably be a mixture of heavy ribosomal precursor and rapidly labelled nuclear RNA (Darnell, 1968). In a second paper, however, Pfeiffer (1968) has made a partial separation of the RNA into nuclear, cytoplasmic ribosomal, transfer and 'messenger' fractions and finds that they all show rate changes similar to those in Fig. 6.2. This is one of the few attempts to follow the synthesis of the various RNA components during the mammalian cell cycle, and it should certainly be extended and developed.

Other evidence from pulse-labelling which supports the concept of rate increases mainly during the S period comes from Crippa (1966), and from cultures synchronised by colcemid arrest (Klevecz & Stubblefield,* 1967; Martin, Tomkins & Granner,* 1969).

This is one side of the present picture, but there is another equally convincing side which shows a continuous increase in the rate of RNA synthesis throughout the cycle. Enger & Tobey (1969) pulsed uridine into synchronous cultures of Chinese hamster cells and showed a steadily rising rate of incorporation without a plateau in G1. They also found a similar result with methyl-labelled methionine which labels the 18S component of ribosomal RNA (Fig. 6.3). Scharff & Robbins (1965) found a continuous increase in the rate of incorporation of uridine in HeLa cells. Its pattern of distribution in sucrose gradients showed that ribosomal and transfer RNA were made at all stages of the cycle and that their relative amounts did not change. Further evidence for continuous increases comes from pulse labels of HeLa cells (Kim & Perez, 1965; Terasima & Tolmach, 1963 a), of mouse L cells (Fujiwara, 1967; Gaffney & Nardone, 1968), and of mouse P815Y cells (Warmsley & Pasternak, 1970).

In many of the experiments, the rate of synthesis of total RNA (unlike that of total protein) appears to increase through interphase by a factor which is substantially greater than 2, at any rate when judged by uridine incorporation. There is nothing which is biologically improbable in this, especially since the rate falls to zero at mitosis. But what is curious is that the factor can vary in one set of experiments according to the label used. In the experiments of Enger & Tobey (1969), the factor is about 2 with a methionine label and about 4 with a uridine label. This is just the kind of effect which might be caused by changes in the internal pools, and it would be as well to bear this in mind in future experimental work with pulse-labels.

Nevertheless, it is hard to believe that factors such as pool fluctuations can explain all the differences between the experiments on synthesis rates. It seems much more likely that there are genuine variations in the patterns of ribosomal RNA synthesis depending on the growth conditions and the cell strains. An example with another type of macromolecule is the change from continuous to periodic synthesis of sucrase in *Bacillus subtilis* with an alteration of the medium (Masters & Donachie, 1966). Quite apart from any gene dosage control, there must also be separate controls which adjust the total amount of ribosomal RNA to the overall rate of protein synthesis.

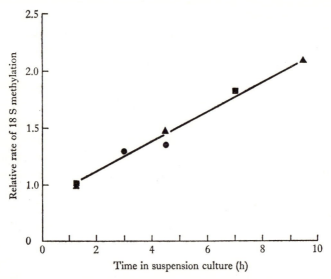

Fig. 6.3. Rate of synthesis of 18 S ribosomal RNA through the cell cycle of Chinese hamster ovary cells. Data from 1 h pulses with methionine-methyl-^{14}C. From Enger & Tobey (1969).

Gene dosage, however, is an easier hypothesis to test and it would be interesting to know whether the sharp rate increase, in those systems where it happens during the S period, is connected with DNA replication by causal relationship or only by temporal coincidence. The fact that the rate increase is stopped by inhibitors of DNA synthesis (Klevecz & Stubble-field,* 1967; Pfeiffer & Tolmach, 1968) suggests a causal relationship, but a further test would be to see whether the rate increase and the S period remained coincident in systems where the position of the S period in the cycle could be varied.

   The position about the non-ribosomal fractions of RNA is far from clear. The existing evidence suggests that they are made throughout interphase and follow the same pattern of synthesis as ribosomal RNA, but this field needs more exploration with modern separation techniques. Questions can be asked not only about the rates of synthesis but also about the variations in composition of some of the fractions through the cycle. The periodic synthesis of enzymes during the cycle (Chapter 8) strongly suggests that there must be some changes in the composition of the true cytoplasmic messenger RNA fraction, and they might also occur in rapidly labelled nuclear RNA. Whether we can detect them with our present techniques is another matter. There is some suggestion in HeLa cells of differences through the cycle in the relative contribution of 'messenger RNA' (Miller,[*] 1967) and in the synthesis of low molecular weight nuclear RNAs (Clason & Burdon,[*] 1969). On the other hand, hybridisation experiments have not revealed changes through the cycle in heterogeneous nuclear RNA (Pagoulatos & Darnell,[*] 1970) or in 'DNA-like RNA' (Bello,[*] 1969; see also, 1968), though these methods are not very sensitive.

It is likely, but not certain, that most of the total RNA content of a nucleus is ribosomal RNA and its precursors which are in transit from their origin on the ribosomal genes to their destination in the cytoplasm. If there was a constant 'processing time' before the RNA was released into the cytoplasm, we might expect the RNA content of the nucleus to increase through the cycle as the rate of synthesis of ribosomal RNA rose. But it does not do so, and remains constant through interphase both in mouse L cells (Zetterberg, 1966b) and in *Vicia* root meristem at 25 °C (though not at lower temperatures; McLeish, 1969), suggesting perhaps that the processing time diminishes. There is, however, evidence for a constant processing time throughout the cycle in HeLa cells (Pagoulatos & Darnell,[*] 1970). When the nuclear membrane breaks down in prophase, the nuclear RNA is released into the cytoplasm (e.g. Davidson, 1964; Newsome, 1966; Prescott & Bender, 1962). The nuclear RNA rises rapidly to its interphase value when the nuclear membrane reforms in telophase. Most of this rise is presumably because RNA synthesis restarts but some of it may be due to the return of RNA from cytoplasm to nucleus, a process which has been shown to happen in *Amoeba* (Goldstein *et al.*, 1969).

Lower eukaryotes

A number of studies have been made on the incorporation of precursor pulses in *Physarum*, and the patterns of RNA synthesis appear to be broadly similar to those in higher eukaryotes. Synthesis stops in mitotic nuclei (Kessler, 1967), but during most of interphase there is continuous synthesis of ribosomal, transfer and rapidly labelled nuclear RNA (Braun *et al.*, 1966*a*; Mittermayer *et al.*, 1964). There is, however, one conspicuous difference from other cells in the presence of a period in mid-interphase, after the S phase, where there is a sharp reduction in the rate of synthesis (Fig. 6.4). This interphase fall in rate, which is also shown in isolated nuclei (Braun *et al.*, 1966*b*; Mittermayer, Braun & Rusch, 1966), is paralleled by a similar fall in the protein synthesis (Mittermayer, Braun, Chayka & Rusch, 1966), but we cannot yet interpret it in terms of other processes in the cell cycle.

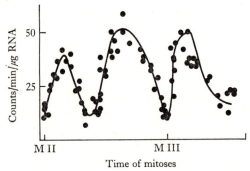

Fig. 6.4. Rate of RNA synthesis in *Physarum*. Incorporation of 10 min pulses of ^{3}H-uridine into RNA at different times in the mitotic cycle. M II and M III are the second and third mitoses after fusion. The time between them is 8–10 h. From Mittermayer *et al.* (1964).

Analysis of sedimentation patterns (Braun *et al.*, 1966*a*) and of the effects of actinomycin (Mittermayer *et al.*, 1964) indicate that there are differences in some of the kinds of RNA made at various points in the cycle, but the clearest demonstration comes from two papers on the base ratios of pulse-labelled RNA (Cummins, Weisfeld & Rusch, 1966; Cummins & Rusch, 1967; see also Cummins, 1969). The rapidly labelled nuclear RNA has a high adenine/guanine ratio in the first half of the cycle and a low ratio in the second half (Fig. 6.5). This has been shown both by *in vivo* labelling and also, in a more satisfactory way, by *in vitro* labelling and 'nearest neighbour' frequency analysis which eliminates the effect of varying nucleotide pools. These experiments are important because they give definite evidence that there are changes through the eukaryotic cycle in the pattern of transcription. They also raise acutely our need to know the function of the rapidly labelled nuclear RNA. If this fraction is composed of a large number of individual RNA species which vary in proportion through the cycle, it is lucky that this variation should be detectable by a measure as crude as the total base ratio.

In the Ciliates, RNA synthesis is continuous throughout the cycle. The rate increases during the last half of the cycle in *Paramecium* and it is possible that this is connected with DNA replication which also happens then (Kimball & Perdue, 1962; Woodard, Gelber & Swift, 1961). This suggestion of gene dosage does not, however, seem applicable to *Tetrahymena*, where, in one set of experiments, the main RNA rate increase was also towards the end of the cycle but the S period was in the first half of the cycle (Prescott, 1960). Nor does gene dosage fit the situation in *Euplotes* where the rate of synthesis increases through G1 to a maximum at the G1/S boundary and then declines through S (Evenson & Prescott, 1970).

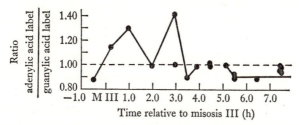

Fig. 6.5. Changes in base ratios of rapidly labelled RNA during the cycle of *Physarum*. From Cummins, Weisfeld & Rusch (1966).

Ciliates differ from most other eukaryotes in showing no gap in RNA synthesis during nuclear division (Prescott, 1964*a*, p. 88; Rao & Prescott, 1967). Since the macronucleus divides amitotically without chromosome condensation, this reinforces the hypothesis that the gap in synthesis in normal mitotic cells is caused by the changes in chromosome structure and not by other processes in nuclear division. We can predict that the Ciliate micronucleus, which divides by mitosis, ought to show the gap in synthesis. This indeed occurs in *Paramecium caudatum* (Rao & Prescott, 1967) though the position is complicated by the fact that the micronucleus has a gap in synthesis which is longer than mitosis and covers G1 and G2 as well. The type of RNA produced by the micronucleus is unknown.

Although RNA synthesis in the macronucleus does not stop during division, it *does* stop in one interesting situation in *Euplotes* (Prescott & Kimball, 1961). In this big Ciliate, DNA synthesis is spatially restricted to two bands (the reorganisation bands), about 4 μm thick, which move through the macronucleus during the S period. RNA synthesis takes place in all regions of the nucleus *except* the bands. There may be a mechanism

5

at the molecular level which stops transcription near a replicating fork, but the situation probably involves higher levels of organisation since there are visible changes in the chromatin both in the light and the electron microscope, and there is also a high level of histidine incorporation near the bands suggesting a simultaneous synthesis of histone. Generalising from this, RNA synthesis in eukaryotes could be switched off by changes in chromosome structure not only during the separative process of mitosis but also during the replicative process. The *Euplotes* nucleus is unusual in concentrating replication spatially, so the switch-off is visible in auto-radiographs. In most other cells, there is neither a spatial concentration which would permit autoradiographic detection nor a temporal one which might show the switch-off as a fall in the total rate of RNA synthesis. Early embryos, however, do concentrate DNA synthesis temporally and they would be the place to look for a diminished rate of RNA synthesis during the S period.

The fission yeast, *Schizosaccharomyces pombe*, was one of the first cells in which RNA synthesis was followed through the cycle. Synthesis is continuous throughout the cycle (Mitchison & Walker, 1959) and there is a fairly steady increase in rate both with ribosomal and with rapidly labelled nuclear RNA, as judged by long and short pulses of labelled adenine (Mitchison & Lark, 1962). These results have been confirmed in a more comprehensive study using four bases and two nucleosides with synchronous cultures as well as autoradiographic methods (Mitchison *et al.*, 1969). Uptake rates and pool sizes were also followed, and it seems that they change through the cycle in such a way that the rate of incorporation is in fact a true measure of the rate of synthesis. There were no signs of base ratio changes in the rapidly labelled nuclear RNA. This differs from the results with *Physarum* though the technique is less sensitive. On the other hand, there was a distinct shift in base ratio of 'step-down' RNA toward the end of the cycle. This kind of RNA is produced for a time after a change from a rich to a poor medium, when ribosomal RNA synthesis is suppressed (Mitchison & Gross, 1965). It was thought at first to be messenger RNA but this conclusion is not justified at present, and its relations to the RNA fractions synthesised in normal growth remain to be established.

Budding yeast, *Saccharomyces cerevisiae*, has a similar pattern of synthesis with ribosomal and transfer RNA being made at a steadily increasing rate through the cycle (Tauro *et al.*, 1969), though Williamson & Scopes (1960) found a gap in synthesis at about the start of the S period.

Neither fission nor budding yeast appear to show any gap in synthesis at the time of nuclear division. It is possible that such a gap does exist for short periods but it has not been detected because of the difficulty of staining nuclei in yeast and associating one stage in division with the absence of a pulse-label. But it is more likely that there is a genuine continuation of synthesis through nuclear division which may be due to the lack of fully condensed chromosomes. Mitotic chromosomes have not been demonstrated is *S. cerevisiae*, and, although they have been stained in some other yeasts, they are often apparent as round bodies differing in shape from the typical chromosomes of higher eukaryotes.

Chlorella shows continuous synthesis through the cycle with a pattern which is approximately exponential (Herrmann & Schmidt, 1965).

RNA synthesis continues throughout interphase in those eukaryotic cycles in which it has been measured, but we should remember that these cycles are all ones in which growth is also continuous. As we shall see in the next chapter, there are cell cycles in which there may be periods with no growth (e.g. *Amoeba, Schizosaccharomyces pombe* at low temperature) and, if so, it is likely that the synthesis of stable RNA species stops during these periods. It would be interesting to know whether the synthesis of unstable species continues and how it compares with the other situations in which ribosomal RNA production is suppressed – early embryos, and cell cultures in stationary phase or after a 'step-down'.

Prokaryotes

The details of RNA synthesis should be easier to follow in prokaryotes than in eukaryotes since there is no cytoplasmic DNA in bacteria and no reason at present to differentiate the rapidly labelled RNA into more than one class. Despite this, the evidence of RNA synthesis during the prokaryotic cell cycle is scanty, particularly by comparison with that on DNA.

Neglecting the earlier work with repetitive temperature shocks, there is good agreement that synthesis is continuous throughout the cell cycle.[1] As with eukaryotes, however, this work has been done in systems where growth is also continuous and the situation may be different in cycles where there appear to be periods of no growth (p. 145). There is less agreement on changes in the rate of synthesis of total RNA. In some cases the

---

[1] For *Escherichia coli*: Abbo & Pardee (1960), Cummings (1965), Cutler & Evans (1967), Manor & Haselkorn (1967), Rudner *et al.* (1965). For *Salmonella typhimurium*: Ecker & Kokaisl (1969). For *Proteus vulgaris*: Cutler & Evans (1966). For *Lactobacillus acidophilus*: Burns (1961). For *Rhodopseudomonas spheroides*: Ferretti & Gray (1968).

rate increases smoothly and in others it shows sharp changes which may be associated with stages in the cell cycle (Burns, 1961; Cutler & Evans, 1966; Ferretti & Gray, 1968; Maruyama & Lark, 1959 for 'sedimentable RNA'). If there is a gene dosage effect for rRNA, it should be detectable as a rate change during the cycle in cases where there is only one replicating fork in the DNA and the ribosomal genes are clustered round one region of the genome, as in *Bacillus subtilis* (Dubnau *et al.*, 1965). It is also possible that a gap in synthesis might occur in a well synchronised culture at the time when the fork was moving through the cluster. But the presence of multiple forks or multiple clusters (as Cutler & Evans, 1967, and Rudner *et al.*, 1965, have suggested in *E. coli*) would make rate changes difficult to detect.

There is evidence in *E. coli* of changes in the composition of the rapidly labelled RNA which is presumed to function as a messenger. Rudner *et al.* (1965) found rhythmic variations in the base ratio during the cycle. Cutler & Evans (1967) used a hybridisation technique and concluded that the genome was being continuously transcribed but that there were changes in the transcriptional activity of different segments of the genome. These results are in general agreement with the concept of ordered patterns of protein synthesis which emerges from the work on enzymes (Chapter 8) but it is much harder at present to isolate and measure a specific messenger than it is to do the same for its product, an enzyme.

Conclusions

RNA synthesis stops during mitosis, presumably because the condensed structure of the metaphase chromosomes prevents transcription. This mitotic gap does not occur in chromosomes which separate amitotically (prokaryotes and Ciliate macronuclei) and may not occur in yeast, which have a somewhat aberrant mitosis, or in cytoplasmic organelles. Total RNA synthesis is continuous through the remainder of the cell cycle and is not periodic like DNA synthesis in eukaryotes. The rate of synthesis increases through the cycle but there is no uniform pattern of increase. In some cases, particularly mammalian cells in culture, there is a pattern consistent with a gene dosage effect – the rate of synthesis doubles when the RNA genes double. In other cases there are smooth increases in rate.

Relatively little work has been done on the synthetic patterns of the various classes of RNA, and this field certainly needs further development. What evidence there is suggests that the main classes (ribosomal, transfer, messenger and rapidly labelled nuclear RNA) follow similar patterns of continuous synthesis through interphase. The composition of rapidly labelled RNA appears to change through the cycle in *Physarum* and *E. coli*.

7 Cell growth and protein synthesis

Volume, dry mass and total protein

One of the basic questions to ask about the cell cycle is what is the pattern of overall cell growth between one division and the next. But precise answers to this question can be given only in terms of certain measurable properties of a cell (Fig. 7.1). The most complete criterion of growth is volume, since it is a measure of all the components of a living cell. Total dry mass, measured either gravimetrically or by interferometry of living cells, covers all components other than water. Fixation or acid treatment removes the pool of low molecular weight compounds (a cellular component which varies considerably but can be as much as a quarter of the total dry mass) and leaves the macromolecular residue. Interferometry, therefore, on fixed cells measures what can be called 'macromolecular dry mass', though exactly which molecular species are removed by the fixative will depend on the treatment. Protein is by far the largest macromolecular component of most growing animal cells, so the total amount of protein will not differ much from the macromolecular dry mass. The difference is much greater in plant cells which have large quantities of carbohydrate polymers in the cell walls. Although total protein is the least complete of the measures of cell bulk, there is the technical advantage that the rate of protein increase can be followed by pulse-labelling with radioactive amino-acids.

CELLULAR CONSTITUENTS

Protein	Other macromolecules – RNA, DNA, carbohydrates	Low mol. wt. pool	Water

Macromolecular dry mass

Total dry mass

Volume

Fig. 7.1. Criteria (italic) which can be used for measuring cell growth.

If we could assume that the pattern of growth through a cycle were the same with all these criteria, it would only be necessary to measure one of them to define the pattern. We shall see, however, that this assumption is not justified and that in some of the cases where, for example, volume and total dry mass have been measured on the same cells, the patterns of increase are not the same. We should therefore consider briefly the type of information we can obtain from the different criteria. Increase in protein (or macromolecules) is the fundamental measure of cell growth, and it is molecular synthesis that distinguishes the growth of biological systems from that of inorganic systems like crystals. No sensible interpretation of cell growth can be made without a knowledge of the overall pattern of protein synthesis. Total dry mass can tell us something about the pool of small molecules and their flow into the cell. Control of this pool by restrictions on entry may well be a mechanism by which a cell alters synthesis or structure during the cycle. Volume gives direct information about the increase in the cell membrane or the cell wall and can also be a measure of the largest constituent of cells – water. Many of the mathematical models of cell growth involve fluctuating concentrations of controlling molecules, and volume is an essential part of concentration.

The more comprehensive of these criteria will yield their full measure of information only when they are combined with the less comprehensive. Total dry mass, for instance, will give the size of the pool only if macromolecular dry mass is known. The very completeness of volume makes it a difficult criterion to interpret if it is used as the sole measure of cell growth. An increase in volume may go hand in hand with molecular synthesis but it may also be due only to an uptake of water (as in many differentiating plant cells). The argument here leads to the conclusion that any comprehensive study of cell growth includes measurements of all the criteria. But studies of this sort are very rare, and we are handicapped at present by the fragmentary state of the information on cell growth during the cycle.

It can be argued that measurements of overall cell growth are unimportant because they add together without distinction a number of disparate components. The pattern of synthesis of individual proteins may indeed be very different from that of the total cell protein, and the pool contains a wide variety of compounds which may be located in a series of separate compartments. There is, of course, every reason for following these components but there are still questions to be answered at the level of overall growth. Cell growth is a surprisingly smooth and regular process and we can hardly fail to believe that there are mechanisms which regulate bulk properties such as total dry mass. Regulation at this level must involve a coordinated control of the synthesis both of RNA, the direct gene product, and of protein, carbohydrate and small molecules which are increasingly far from direct genetic influence. These mechanisms are as yet unknown, but we could guess that an essential part of them will lie in the fine control that can be exerted in a cell over the rate of uptake of nutrients.

A small, but not unimportant point is that volume and total dry mass are the only measures of growth which can be made without damage on living cells. With the right material it is possible to make successive measurements on single growing cells and to achieve a precision about the finer details of the cycle which cannot be equalled with synchronous cultures or other methods of cycle analysis.

Mammalian cells

Some of the best information about the overall growth pattern of mammalian cells comes from the work of Zetterberg & Killander. They measured the macromolecular dry mass of mouse L cells by interferometry and fitted the data to the cell cycle both by cell histories from films (Killander & Zetterberg, 1965a) and by frequency distributions (Zetterberg & Killander, 1965a). The results by the two methods are shown in Fig. 7.2. The dry mass increases continuously and doubles by the end of the cycle. There is an increase in rate through the cycle but the pattern of this increase differs slightly between the two curves. Most of the increase happens about half way through the cycle with the first method, and in the last quarter of the cycle with the second method. Points of rate change are difficult to determine in absolute measurements and their appearance at different points with the two methods suggests that they should not be given any great weight. Protein synthesis seems to follow a very similar pattern to dry mass increase in these cells. Pulse-labels with three amino-acids show continuous incorporation through the cycle with a steadily increasing rate (Zetterberg & Killander, 1965b).

Protein synthesis has been followed in other mammalian cell cultures by pulse-labelling with amino-acids. There is general agreement both that protein synthesis is continuous and that its rate increases through the cycle, but there are differences in the patterns of rate increase. HeLa cells (Scharff & Robbins, 1965) and Chinese hamster lung cells (Robbins & Scharff, 1967) show a steady increase in rate similar to that in L cells. Mouse

P815Y cells show a rate increase during G1 and S and a slight decrease during G2 (Warmsley & Pasternak, 1960). After colcemid arrest of rat hepatoma cells (Martin, Tomkins & Granner,* 1969) and Chinese hamster Don C cells (Stubblefield *et al.*,* 1967), there is constant rate for a period which is then followed by a sharp increase during the S period, reminiscent of some of the RNA synthesis patterns discussed in Chapter 6.

A sophisticated analysis of the volume growth of CHO cells has been made by Anderson *et al.* (1969). In many cases the rate of volume increase

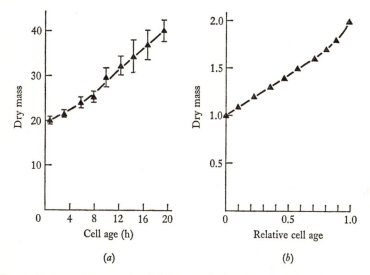

(a)   (b)

Fig. 7.2. Dry mass (macromolecular) increase during the cell cycle of mouse L-929 cells. (*a*) From cell histories. From Killander & Zetterberg (1965 *a*). (*b*) From frequency distributions. From Zetterberg & Killander (1965 *a*).

was proportional to cell volume at a particular stage of the cycle, giving an exponential curve of volume growth through the cycle with a doubling in rate between the beginning and the end. In some cultures, however, the rate increase was much greater than a doubling and the pattern of volume growth more curved than an exponential. Sinclair & Ross (1969) used a similar technique with two sublines of Chinese hamster cells. The volume increased smoothly with time and there were no sharp changes in rate. Both linear growth and exponential growth were acceptable as models, though an exponential gave a better fit to the data from one of the sublines. Terasima

* This asterisk, here and later in this chapter, marks experiments with cultures synchronised by DNA inhibitors or colcemid arrest. The reasons for this are explained on p. 33.

& Tolmach (1963a) also made estimates of cell volume in synchronous cultures of HeLa cells and found that there was an increase in the rate of volume growth during the cycle. The density of living CHO cells, measured in Ficoll gradients, fluctuates by less than 2 per cent during the cycle (Anderson *et al.*, 1970). This indicates that volume measurements on these cells are a fairly accurate measure of total dry mass. It may be that this is a general rule for many animal cells and that the fluctuations of density which, as we shall see, occur in cells with rigid walls such as yeast or bacteria are only tolerated because such cells can more easily resist the swelling action of osmosis.

The pattern that emerges from this evidence is one of continuous cell growth during interphase in cultured mammalian cells. Growth also seems to be non-linear with an increasing rate of protein synthesis and of growth in volume and macromolecular dry mass. Beyond this, we can say nothing with certainty. The apparent diversity in the patterns of rate increase may be a measure of true biological variation. Different cell strains in different media could well vary in the finer details of their growth, particularly since Anderson *et al.* (1969) have shown that cultures of the same cell line in the same medium can show different volume growth curves.

Protein synthesis is similar to RNA synthesis in being continuous through interphase and increasing in rate. Is it also similar in being shut off during mitosis? Most synchronous cultures show a sharp reduction in amino-acid incorporation but it is not cut off completely (e.g. Martin *et al.*,* 1969; Scharff & Robbins, 1965). Because of imperfections of synchrony, however, this residual incorporation at the time of mitosis may be due either to some continuing protein synthesis in mitotic cells or to the presence of contaminating interphase cells. Autoradiography gives a more precise picture. In the Chinese hamster cells of Prescott & Bender (1962), histidine incorporation during mitosis (at its minimum level in early telophase) was about a quarter of the interphase value. This is a definite reduction, though much less marked than it is with uridine incorporation into RNA (about one fiftieth). Konrad (1963) found a reduction of phenylalanine incorporation to about the half of interphase value in human amnion cells, but no reduction with hamster cells. There is also no reduction in the Chinese hamster cells of Taylor (1960a) and in sea urchin eggs (Gross & Fry, 1966).

The answer to the question about protein synthesis during mitosis is that the position seems more variable than it is with RNA synthesis. In some cells protein synthesis continues at an unaltered rate through mitosis and in others it is reduced. This is not a surprising conclusion. The mitotic cell is enucleate from the point of view of communication by RNA from the nucleus, but there is ample evidence that protein synthesis can continue in other cells that have been physically enucleated or chemically enucleated by actinomycin, presumably because long-lived messages are available in the cytoplasm. The problem here is to explain the cases where protein synthesis is reduced rather than those where it continues. One explanation is the presence of a large proportion of short-lived messages (a proportion which might be enhanced at mitosis compared with interphase) but another is that mitosis causes a block to protein synthesis in the cytoplasm. This type of translational inhibition has been invoked as a reason for the inability of Newcastle disease virus to multiply in HeLa cells arrested in mitosis by vinblastine (Marcus & Robbins,* 1963). These experiments can be criticised on the grounds that the cells were exposed for long periods (up to 20 hours) to the blocking agent and might be abnormal. Also, poliovirus will multiply in similar cells (Johnson & Holland,* 1965), which argues against the presence of a specific lesion in mitotic ribosomes (Salb & Marcus,* 1965). Fan & Penman* (1970*b*) find that the rate of protein synthesis is reduced to 30 per cent of the interphase level in CHO cells after colcemid arrest and they have evidence that this is due to a lowered rate of attachment of messenger RNA to the ribosomes. The amount of polyribosomes drops sharply at mitosis (Johnson & Holland,* 1965; Scharff & Robbins, 1966), but it is uncertain whether this is caused by the normal breakdown of messenger RNA which is not replaced from the nucleus or by a specific mechanism operating at this time in the cycle. In the latter case, a message which was being translated before mitosis might survive through the period and be used again after mitosis. Hodge, Robbins & Scharff (1969) have some evidence that this may occur, including the fact that protein synthesis recommences after mitosis in cells in the presence of actinomycin which should stop the production of new messages from the nucleus. But Buck *et al.** (1967) have contrary evidence and argue that new messages are essential for starting off protein synthesis after mitosis. The present situation is confused, particularly because of uncertainty about the possible side effects of colcemid, and we cannot tell for certain whether or not there is a mechanism of translational inhibition at mitosis. But if such a mechanism exists it is unlikely to be a universal phenomenon, even in mammalian cells in culture. This does not remove its interest, only its generality.

Lower eukaryotes

The work of Prescott (1955) on *Amoeba proteus* was a pioneer study of cell growth. He used a Cartesian diver balance to measure the 'reduced weight' of single living amoebae at intervals through a cell cycle of about one day. Reduced weight, or weight of the cell in water, is proportional to total dry mass provided there is no change in the density of the dry material. The reduced weight follows a growth curve with a fairly steady decrease in rate through the cycle and with virtually no growth in the four hours

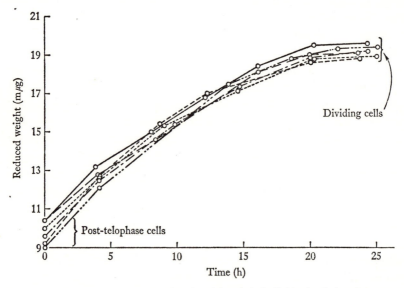

Fig. 7.3. Growth curves in reduced weight of six individuals of *Amoeba proteus* through the cell cycle. From Prescott (1955).

preceding division (Fig. 7.3). Total protein content gives a similar curve which makes it unlikely that there are large fluctuations in the pool. So also does cell volume, which indicates that the density of the amoebae does not change (Fig. 7.12). This work, especially with the diver balance, was a technical *tour de force* and provided a comprehensive picture of the growth of a well-known cell. It has been criticised on the grounds that amoebae have to be fed on solid food (*Tetrahymena*), which spends an appreciable time in food vacuoles before digestion and will be included in bulk measurements such as reduced weight. One does not know therefore how much of the total protein is undigested food and how much is amoeba protein. There is certainly truth in this criticism, but we should remember that it also applies in some measure to other cell systems. The small molecule food of other cells can remain for some time in the pool before it is incorporated, and mammalian cells may contain protein that comes from the culture medium.

The big multinucleate amoeba *Chaos chaos* has a similar pattern of increase in reduced weight, with minimum growth rates generally occurring before division (Satir & Zeuthen, 1961). In some cases, however, the cells do not divide at this point and one amoeba can show several successive periods of low growth rate before it finally divides. The mechanism that controls the growth cycle is in this case dissociated from the division cycle.

Turning to Ciliates, we find growth patterns that resemble those in mammalian cells. *Paramecium aurelia* has an exponential curve of increase of dry mass, as measured by interferometry and X-ray absorption (Kimball *et al.*, 1959). There is, however, a small but significant loss of weight during division. What was measured in these experiments was probably the total dry mass although some of the pool may have been extracted by the glycerol mounting medium. The solid food criticism also applies here since the cells were fed on bacteria. Similar results were found in *P. aurelia* by Woodard, Gelber & Swift (1961) using a cytochemical stain for measuring total protein. The curve for growth in total protein was one of increasing rate but there were indications of a particularly sharp rate increase during the S period. Volume growth also follows a roughly exponential curve (Schmid, 1967).

Tetrahymena appears to have a linear growth in volume except during division where there is a sharp increase (Cameron & Prescott, 1961). A similar result was obtained by Summers (1963). He did not find the increase at division but the frequency histogram method which was used is not very sensitive to small rate changes during a limited part of the cycle. On the other hand, Schmid (1967) found an increasing rate curve using the same type of technique. The increase in total protein was found to be linear as judged by the accumulation of incorporated methionine (Prescott, 1960). Lövlie (1963) performed the technically difficult feat of weighing single cells of *Tetrahymena* on a diver balance and of growing single cell clones in a diver respirometer (a method first developed and exploited by Zeuthen, 1953). Unlike the experiments of Prescott (1955) on amoeba, the cells were fixed with formalin before weighing, so the reduced weight was proportional to macromolecular dry mass. The respirometer measured oxygen consumption, which was called 'respiratory mass'. The results varied in

different experiments but approximated to one of the three main patterns
of growth curve – exponential, linear, and linear with a terminal plateau
(Fig. 7.4). Lövlie makes the suggestion that the fundamental growth
pattern is the exponential one but where growth is 'unbalanced' in the
sense that there is less than a doubling over the cycle, the cell is affected
in a sensitive phase at the end of the cycle and either slows down to produce
the apparent linear pattern or stops to produce the linear plus plateau
pattern. Whether or not this suggestion is valid (and there is some support
for it in the fission yeast results which are discussed later) these experiments

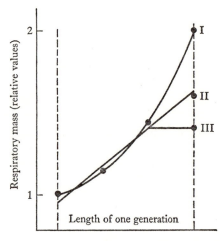

Fig. 7.4. Three possible growth curves of reduced weight or respiratory mass during a cell
cycle of *Tetrahymena*. I, Exponential. II, Linear. III, Linear with plateau around division.
From Lövlie (1963).

give a clear warning. Growth patterns may be much more labile than we
might imagine. They may alter not only with the slight differences between
one experiment and the next but also from cell to cell so that bulk measurements on synchronous cultures may give a misleading picture of the individual cell cycle. Growth may also be altered by the technique of measurements, as is apparent in some of the respirometer experiments. One other
warning is appropriate at this point. It needs very accurate measurements
to distinguish an exponential from a linear pattern. Fig. 7.4 exaggerates
the position since the maximum difference between an exponential curve
which doubles in rate over the cycle and the line of best fit is less than
3 per cent.

There is an interesting situation in the Ciliate *Bursaria truncatella* (Zech,

1966). Division is unequal with the 'anterior animal' being always smaller than the 'posterior animal'. The growth of the posterior animal in macromolecular dry mass (measured by X-ray absorption) is fairly steady through the cycle though there is a fall-off in rate towards the end. The anterior animal, however, has a period of three hours without any growth at the start of the cycle and then grows faster than the posterior animal for the rest of the cycle. Both animals finish up with the same dry mass at the end of the cycle. The anterior animal also has a shorter G1 in its DNA cycle than the posterior animal. What is unusual about this growth pattern is the

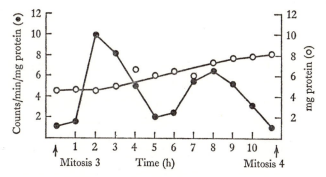

Fig. 7.5. Protein synthesis through the cycle of *Physarum*. Open circles show the total protein measured with the Biuret method. Closed circles show the incorporation of 10 min pulses of ³H-lysine. From Mittermayer, Braun, Chayka & Rusch (1966).

presence of a period of no growth at the beginning of the cycle. In most other cases, if there is such a period it occurs at the end of the cycle.

Physarum is another organism which shows little or no growth in total protein for 1–2 hours after mitosis (Mittermayer, Braun, Chayka & Rusch, 1966). Thereafter growth is fairly steady for 7 hours and then falls off slightly for the last 2 hours of the cycle (Fig. 7.5). Lysine pulses are in fair agreement with this. They show a fall in rate at the start and end of the cycle, but they differ from the total protein curve in showing another fall in rate in mid cycle. The resulting premitotic and postmitotic peaks in rate parallel those in RNA synthesis except that the postmitotic rise in rate is delayed. Another interesting result from this work is that the proportion of polyribosomes drops not at mitosis (as in mammalian cells, see earlier) but a little later to reach a minimum at 40 minutes after mitosis. Whatever mechanism restricts protein synthesis during mitosis must operate for an appreciable time after the very short metaphase–anaphase (about 5 minutes) in this organism.

A comprehensive survey of growth patterns has been made by my colleagues and myself on the fission yeast *Schizosaccharomyces pombe*. Measurements with an interference microscope on single growing cells show that the growth in total dry mass is linear over the cycle (Mitchison, 1957). Volume follows an approximate exponential curve for the first three quarters of the cycle and then remains constant for the last quarter (Fig. 7.6(*a*), see also Knaysi, 1940). During this 'constant volume phase', the cell plate or septum forms across the cells. Since this cell plate is made of cell wall material and there is no apparent change in the thickness of the main cell wall through the cycle, it is likely that the curve for cell wall growth is roughly exponential over the whole cycle. This point is discussed later (p. 190). The differing shapes of the volume and dry mass curves show that the cell density or concentration fluctuates through the cycle with a minimum at the beginning of the constant volume stage and a maximum at the end of the cycle. Although the total dry mass increase is linear, the pattern of synthesis of each of the main classes of macromolecules (protein, carbohydrate and RNA) is an increasing rate curve which is nearly exponential, as judged by pulses of the appropriate precursors (Mitchison & Wilbur, 1962; Mitchison & Lark, 1962). This apparent paradox is easy to resolve if we remember that the total dry mass includes the pool as well as macromolecules, so changes in pool size would account for the difference in the curves. These changes were demonstrated in a series of interference microscope measurements on cells before and after extraction of the pool (Mitchison & Cummins, 1964). The pool fluctuates by a factor of about 2 with a maximum size in mid cycle (Fig. 7.7). This fluctuation is out of phase with respect to the changes in cell density in which the peak value is later in the cycle.

There are other ways of describing these growth patterns. The cell nutrients are built into macromolecules at an increasing rate through the cycle yet they are accumulated at a constant rate. Note here that this is net accumulation and not uptake since there is no measure of the outflow of materials (including CO_2). Neither nuclear division nor cell division cause any cessation of synthesis or accumulation. Cell wall synthesis follows the same pattern of increasing rate as the other macromolecular classes, yet there is a constant volume stage near the end of the cycle because the growing point moves from the tip of the cell where it produces volume increase to the site of the cell plate where it does not. During this stage there must be a sharp drop in the rate of entry of water. The facts here are fairly clear but the deeper meaning is still elusive. Interpretation is also

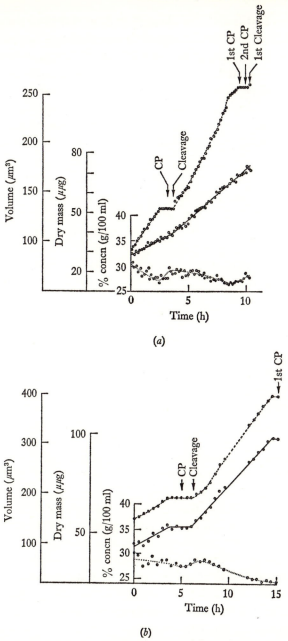

Fig. 7.6. Growth curves of cells of *Schizosaccharomyces pombe* from interference micro-scope measurements. Top curve, volume; centre curve, total dry mass; lower curve, concentration. (*a*) Growth at 23 °C. One cell is followed through division to two daughter cells and then to the division of one of these daughters. This pattern of growth is shown by all cells in the range 23 °C to 32 °C. CP, first appearance of cell plate. (*b*) Growth at 17 °C. One cell is followed through division to two daughter cells and then to the appearance of a cell plate in one of these daughters. From Mitchison (1963).

made no easier by the fact that these growth patterns are not fixed attributes of this cell and can change with alterations in temperature and medium. Total dry mass increases linearly throughout the cycle at temperatures between 23 °C and 32 °C, but there is a change in pattern at lower temperatures and at 17 °C nearly all cells have a plateau with little or no dry mass increase during the last quarter of the cycle (Mitchison *et al.*, 1963; Fig. 7.6(*b*)). The volume growth curve, however, remains unaltered in shape. The appearance of the dry mass plateau at the end of the cycle

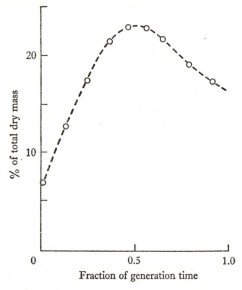

Fig. 7.7. Change in the total acid-soluble pool (as % of total dry mass) during the cell cycle of *Schizosaccharomyces pombe*. From Mitchison & Cummins (1964).

resembles the situation in *Tetrahymena* where it is this stage of the cycle that appears to be most liable to changes of pattern (p. 136). The growth curves which have been described are for cells growing in a rich medium (wort or malt extract broth) but in a minimal medium there appears to be no pool fluctuation and the total dry mass curve is exponential rather than linear (Stebbing, 1969). The pool in minimal medium is small and it may be that the larger pool in rich medium has an added fluctuating component which corresponds to the 'expandable pool' of amino-acids described in budding yeast by Cowie & McClure (1959).

The budding yeast, *Saccharomyces cerevisae*, is similar in many respects to fission yeast. Volume growth shows a plateau towards the end of the cycle (where the start of the cycle is taken as bud initiation), though the plateau is not as sharply defined as in fission yeast and the growth curve is more sigmoid in shape (Bayne-Jones & Adolph, 1932; Lindegren & Haddad, 1954; Mitchison, 1958; Scopes & Williamson, 1964). Total dry mass increases with the same linear pattern as in fission yeast. There is no pause at the end of the cycle and in many cases a sharp increase in rate (Mitchison, 1958; Scopes & Williamson, 1964). Total cell density therefore fluctuates with a maximum at the start of the cycle. Unlike fission yeast, total protein seems to follow the same linear pattern as total dry mass, though the data are not accurate enough to make a clear-cut distinction between linear and exponential patterns (Williamson & Scopes, 1961 *b*; Gorman *et al.*, 1964; Hilz & Eckstein, 1964). There is some evidence of pool fluctuations, since total nitrogen follows a different time course through the cycle from total protein (Williamson & Scopes, 1961*b*). The results are somewhat variable but they suggest that the pool nitrogen may reach a maximum roughly in mid cycle – the same point where the fission yeast pool is also at a maximum.

The volume growth of *Euglena gracilis* has been determined from the volume distribution in an asynchronous culture measured with a Coulter counter (Kempner & Marr, 1970). The curve of increase is sigmoid, similar to that in budding yeast. *Chlorella* shows an exponential increase through the cycle in total dry mass and protein nitrogen (Schmidt, 1969).

We can ask whether the reduction in protein synthesis which occurs during mitosis in some higher eukaryotes also occurs in the lower ones, but we will not get a clear answer because of the scarcity of evidence. The most sensitive test is to pulse-label with amino-acids but this has only been done on two of the lower eukaryotes. To recapitulate, there is reduction in lysine incorporation during and, unlike higher cells, after mitosis in *Physarum* (Mittermayer, Braun, Chayka & Rusch, 1966). Fission yeast, on the other hand, shows no reduction in amino-acid incorporation (Mitchison & Wilbur, 1962) – a situation which is similar to that with RNA precursors and may be due to the unusual chromosomes in yeast (p. 125). As far as one can tell from dry mass curves, there appears to be a plateau (or fall) at division in *Paramecium* (Kimball *et al.*, 1959) and in some cells of *Tetrahymena* (Lövlie, 1963). There is also a plateau in *Amoeba* (Prescott, 1955) but this starts several hours before division. The position in *Amoeba* underlines one difficulty in interpreting a reduction in protein synthesis at division. It may be that this reduction is caused by changes in the cell at mitosis, as has been suggested for mammalian cells. But this cannot explain the plateau in *Amoeba* which starts much earlier and is the culmination of a curve of decreasing rate rather than an abrupt cessation. Nor would it explain the appearance of the plateau only in some cells of *Tetrahymena* or only at some temperatures in fission yeast. In these cases it is easier to interpret the plateau as marking a sensitive stage at the end of the cycle rather than having a close causal connection with mitosis.

Prokaryotes

Considering the great importance for molecular biology of bacterial rods in general and *Escherichia coli* in particular, there is surprisingly little information about their overall growth patterns. There is general agreement that volume (and length) growth is continuous over most of the cycle in fast-growing rods, but the detailed pattern is uncertain. Microscopic measurements on single living cells suggest roughly exponential growth for *Bacillus megaterium* (Adolph & Bayne-Jones, 1932), for *Bacillus cereus* (Collins & Richmond, 1962) and for *Salmonella typhimurium* and *E. coli* B/r (Fig. 7.8, Schaechter *et al.*, 1962), but the fit is only approximate and there is an appreciable possibility of error in measuring these small objects. Collins & Richmond (1962) used an intricate analysis of length distributions of *B. cereus* and showed that this gave the same pattern of an increasing rate curve through the cycle as their measurements on single living cells. The Collins–Richmond analysis was applied by Harvey *et al.* (1967) to Coulter counter measurements of the volume of *E. coli* and *Azotobacter agilis*. The growth rate increased during most of the cycle but the curves differed in detail from an exponential model. This technique, however, has been criticised by Koch (1966) and Kubitschek (1969*b*) and there is some doubt as to whether present methods of analysing cell size distributions in asynchronous cultures have sufficient resolution to distinguish linear from exponential growth. Linear growth (for at any rate most of the cycle) has been found for *E. coli* by Kubitschek (1968*a*) with a more sensitive technique in which size distributions are measured on synchronous cultures. It is clear at the present time that the volume growth pattern is fairly near an exponential or a linear model (which are of course very similar) but we do not know which is the best fit or whether any model is a universal one which can be applied to all bacterial rods in a variety of media. One other facet of volume growth is that there is a short plateau at the time of division which has been detected by time-lapse photography of single cells of *E. coli* (Hoffman & Frank, 1965). This constant volume

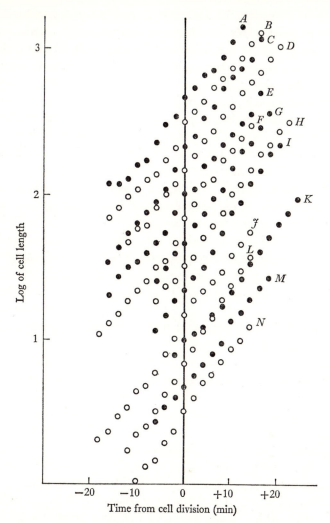

Fig. 7.8. The growth in length of individual cells of *E. coli* B/r, measured every 2 min. The lengths of resulting daughter cells are added. The lengths at division of individual cells are here spaced out equally along the ordinate which is in arbitrary units. In the experiments, the length at division varied between 4.1 and 6.0 μm. From Schaechter *et al.* (1962).

period lasts for 1.5 minutes in a generation time of 21 minutes, and though relatively shorter, resembles the equivalent stage in the fission yeast *S. pombe*. In both cases, the formation of a septum interrupts the growth in length.

Growth in total protein is somewhat similar to growth in volume. There is an approximately exponential pattern in *E. coli* (Abbo & Pardee, 1960) and in *Rhodopseudomonas spheroides* (Ferretti & Gray, 1968), whereas there is a linear pattern in *Lactobacillus acidophilus* (Burns, 1961). These measurements, however, are all on total protein in synchronous cultures where it is difficult to distinguish with certainty between linear and exponential patterns. With total protein, however, it is possible to get a more sensitive measure of rate by using pulses of precursors. Here again the results are equivocal. The rate of incorporation increases during the cycle, indicating an exponential pattern, with leucine pulses in *S. typhimurium* (Ecker & Kokaisl, 1969). In *E. coli*, however, the rate remains constant, indicating a linear pattern, with leucine and glycine pulses (Kubitschek, 1968*b*). These experiments with *E. coli* were discussed in terms of uptake, but it is likely that what was measured was incorporation, since the cells were treated with formalin before counting. They also show that there is a difference between adding a labelled precursor to cultures previously grown in the presence of the same unlabelled precursor and adding the precursor *de novo*. The former procedure, which is least likely to disturb the synthetic patterns, gives the linear pattern while the latter gives an approximately exponential one. This is an important point which is often neglected in precursor pulse experiments.

Nishi & Kogoma (1965) found an interesting situation in two strains of *E. coli* synchronised by filtration. There was a plateau at the end of the cycle in the net increase in total protein and in constitutive β-galactosidase activity. During this period, however, there was continued incorporation of labelled leucine and sulphate, and no reduction in the capacity to form induced enzymes. In prelabelled cells, also, there was a temporary decrease in the activity of the protein fraction at this time in the cycle. The implication here is that there is a partial breakdown and turnover of protein at the end of the cycle which might perhaps be correlated with the increase in proteolytic activity which is also found at this time (Kogoma & Nishi, 1965). Although the total protein turnover may be small in growing cells (Koch & Levy, 1955), the possibility of a restricted period of breakdown should be borne in mind both in *E. coli* and in other cells.

So far we have been almost entirely concerned with bacterial rods, and we can now ask whether the same patterns are shown by bacteria with a different morphology – the Cocci. The information is scanty, but what there is suggests that there may be quite different patterns. Knaysi (1941) found an S-shaped curve for the volume increase of *Streptococcus faecalis*. This was confirmed by Mitchison (1961) who made interference microscope measurements of volume and dry mass on the same organism. The total dry mass curves are somewhat different and resemble the decreasing rate curves of *Amoeba proteus* (Fig. 7.9). The relation between the dry mass curve and the volume curve appears to vary from cell to cell, and also on average between the two growth temperatures (40 °C and 17 °C) that were used. There are concomitant variations in cell density. It may be that the

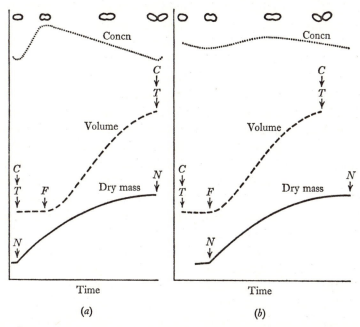

Time   Time

(*a*)   (*b*)

Fig. 7.9. Diagrammatic growth curves of individual cells of *Streptococcus faecalis* from interference microscope measurements. (*a*) The majority pattern at 17 °C. (*b*) The majority pattern at 40 °C. Top curve, cell concentration or density; middle curve, cell volume; lower curve; total dry mass. *T* is the 'turning' point when the two daughter cells are finally separate and usually turn at angle to each other. *F* is when a furrow first appears at the surface. *C* may mark the start of the cell cycle and *N* of the nuclear cycle. From Mitchison (1961).

volume curve is linked to the cell cycle and the dry mass curve is linked to the nuclear cycle. If so, the phase relation between these two cycles varies between different cells at the same temperature, and also on average between the two growth temperatures. This variable relation between the nuclear and the cell cycle also occurs in *E. coli* and *S. typhimurium* (Schaechter *et al.*, 1962).

One conclusion from this short survey of growth patterns in bacteria is that they seem to vary as much as those of eukaryotes. But what we know is still very limited and there are large gaps – there is, for example, no information about the variation of pool size during the cycle or of the growth patterns of slow-growing bacteria where the DNA cycle is certainly different from that in fast growth (p. 106).

Linear patterns

Although linear growth is not the dominant pattern, it does occur both in higher and lower eukaryotes as well as in prokaryotes, and it is worth considering briefly one of its implications (see also Kubitschek, 1970). Growing cells increase in a multiplicative manner (1 to 2 to 4 to 8) and so an exponential growth curve will increase smoothly through a series of cycles and will show no discontinuities of rate. With a linear pattern, on the other hand, there must be discontinuities with a doubling of rate once per cycle (e.g. Fig. 7.6). This raises the question of what cellular event is associated with the point of rate doubling. This point seems to be in mid-cycle and associated with the S period in those mammalian cells which show linear patterns, and it is tempting to believe that the doubling in rate might be due to the doubling in DNA, though there is no firm proof of this. But in lower cells (*Tetrahymena*, yeasts and bacteria) the rate doubling point is at the end of the cycle. This is not at the time of DNA synthesis in budding yeast and *Tetrahymena*, or in bacteria where the S period occupies the whole cycle. It is at the same time in fission yeast but there is evidence that the 'functional' replication of the genome occurs later in the cycle (p. 167). What else then doubles at the end of the cycle? Clearly the number of cells does this, but it is difficult to see why splitting a cell in two (a separation event and not a synthetic event) should double the growth rate. It is more plausible to implicate the nucleus in eukaryotes since division produces two complete nuclei each with a full set of chromosomes.

Another alternative is control by the number of growing points. Fission yeast cells, for example, usually have a single growing point (p. 190) so two growing points are formed after division. Even so the situation here is not a simple one since, although the total dry mass shows a rate doubling at division, the rate of wall growth (which one might suppose would reflect the number of growing points) does not show a doubling at that point and follows an exponential curve. A third possibility is a doubling in the uptake mechanism, possibly by a sharp increase in the number of carrier molecules in the cell membrane. It is impossible now to decide between these or other alternatives since there is very little relevant evidence, but it remains an intriguing problem.

Conclusions

The commonest pattern for growth in volume, dry mass or total protein is one of continuous increase through the cycle. The rate of growth may increase smoothly and double through the cycle, giving a roughly exponential curve, or it may stay constant giving a linear pattern until it doubles sharply at one point in the cycle. When there is a rate doubling in mammalian cells, it seems to occur in mid-cycle. But in lower eukaryotes and bacteria, it occurs at the end of the cycle. A different pattern with a decreasing rate through the cycle and a plateau at the end with no growth has been found in two cells – *Amoeba* and *Streptococcus*. A terminal plateau also occurs in the volume or dry mass growth of some eukaryotes and prokaryotes which have linear or exponential growth for the rest of the cycle. There are no cells in which there is a marked burst of growth before division (as in nuclei) or a growth curve with a rate increase much greater than a factor of 2.

In the few cases where growth parameters have been measured on the same cell, these parameters do not necessarily follow the same pattern of increase. As a result, low molecular weight pools and cell density may fluctuate during the cycle.

In some higher cells, the rate of synthesis may be reduced during mitosis, but in others it remains unaltered. It is uncertain whether or not there is a translational inhibition of protein synthesis.

A major problem in the interpretation of growth curves is that the evidence is not only scanty in coverage but also lacks in many cases the precision needed to distinguish patterns as similar as exponential curves and lines. In addition, there are cases which show that the shape of growth curves can vary with changes of medium and temperature, between different cultures, and from cell to cell so that synchronous cultures may give only an average pattern.

Nucleus and cytoplasm

We have followed in the preceding section the patterns of growth of the whole cell. It is now time to see how far these patterns are mirrored in the growth of the components of the cell. Individual protein components will be considered later, and this section will concentrate on the growth of the nucleus, the one component organelle which has been studied in a number of different systems. Other organelles will be considered in Chapter 9.

The work of Zetterberg (1966*a*, *b*, *c*) provides a detailed picture of the growth relationships of nucleus and cytoplasm in mouse L cells. Most of the *net* nuclear growth, in terms of macromolecular dry mass, happens in the second half of the cycle (Fig. 7.10(*a*)). Since the dry mass of the whole cell follows an approximately exponential curve (e.g. Fig. 7.2), by subtraction the cytoplasmic dry mass shows most growth in the first half of the

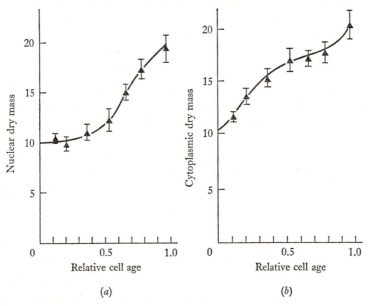

Fig. 7.10. Growth in dry mass (macromolecular) during the cycle of mouse L-929 cells. Vertical bars are standard errors of mean. (*a*) Nuclear dry mass. (*b*) Cytoplasmic dry mass. From Zetterberg (1966*a*).

cycle (Fig. 7.10(*b*)). In this cell line, incidentally, the nuclear and cytoplasmic dry masses are about equal. The pattern of protein synthesis in the cytoplasm appears to be different from the dry mass. Short (and long) pulses of leucine incorporated in the cytoplasm show a constant *specific* activity per unit mass of cytoplasm throughout the cycle (Fig. 7.11, see also Zetterberg, 1966*b*). Since the cytoplasmic mass increases through the cycle, the rate of incorporation and, by assumption, the rate of protein synthesis in the cytoplasm also increases. There is therefore a discrepancy between accumulation of cytoplasmic dry mass which has a lower rate in the second half of the cycle than in the first half, and cytoplasmic protein synthesis which has a higher rate in the second half. Unless there is considerable protein turnover (and there is evidence against this), the implication is that a substantial proportion of the protein synthesised in the cytoplasm in the second half of the cycle is transported to the nucleus and accumulated there. It can be calculated that 65 per cent of the final nuclear dry mass has come from the cytoplasm during this period. It is likely that

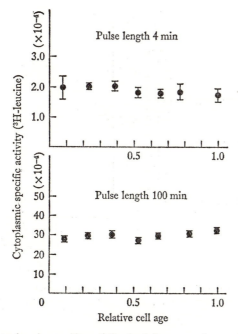

Fig. 7.11. Cytoplasmic specific activity (activity per unit cytoplasmic dry mass) of incorporated ^{3}H-leucine (mean + standard error) after 4 and 100 min pulses during the cycle of mouse L-929 cells. From Zetterberg (1966a).

much of this material is the non-chromosomal protein which leaves the nucleus at mitosis and which can make up to 75 per cent of the mass of a prophase nucleus (Richards & Bajer, 1961).

So far, the evidence points to transport of protein from cytoplasm to nucleus, but there are further data which suggest that there is also transport in the opposite direction. After a pulse of leucine the specific activity of the incorporated label in the cytoplasm falls by about 30 per cent in the first hour and then stays fairly constant (Fig. 7.12(a)). The nuclear activity shows a reverse pattern with a rise followed by a plateau (Fig. 7.12(b)).

This is what would be expected with transport from cytoplasm to nucleus, but there is the interesting additional observation that these labelling kinetics exist throughout the cycle *including* the first half (G1). Since there is little or no accumulation of nuclear protein during this first half (Fig. 7.10(*a*)), there must be outward transport of protein from the nucleus to balance the inflow. It is not, however, clear whether this outflow also occurs in the second half of the cycle.

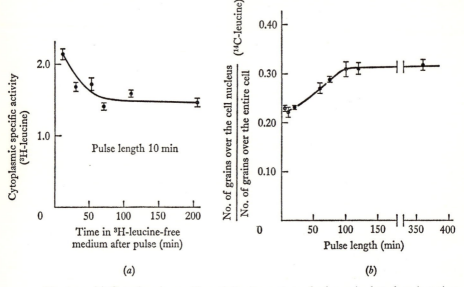

Fig. 7.12. (*a*) Cytoplasmic specific activity (mean + standard error) plotted against time in non-radioactive growth medium after a 10 min pulse of ³H-leucine. (*b*) Relative nuclear grain counts (mean + standard error) plotted against pulse time in ¹⁴C-leucine. From Zetterberg (1966*c*).

The picture that emerges from this work on L cells is of nuclear proteins moving to and fro across the nuclear membrane, at any rate in the first half of the cycle. These proteins are largely made in the cytoplasm but they accumulate in the nucleus during the second half of the cycle and give a net increase of dry mass. The shuttling action and the cytoplasmic site of synthesis are very much in accord with the results from the important experiments of Goldstein & Prescott (1967; 1968) on the proteins of the *Amoeba* nucleus. The functions of these proteins is at present unknown, though we may guess that they are concerned with the passage of information from nucleus to cytoplasm and vice versa.

The macromolecular dry mass of mammalian nuclei has also been measured by Seed (1962; 1964; 1966). In fresh embryonic cells (mouse and man), the nuclear dry mass followed the same growth curve as DNA giving a pattern similar to that of the L cells in Fig. 7.10(a). In tumour cell lines (mouse Ascites and HeLa), however, there is some evidence that the growth curve of nuclear dry mass differs from that of DNA and is more continuous. This difference between embryonic cells and mouse Ascites tumour cells was not confirmed in similar measurements made by Bassleer (1968). He found a continuous increase in nuclear dry mass through the cycle in both types of cell, but he points out the difficulty of deriving accurate synthesis curves from his data.

Macromolecular dry mass has been followed in the nucleus of *Paramecium aurelia* by Kimball *et al.* (1960). Its curve of increase is similar to that of the L cells with a doubling in the later part of the cycle. This is like the DNA, which also has a late S period, but is unlike the dry mass of the whole cell which increases exponentially.

Nuclear volume also follows a growth curve with the majority of the increase happening in the later stages of the cycle and a rapid rise before division. This volume has been measured in liver cells (Carrière *et al.*, 1961), in bean root meristem (Woodard, Rasch & Swift, 1961), in *Amoeba* (Fig. 7.13; Prescott, 1955), and in the Ciliates *Tetrahymena* (Summers, 1963), *Frontonia* (Popoff, 1908) and *Paramecium caudatum* (Popoff, 1909). Cameron & Prescott (1961) found a more linear increase in *Tetrahymena* though there was a sharp rise at division. In the dividing cells of pea roots, nuclear volume increases five times through the cycle but the dry mass only doubles, so there must be a considerable increase in hydration towards the end of the cycle (Lyndon, 1967).

There is a remarkable congruity about the growth curves for nuclear dry mass and volume over a wide range of cells. With a few exceptions, the major increases all take place in the later parts of the cycle before division. It would be tempting to associate this with the S period, particularly since a marked swelling occurs at or just before DNA synthesis in nuclei activated by injection into egg cytoplasm (Gurdon & Woodland, 1968) or by virus-induced fusion (Harris, 1967). But such an association is ruled out in the liver cells where the volume increase starts after the beginning of S and continues through G2, and even more strikingly in *Amoeba* where the S period is at the start of the cycle. Nuclear increase certainly seems to be one of the preparations for division, and it would be interesting to know whether it was due to the accumulation of a particular type of protein – perhaps microtubular protein.

This raises a point about nuclear division. The contents of a whole cell are divided more or less equally between the two daughter cells. The chromosomes are also divided equally – indeed this is the whole purpose of mitosis – but they are only a portion of the nuclear contents. The rest of the nucleus (up to 75 per cent of the nuclear dry mass) escapes into the cytoplasm when the nuclear membrane breaks down in prophase and we do not know how much of it returns to the two daughter nuclei when they reform in telophase. The contents of a telophase nucleus are not necessarily

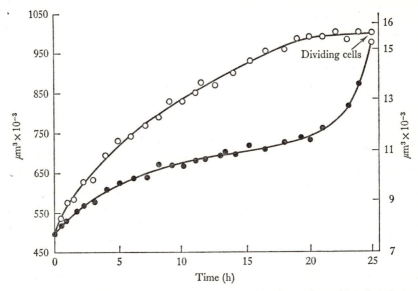

Fig. 7.13. Cytoplasmic volume (open circles) and nuclear volume (closed circles) during the cell cycle of *Amoeba proteus*. From Prescott (1955).

an exact half of the contents of a prophase nucleus. Nor is there any obvious reason why the dry mass or volume should halve at mitosis, though in practice this does seem to happen, at any rate approximately. A particular protein could well be accumulated in large quantities at the end of a cycle, lost to the cytoplasm in mitosis and not reappear in the telophase nuclei – though it might have to be replaced by a substitute to keep the dry mass up. There is less opportunity of loss in those lower eukaryotes where the nuclear membrane does not break down at mitosis, though even here there is the possibility of a substantial change in composition through protein exchange across the nuclear membrane. A conclusion from this argument is that we may find much larger changes through the cycle in the composition of nuclei than in that of whole cells. The amount of a stable macro-molecule can only increase by a factor of 2 in a whole cell whereas the increase in a nucleus could be much greater.

In summary, there appears to be a nearly universal rule that nuclear growth in dry mass and volume differs from the growth of the whole cell in showing most increase in the later parts of the cycle and especially just before division. Increase in nuclear dry mass also differs from that of the whole cell in being mainly a net rather than a gross accumulation of protein, since protein molecules are probably moving in both directions across the nuclear membrane.

Histones

Histones are proteins which are important but difficult to define precisely. Murray (1964) calls them 'basic proteins that at some time are associated with DNA'. This leads to an operational definition as those proteins which are extracted at low pH (or by salt solutions) from 'chromatin' or washed nuclei. Some but not all histones are rich in lysine or arginine and the ratio of these amino-acids is often used as basis of classification. They also lack tryptophan in many cases, and have molecular weights which are relatively low for proteins. Histones are absent in bacteria, and do not have a homo-geneous pattern through the eukaryotes although there are regularities within a group such as the mammals, and a remarkable resemblance between the histones of calf thymus and pea-buds. Their association with DNA makes it likely that they are integral parts of chromosomes and are important as structural components and perhaps as regulators of gene expression.

A good deal of work has been done on the synthesis of histones through the cell cycle, and the results can be divided broadly into those stemming from cytological techniques and those from biochemical ones. The early cytological work was photometric using alkaline fast green as a stain for histones. This showed that nuclear histones doubled over the same period in which the DNA doubled, (the S period) in mammalian cells (Bloch & Goodman, 1955), in plant cells (Woodard, Rasch & Swift, 1961) and in the reorganisation band of *Euplotes* (Gall, 1969; see also Prescott & Kimball, 1961). Tracers have been used in recent years, especially labelled arginine and lysine which should be incorporated preferentially in some of the histones. Histone labelling of the macronucleus of *Euplotes* is absent in GI but present throughout S at a constant rate (Fig. 7.14; Prescott, 1966*b*). The spermatocytes of *Urechis* incorporate 2–4 times more arginine, lysine and phenylalanine during S than during GI or G2. This increased incorporation occurs mainly in histones which do not incorporate tryptophan and which are acid soluble (Das & Alfert, 1968). Bloch *et al.* (1967), using the

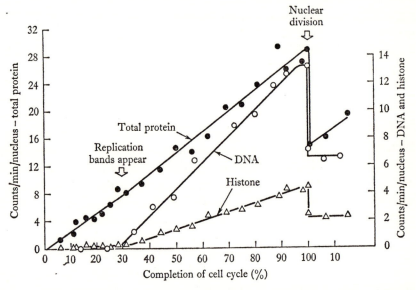

Fig. 7.14. Accumulated incorporation of ³H-thymidine into DNA and ³H-amino-acids into total protein and histone during the cell cycle of *Euplotes eurystomus*. From Prescott (1966*b*).

labelled mitoses method, showed that the S period was the main time in which lysine and arginine were incorporated into the chromosomal proteins of onion root cells, though there was also a subsidiary time in G2. Bloch & Teng (1969) found two classes of histone in the X chromosome of grasshopper spermatocytes, one of which increased in parallel with the DNA whereas the other increased later. Histone labelling also occurred during the S period.

There is a different picture in cultured leucocytes stimulated by phytohaemagglutinin where lysine and arginine are incorporated into chromosomal protein at all stages and not solely in the S period (Cave, 1966; Schneider & Rieke, 1967; Shapiro & Levina, 1967). This difference, however, may be more apparent than real since the cells were fixed with acid fixatives. Swift (1964) found that acid fixatives did not remove appreciable quantities of the histones of *Drosophila* salivary chromosomes, but Dick & Johns (1968) found the reverse to be true with biochemical assays of calf thymus, a tissue which is nearer to leucocytes. It is quite possible, therefore, that what was being followed in the leucocytes was the acidic protein of chromosomes and not the histones.

On balance, the cytological work suggests that many of the histones are synthesised at the same time as DNA. But several warnings should be made. The first is that arginine and lysine are not specific labels for histones, nor is tryptophan a satisfactory 'negative' label with lower eukaryotes. The second is that tritium from labelled arginine can be incorporated into DNA if there is a long period between labelling and fixation (see the criticism of Chernick, 1968 by Comings, 1969). The third, and far the most serious warning is that although the results that have been described (apart from those on leucocytes) are consistent with histone synthesis being restricted to the S period, they are also consistent with a different hypothesis – that histones are synthesised continuously through the cycle in the cytoplasm but only move into the nucleus and onto the chromosomes when new DNA is being made during S. The one set of results that would not fit this second hypothesis is that from *Urechis*.

This last criticism also applies to most of the biochemical work where the histones have been extracted by acid from isolated nuclei in synchronous cultures and have been found to be synthesised concomitantly with DNA.[1] Two papers suggest that histones are synthesised in the cytoplasm but only during the S period (and another recent paper by Pederson & Robbins (1970*a*) indicates that the control is not translational). Robbins & Borun (1967) found that there were small cytoplasmic polyribosomes in synchronised HeLa cells which contained nascent protein with a higher lysine/tryptophan ratio during the S period than during G1, and that this high ratio did not occur when DNA synthesis was stopped by inhibitors. Gallwitz & Mueller* (1969) showed that a microsomal preparation from HeLa cells would incorporate lysine, arginine and leucine into an acid-soluble protein only when the microsomes were prepared from cells in S. In neither case, however, was there a clear demonstration that the proteins

---

[1] For mammalian cells: see Gurley & Hardin* 1968; Hancock 1969, Littlefield & Jacobs* 1965, Robbins & Borun 1967, Sadgopal & Bonner* 1969, Spalding *et al.** 1966. For regenerating liver: see Niehaus & Barnum 1965. For heat-synchronised *Tetrahymena:* see Christensson 1967, Hardin *et al.* 1967. For *Physarum:* see Jockusch *et al.* 1970, and discussion on p. 157. For *Schizosaccharomyces:* see Duffus 1971.

involved were nuclear histones and not another type of protein with similar properties (e.g. ribosomal proteins).

Even if the major increase in histones takes place during S, there may be minor net increases or turnover during the rest of the cycle. The evidence is mixed. There is definite turnover of at least some histones in HeLa cells synchronised by DNA inhibitors (Spalding *et al.*,* 1966; Sadgopal & Bonner,* 1969), in *Vicia* root synchronised by aminouracil (Prensky & Smith,* 1964), in unsynchronised Chinese hamster cells (Gurley & Hardin, 1969) and in calf endometrium (Chalkley & Maurer, 1965). On the other hand, Hancock (1969) found no turnover during eight generations of mouse mastocytes and suggested that the different results with other mammalian cells might be due to turnover being caused directly by the DNA inhibitors used for synchronisation. There is indeed direct evidence of this in the results of Chalkley & Maurer (1965), but it would not explain the turnover in unsynchronised cells, nor would it account for the situation in *Amoeba* where the amount of stable nuclear protein is negligible (Goldstein & Prescott, 1968). At present the best way to resolve these differences is to suggest that there is real variation in the degree of histone stability and in the proportion which is synthesised concomitantly with DNA.

To summarise, many of the nuclear histones double in amount during the S period. This doubling may be associated with histone synthesis, but the evidence is not clear cut. In some cases, there is histone synthesis outside the S period which is manifest as turnover or net accumulation.

Other proteins

Leaving enzymes and 'division proteins' to later chapters, we can now examine the few cases in which other proteins have been followed through the cycle.

In HeLa cells synchronised by wash-off, the protein that binds colchicine appears to be synthesised throughout the cycle (Robbins & Shelanski, 1969). This protein, which separates as one peak on acrylamide gel electrophoresis, is believed to be the main component of the microtubules of the mitotic apparatus and perhaps of some other structures. Similar results were found with vinblastine precipitation which may be another way of identifying microtubular protein.

Jockusch *et al.* (1970) have pulsed a *Physarum* plasmodium with labelled leucine and lysine and then electrophoresed the acid extract from isolated nuclei on acrylamide gels. There are a set of peaks identified as the main histones which have a higher radioactivity with pulses in S than with pulses in G2, and another peak, migrating more slowly, which behaves in the reverse way and has the highest activity in G2. This latter peak is regarded as a non-histone protein though there does not seem to be a very good reason for this in view of the extraction of this protein from nuclei by acid. The evidence is consistent with the preferential synthesis of the histones in S and of the other protein in G2, but it is also possible that the proteins are made at all times and only accumulate in the nucleus during S or G2.

Immunoglobulins appear to be synthesised during a restricted period of the cycle, from late G1 until nearly the end of S, in human lymphoid cell lines (Buell & Fahey, 1969) and in mouse myeloma cells (Byars & Kidson, 1970). They also reach their maximum concentration in early S in human lymphoid cells (Takahashi *et al.*, 1969).

Pulse-labelling and acrylamide gel separation was used by Kolodny & Gross (1969) to compare the times of synthesis of components of the total soluble protein of HeLa cells. There is a marked difference between the protein components made in G2 and those which on average are made throughout the cycle. Fractionation of HeLa cell soluble proteins has also shown one component which is synthesised at a maximum rate during the S period (Hodge, Borun, Robbins & Scharff, 1969). A pattern of continuous increase through the cycle similar to that for total protein has been found for some more specialised proteins – microsomal and mitochondrial proteins in mouse P815Y cells (Warmsley *et al.*, 1970); glycoproteins in mouse L5178Y cells (Bosmann & Winston,* 1970); and collagen in mouse, rat and human embryo cells (Davies *et al.*,* 1968).

Conclusions

Separate summaries of the sections of this chapter have been made earlier, and we only need here to note one point. The dominant pattern of growth for the whole cell is a continuous increase through most of the cycle, following a linear or exponential course. But when we look at the components of the cell, this is not the dominant pattern. Nuclei have a burst of growth at the end of the cycle. Many of the histones are probably synthesised and certainly accumulated in the nucleus discontinuously, as is DNA. Immunoglobulin has a restricted period of synthesis: so also have many enzymes, as we shall see in Chapter 8. These patterns of ordered synthesis are a measure of the chemical differentiation that continues through the whole cell cycle.

8 Enzyme synthesis

Patterns of enzyme synthesis

One of the most active fields of cycle research in recent years has been the mapping of the changes in enzyme activity through the cell cycle in a wide variety of synchronous cell systems (reviewed by Donachie & Masters, 1969; Halvorson *et al.*, 1971; Mitchison, 1969*a*). If it is assumed that an increase in enzyme activity represents an increase in enzyme protein (p. 174), then measurements of activity provide a simple and sensitive way of following the synthesis of specific proteins. As with the other proteins considered at the end of Chapter 7, the synthesis of many enzymes appears to be restricted to regions of the cycle characteristic for each enzyme and does not follow the common pattern of continuous increase of *total* cellular protein. This emphasises the fact that the composition of a cell is changing continuously through the cycle and encourages us to speculate about the mechanisms that may control this process of differentiation at the chemical level.

The patterns of enzyme synthesis can be classified into two broad groups, depending on whether synthesis is continuous or periodic during the cycle, and each of these groups can be subdivided again (Fig. 8.1). The majority of enzymes that have been examined are synthesised periodically during one or more stages of the cycle. A stable enzyme which doubles during one stage of the cycle gives a 'step' pattern which is similar to that for DNA synthesis in many eukaryotes (Fig. 8.2). Each step enzyme, therefore, has its own characteristic G1, S and G2. An alternative form of periodic synthesis is shown by 'peak' enzymes which are presumably unstable. The activity rises during the period of synthesis but then falls as the enzyme breaks down or is inactivated. The other group of enzymes are synthesised continuously and may show a variety of patterns. Two of the simplest, illustrated in Fig. 8.1(*c*) and (*d*), are an exponential curve of increase and a linear pattern with a constant rate of synthesis until a particular point in the cycle is reached, after which the rate doubles (Fig. 8.3).

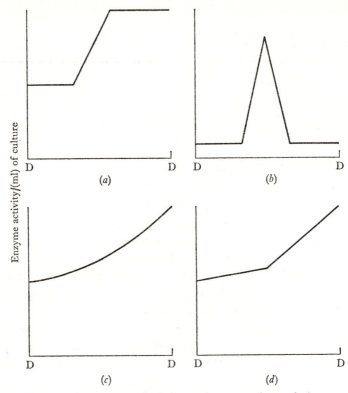

Fig. 8.1. Patterns of enzyme synthesis in synchronous cultures during one cell cycle. (*a*) Step. (*b*) Peak. (*c*) Continuous exponential. (*d*) Continuous linear. From Mitchison (1969*a*).

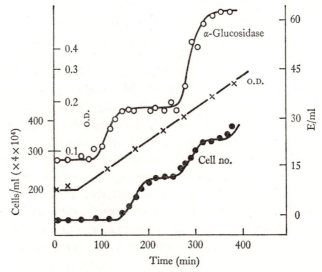

Fig. 8.2. Enzyme synthesis ('step') in a synchronous culture of *Saccharomyces cerevisiae*. From Tauro & Halvorson (1966).

This simple classification is of course somewhat artificial and does not fit all the cases. Periodic enzymes may be synthesised during more than one period of the cycle, giving multiple steps or peaks. Continuous enzymes may also show patterns which are different from those in Fig. 8.1. If, for example, a continuous enzyme is unstable, it could follow a step

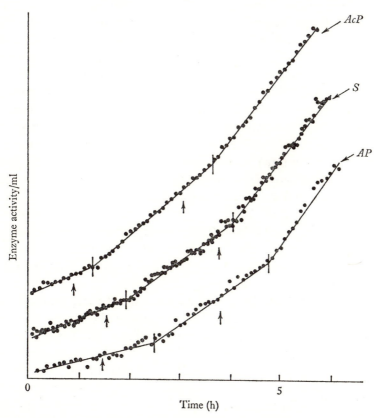

Fig. 8.3. Enzyme synthesis ('continuous linear') in three different synchronous cultures of *Schizosaccharomyces pombe. AcP*, acid phosphatase. *S*, sucrase. *AP*, alkaline phosphatase. Arrows mark peaks of cell plate index (see Fig. 3.11). Vertical lines mark points of rate change calculated by a statistical method. From Mitchison & Creanor (1969).

pattern (p. 174). The linear pattern in Fig. 8.3 is difficult to distinguish from an exponential since there is a maximum difference of only 3 per cent between an exponential curve that doubles in rate over a cycle time and the straight line of best fit. The enzyme assays must be frequent and accurate and the results should be tested statistically against controls (Mitchison & Creanor, 1969). It may be worth doing this, since there is an interesting interpretation that can be made of the linear pattern (p. 167). It could be that a number of enzymes which have been reported as exponential are, in fact, linear. For example, fully induced β-galactosidase in *Escherichia coli* B/r was first described as exponential (Cummings, 1965) but, on closer examination, appears to be linear (Donachie & Masters, 1969).

Most of the results of enzyme assays through the cycle are summarised in Table 8.1 for prokaryotes and Table 8.2 for eukaryotes. About two-thirds of the enzymes are periodic in the three major groups – prokaryotes, lower eukaryotes and mammalian cells. This statement, however, conceals a difference between lower eukaryotes and mammalian cells which may be important. In both groups, the enzymes concerned with DNA synthesis are almost always synthesised periodically at a stage in the cycle which is often during or just preceding the S period. In the case of other enzymes, however, a distinctly smaller proportion of them are continuous in lower eukaryotes (*c.* 25 per cent) than in mammalian cells (*c.* 50 per cent). More information, particularly about mammalian cells, is needed before we can be sure that this trend indicates a real difference between the two groups of eukaryotes. Halvorson *et al.* (1971) argue that such a difference might be expected since the lower eukaryotes are mostly free-living single cells in what may be a rapidly changing environment whereas the higher eukaryotic cells are derived from tissues and organs which normally exist in a comparatively stable environment.

Most of the difference between Tables 8.1 and 8.2 and similar tables which I drew up two years ago (Mitchison, 1969*a*) lies in the increased information about enzyme synthesis in mammalian cells. This is all to the good, if for no other reason than to raise the question in the preceding paragraph. But much of the information is difficult to interpret. One reason for this and for many of the question marks put after the pattern classification in Table 8.2, is that the enzyme activity is expressed as specific activity (for example, with respect to protein N) and not as activity per ml of culture.[1] Another problem is that some of the results do not span a complete cycle so it is impossible to apply the simple check that specific activity should return to the same value after one cycle and that activity per ml should double. The most serious difficulties, however, arise because of the use of DNA inhibitors and colcemid in synchronising the cultures.

---

[1] Although the initial measurements in enzyme assays may be in terms of specific activity, it is much better to convert them to activity per ml of culture. The arguments against specific activity are well illustrated and discussed in Duynstee & Schmidt (1967). Consider also how difficult it would be to interpret a curve for DNA synthesis expressed as mg DNA/mg protein.

Changing the chemical composition of the medium by adding and subtracting the inhibitor is in itself a dubious procedure since it may affect the synthetic patterns; but there is a more important argument, developed elsewhere (pp. 32, 244), which suggests that most of the enzymes (except those concerned with the 'DNA-division cycle') are associated with the 'growth cycle' and that this cycle is not synchronised either by a DNA block or, perhaps, by a metaphase block. It is a possibility, therefore, that the 'continuous' enzymes measured in cultures synchronised by these techniques are in reality periodic enzymes and would appear as such with selection synchrony. This might account for the relatively higher proportion of cases where there appears to be continuous enzyme synthesis in mammalian cycles as compared to lower eukaryote cycles.

There appear to be differences in the pattern of some enzymes both between different strains of mammalian cells and also within the same strain. Glucose-6-phosphate dehydrogenase and lactate dehydrogenase show three peaks in the diploid Don line of Chinese hamster cells, whereas lactate dehydrogenase is a continuous enzyme in the heteroploid G_3 line. Both these enzymes are probably continuous in rat HTC cells, and Warmsley *et al.* (1970) found that they were sometimes continuous in mouse P815Y cells and sometimes showed peaks similar to those in the Chinese hamster cells but less marked. Klevecz (1969*b*) suggests that the difference between the patterns of lactate dehydrogenase (and also thymidine kinase) in the diploid and heteroploid Chinese hamster cells may be related to different modes of DNA synthesis in these two lines. Halvorson *et al.* (1971), however, point out that this is unlikely, at any rate for the lactate dehydrogenase, since the enzyme peaks continue after inhibition of DNA synthesis (Klevecz, 1969*a*).

Although the majority of enzymes appear to be synthesised periodically in the cycles of lower eukaryotes, seven enzymes have recently been reported as continuous in *Physarum* (Hüttermann, described in Rusch, 1969). These may indeed be continuous, but there is the possibility that the same dissociation of the DNA-division cycle and the growth cycle may take place after starting a synchronous culture of *Physarum* as may also happen after a DNA block (p. 248). These enzymes might therefore be periodic in a normal cycle even though they were continuous in the experimental conditions that were used.

Certain results on enzyme activity through the cycle have been omitted from Tables 8.1 and 8.2 because they involve synchrony procedures where there is one shock per cycle during the course of the enzyme assays and there is therefore some uncertainty as to how far the changes in activity might be caused by the shocks. These include experiments with cyclic illumination on *Chlamydomonas* (Kates & Jones, 1967) and *Euglena* (Walther & Edmunds, 1970), and with cyclic additions of limiting nutrient on *Escherichia coli* (Goodwin, 1969*a*, *b*). On the other hand, the experiments of Knutsen, and of Schmidt and his colleagues on *Chlorella* have been included since, although the synchrony was induced by cyclic illumination, the cells were left in continuous light for the period of the assays and, in some cases at any rate, measurements were continued into a second cycle and were similar to those in the first cycle. In their latest experiments, Schmidt and his colleages have used both cyclic illumination and selection techniques and have found similar results with the two methods (Molloy & Schmidt, 1970); or they have used cyclic illumination followed by selection (Baechtel *et al.*, 1970).

An important question that can be asked about periodic enzymes is whether or not their times of synthesis are concentrated at particular regions of the cycle. The association of the enzymes of DNA synthesis with the S period has been mentioned earlier. With the other enzymes, the best information comes from yeast (particularly Tauro *et al.*, 1968) and indicates, to the extent of our present knowledge, that the periods of synthesis are not concentrated at one time in the cycle but are spread through the cycle in a fairly uniform way. It remains to be seen whether this is also true in mammalian cells.

Changes in enzyme potential

Before discussing the possible mechanisms that underlie the patterns of enzyme synthesis, it is better to consider another way of using enzyme assays with synchronous cultures. This is to remove samples from a culture and challenge the cells to synthesise an enzyme under conditions of induction or derepression. The rate at which the enzyme is then made is the inducibility or 'potential' (Kuempel *et al.*, 1965) for that enzyme at that stage of the cycle. In most cases in prokaryotes, the pattern of change in potential through the cycle is very similar to the pattern of synthesis of a step enzyme (Fig. 8.4, Table 8.3). The potential doubles fairly sharply at a particular point in the cycle and then remains constant until the same point is reached in the next cycle.

There is good reason in prokaryotes for supposing that this point where the potential doubles is also the point where the structural gene for that enzyme doubles. The most direct evidence is that the sucrase potential in *Bacillus subtilis* doubles at the same time as the sucrase-transforming ability of the DNA (Masters & Pardee, 1965). A prediction would be that the points of potential doubling should occur during the cycle in the same order and with the same spacing as the appropriate genes on the chromosome – provided there was only one replicating fork and that DNA

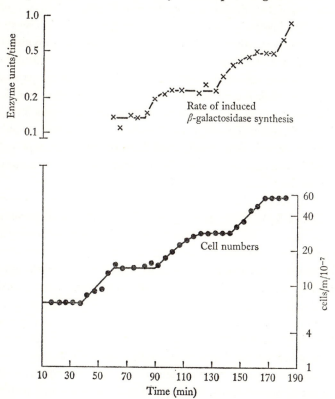

Fig. 8.4. Changes in rate of induced synthesis (potential) of β-galactosidase in a synchronous culture of *Escherichia coli* B/r. Induced by thiomethyl galactoside. From Donachie & Masters (1969).

synthesis lasted the whole cycle. This has been shown to be approximately true for six enzymes in *E. coli* B/r (Donachie & Masters, 1969) and to be accurately true for three of them using membrane elution (Helmstetter, 1968; Pato & Glaser, 1968). We would also expect that there should be no increase in potential if DNA synthesis were blocked yet growth continued. This happens with D-serine deaminase in *E. coli* 15 T⁻ deprived of thymine (Donachie & Masters, 1966) and with β-galactosidase in *E. coli* B/r treated with fluorodeoxyuridine or nalidixic acid (Pato & Glaser, 1968). Finally, there is the general supporting evidence of gene-dosage effects on enzyme concentration. Enzyme quantities per cell increase as the number of copies of the appropriate gene increases, both in *E. coli* (Pittard & Ramakrishnan, 1964) and in *Neurospora* (Donachie, 1964).

All the bacterial data, except for two cases, agree with a model in which potential is not restricted during the cycle and doubles when the gene doubles. One of these cases is the continuous rise of potential in one strain of *E. coli* K12, whereas other strains showed the normal stepwise rise (Nishi & Horiuchi, 1966). Donachie & Masters (1969) and I (Mitchison, 1969*a*) have already commented on this, and it is sufficient here to say that this interesting work merits further exploration. The other case is in the germinating spores of *B. subtilis* where two enzymes are inducible, but only at particular times during outgrowth (Steinberg & Halvorson, 1968). This restriction of potential is quite different from the continuous potential which is normal in the bacterial cycle. The controls in spore germination, however, may not be the same as those in the cell cycle.

There is a small but growing amount of information on potential changes in eukaryotes (Table 8.3). There are three cases (two in *Schizosaccharomyces* and one in *Chlorella*) where the pattern of potential changes is very like that of bacteria – potential is unrestricted and increases in a stepwise manner at one point in the cycle. In the yeast, however, this point is later than the S period, a result which will be discussed later. With the other three enzymes in *Chlorella*, the curve for potential rises to a peak in mid-cycle and then falls off. This restriction of potential may have some connection with an inhibition of transcription during mitosis but it is in any case a different situation from bacterial cycles. The only enzyme whose potential has been measured in mammalian cells is the tyrosine amino-transferase of rat hepatoma cells. Its regulation is complex and seems to involve controls both at a transcriptional level and at a post-transcriptional level. There does, however, appear to be a restriction of inducibility at the gene level during the period of G2, M and the first 3 hours of G1. The remaining part of the cycle, when the enzyme is inducible, lasts for 16 hours or 65 per cent of the cycle.

Linear enzymes

Returning now to the normal patterns of enzyme synthesis, we can find a connection between linear enzymes and potential changes. In bacteria, there are linear patterns in repressed alkaline phosphatase in *B. subtilis*, in fully induced *β*-galactosidase in *E. coli* and in repressed *β*-galactosidase and alkaline phosphatase in *E. coli* (Table 8.1). These have been interpreted in the same way as the potential changes – that the rate of enzyme synthesis doubles at the point when the structural gene for that enzyme doubles. The evidence for this is thinner than it is for the potential changes. There is the general argument about gene dosage that has been mentioned earlier, and there is also the evidence that the points of rate change of two linear enzymes in *E. coli* (fully repressed *β*-galactosidase and alkaline phosphatase, Kuempel *et al.*, 1965) are at the same place in the cycle as the points of potential doubling.

Linear enzymes have been clearly demonstrated in only one eukaryotic cell, *Schizosaccharomyces pombe*. There are three enzymes (sucrase, alkaline phosphatase and acid phosphatase) which have linear patterns and they all show a doubling in rate of synthesis at a point in the cycle which has been called the 'critical point' (Mitchison & Creanor, 1969). This is also the point where the sucrase potential doubles. It occurs, however, not during the short S period but a third of a cycle later, during G2. This suggests that there is a delay between the formation of new genes and the time when they come into use at the critical point – in other words, a delay between chemical replication of the genome and 'functional replication'. There are other physiological events which take place at the critical point and which suggest that there may be a change in chromosome configuration, perhaps a separation of the chromatids. It is possible that there may be the same delay between chemical and functional replication in *Chlorella* since the stepwise rise in the potential of isocitrate lyase starts an hour after the beginning of the S period. In this case, however, the delay could also be caused by late replication of the lyase gene. Whatever may be the final explanation of this delay, it does raise an interesting question for other eukaryotes. We know that an extra genome is made during the S period. When is it first used – at once (as in bacteria), during G2 (as in *Schizosaccharomyces*), or only after it has separated as a new nucleus after mitosis?

Step and peak enzymes

Relation to DNA synthesis

The changes in potential and in the rate of synthesis of linear enzymes appear to be closely related to the replication of the genome. The reverse, however, seems to be true for most step and peak enzymes. There are three lines of evidence for this statement. One, which has been mentioned earlier, is that in yeast (both *Saccharomyces cerevisiae* and *Schizosaccharomyces pombe*) enzyme steps are spread through the cycle, whereas the S period is restricted to one part of it. Second, the steps in aspartate transcarbamylase in *E. coli* occur at a point in the cycle different from that for potential doubling (Kuempel *et al.*, 1965). The third and most cogent evidence is that enzyme steps can continue after DNA synthesis has been blocked by inhibitors, the inverse of what happens with steps in potential. Steps of ornithine transcarbamylase continue in *B. subtilis* after DNA replication has been blocked with fluorodeoxyuridine (Masters & Donachie, 1966). Although not done with synchronous cultures, the experiment of Fangman *et al.* (1967) shows a sixfold increase in two enzymes after stopping DNA synthesis in a temperature-sensitive mutant of *E. coli*. Steps of ornithine transcarbamylase, aspartate transcarbamylase and alcohol dehydrogenase in *S. pombe* continue after inhibition of DNA synthesis by hydroxyurea, and after inhibition of both division and DNA synthesis by mitomycin C (Robinson, 1971). Steps in alcohol dehydrogenase and glyceraldehyde dehydrogenase, and peaks in DNA polymerase continue in *S. cerevisiae* after division has been blocked and DNA synthesis considerably delayed by X-irradiation (Eckstein *et al.*, 1966; 1967). The patterns, probably steps, of acid and alkaline phosphatase in HeLa cells continue unaffected after DNA inhibition by excess thymidine (Churchill & Studzinski, 1970). Peaks of lactate dehydrogenase in Chinese hamster cells (Klevecz, 1969*a*) and of deoxycytidine-P deaminase in HeLa cells (Gelbard *et al.*, 1969) continue after DNA inhibition by thymidine and by hydroxyurea respectively. This evidence has been spelt out in some detail because it is important to show that DNA replication can be dissociated from what is probably periodic transcription. It is the basis of the argument put forward elsewhere in this book that there may be separable 'DNA-division cycles' and 'growth cycles' (pp. 32, 247). The question then arises whether the periodic enzymes concerned with DNA synthesis should be affected by a DNA block since they are presumably associated with the DNA-division cycle. The answer might depend on whether the signal for their synthesis came before or after the start of the S period when the block becomes effective, and we should therefore expect a variable response. As far as the evidence goes, this is what happens. Alkaline DNAase in HeLa cells is affected by the block (Churchill & Studzinski, 1970) whereas deoxycytidine-P deaminase in the same cells is not (see above).

We can conclude then that the periodic synthesis of step and peak enzymes (except those concerned with DNA synthesis and other aspects of the DNA-division cycle) is *not* closely and causally related to DNA replication. We have therefore to look elsewhere for possible control mechanisms and here there are, at present, two rival theories.

Oscillatory repression

The first model for the control of periodic enzyme synthesis can be called 'oscillatory repression'.[1] It has been developed primarily for prokaryotes, though there is no inherent reason why it should not also apply to eukaryotes. The essential idea is simple and attractive. A system where an enzyme product can repress the synthesis of that enzyme has negative feedback and, with the right choice of constants, will produce stable oscillations. When the pool of end product is high, the synthesis of the enzyme will be repressed: when the pool is low, the enzyme will be produced. The system generates its own steps, so they have been called 'autogenous'. There is no obvious reason why these oscillations should be at the same frequency as the cell cycle so that an enzyme step occurs at the same point in successive cycles. In order to meet this objection, it has been suggested that the oscillations are entrained by an event which is dependent on the cycle, for instance a pulse of messenger RNA produced at the time of gene replication (Goodwin, 1966). Although the steps for different enzymes could come at varying and relatively long intervals after the entraining pulse, there is no reason why they should bear any relation to the order of their structural genes on the genetic map.

Oscillations should occur when the enzyme is being controlled by end-product repression and is partially de-repressed. If the enzyme is completely repressed and at basal level, or is fully induced or derepressed, the oscillations should cease and the enzyme would be synthesised continuously, probably following the linear pattern. Potential should be present throughout the cycle.

This is a persuasive theory for prokaryotes and fits all the data in Table 8.1. In particular, it provides a neat explanation of why sucrase and alkaline phosphatase in *B. subtilis* are linear (or exponential) enzymes when repressed and step enzymes when derepressed. Support for it in eukaryotes comes from a similar result in *Chlorella* (Molloy & Schmidt,

---

[1] This theory has been developed in a number of papers, e.g. Kuempel *et al.* (1965), Masters & Pardee (1965), Masters & Donachie (1966), Goodwin, (1966; 1969c), and is expounded well by Donachie & Masters (1969). Its name here comes from Mitchison (1969a).

1970). Ribulose-1,5-diphosphate carboxylase is a continuous enzyme at high growth rates when it may be fully derepressed, and a step enzyme at low growth rates when repression is likely to be operating. The theory, however, does not account for the restricted potential in some of the eukaryotic cases in Table 8.3 unless an additional mechanism for preventing transcription is invoked. Nor does it fit the observation that induced α- and β-glucosidases in yeast are step rather than continuous enzymes, with the steps at the same place in the cycle as in uninduced cultures (Halvorson *et al.*, 1966). But these steps might be controlled by a general catabolite repression which is less specific than the end-product repression of bio-synthetic enzymes. Some doubt has also been thrown on the theoretical side of this theory since it is not clear that single step oscillations can be produced from a system of one gene and one enzyme (Griffith, 1968; Walter, 1969; see also Morales & McKay, 1967).

It is not easy to find a critical and clear-cut test of this theory. If the end-product pool was found to fluctuate, it would agree with the theory but would not prove that the pool fluctuations caused the enzyme steps since the converse might be equally true. Enzyme synthesis might be switched on and off by another mechanism and these changes would cause pool oscillations. If, on the other hand, the pool was found not to fluctuate during the cycle, it could be argued either that this pool was not the controlling one or that this pool was partitioned spatially in the cell and the measurements could not resolve the controlling oscillations in one of the compartments.

Linear reading

The second theory of the control of periodic enzymes has been developed by Halvorson and his colleagues as a result of their work on budding yeast.[1] It suggests that genes are transcribed in the same order as their linear sequence on the chromosomes. The image is of an RNA polymerase which moves along the chromosome transcribing the genes in sequence. As a result a gene is only transcribable, and therefore only inducible, for a short time in the cycle. This has been called 'linear reading' or 'sequential transcription'.

Several experimental results support this theory. *Saccharomyces dob-zhanskii* has one β-glucosidase step per cycle. When it is crossed with

---

[1] Halvorson *et al.* (1964, 1966), Gorman *et al.* (1964), Tauro & Halvorson (1966), Tauro *et al.* (1968). Recently reviewed by Halvorson *et al.* (1971).

Saccharomyces fragilis, the hybrid produces two species of this enzyme which are antigenically distinct but are subject to the same regulatory control. With oscillatory control, we would expect one step per cycle. But instead, there are two steps, suggesting that there are two non-allelic genes which are transcribed at different points in the cycle (Gorman *et al.*, 1964; Halvorson *et al.*, 1966. See also a criticism by Fleming & Duerksen, 1967). This has been followed up by a study of the multiple M genes for α-gluco-sidase in *S. cerevisiae* (Tauro & Halvorson, 1966). The homozygote M_1M_1 and the heterozygote M_1m_1 both produce a single enzyme step per cycle, and always at the same point in the cycle, showing that an increase in the gene dosage at a particular locus does not affect the step timing. But the introduction of other non-allelic structural genes (M_2 and M_3) produces additional steps – two per cycle for two non-allelic genes and three per cycle for three genes. In a later paper, Tauro *et al.* (1968) have compared the step timings of nine enzymes with the position of their genes on the genetic map of *S. cerevisiae*. The timings are consistent with linear reading of the chromosomes from end to end, though not with a reading mechanism starting at the centromere and moving simultaneously along each arm. The most convincing evidence comes from the four enzyme genes located on the fifth chromosome. The other chromosomes have only two or one enzyme genes, and it is difficult to draw any conclusions from their timing especially as there is no reason to suppose either that transcription would start at the same time on each of them or that it should move in the same direction. An interesting recent support for linear reading comes from the experiments of Cox & Gilbert (1970) with two strains of *S. cerevisiae*. The distance between two enzyme genes on the second chromosome is much greater in one strain than in the other and so also is the distance between the two steps in the cell cycle.

Linear reading in its simplest form does not allow for continuous syn-thesis or for continuous and unrestricted potential through the cycle. It cannot therefore be universally applicable to eukaryotes, since there is now clear evidence in Tables 8.2 and 8.3 that these two continuous patterns do occur in eukaryotes and even in a close relative of budding yeast, the fission yeast *S. pombe*. The argument that a continuous enzyme pattern might be in reality a series of steps caused by the presence of multiple genes and so not resolvable in experimental curves, is not convincing since it would not account for the rate changes in the linear patterns or the steps in potential. Although tyrosine aminotransferase in rat cells has a period of the cycle when its potential is restricted, the time when this potential is *not* restricted is 65 per cent of the cycle and this is much too long to be explained in terms of the passage of a polymerase over a gene.

Halvorson and his colleagues do not suggest that this mechanism would apply to bacteria, where there is also good evidence of continuous synthesis and potential, but there is an important paper which does support linear reading in prokaryotes. Using sucrase potential as a marker for aligning the enzyme step map with the genetic map of *B. subtilis*, Masters & Pardee (1965) showed that the order and timing of the steps of three enzymes during the cycle was the same as the order and spacing of the relevant

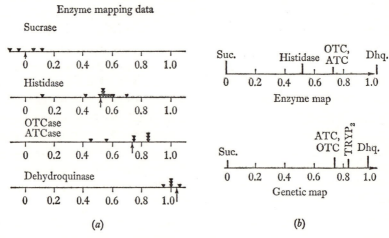

Fig. 8.5. Enzyme synthesis in synchronous cultures of *Bacillus subtilis*. (a) Each triangle represents the half-doubling time (HDT) of an enzyme in one cycle, taking the average HDT for histidase in that entire experiment as 0.54. Arrows indicate the mean HDT for each enzyme. Ornithine transcarbamylase (OTCase) and aspartate transcarbamylase (ATCase) are plotted on the same line since their HDTs were indistinguishable in all experiments. (b) The points on the enzyme map are the average HDTs from (a). The genetic map was compiled from other data. From Masters & Pardee (1965).

genes on the genetic map (Fig. 8.5). This is a somewhat awkward observation to accommodate within the theory of oscillatory repression since, as was mentioned earlier, this theory would not predict such a result. It would if there was a constant time between the replication of a gene and its step, but this extra hypothesis would then not fit the evidence from yeast where there are variable intervals between steps and the S period. It would certainly be worth extending this work and seeing whether this gene-step identity was a matter of chance or whether it holds good with a larger number of genes and enzymes. The obvious cell to use would be *E. coli*. It has a single well-mapped chromosome, and in addition enough is known about its modes of DNA replication to make it very desirable to follow the way that growth rate affects enzyme patterns. If linear reading does apply to bacteria, it will be much easier to prove its existence in a genome with one chromosome than in a genome with many. In a eukaryotic cell, there is no reason why the time when transcription starts and its speed and direction should be the same in all the chromsomes.

Comments

It is too early to come to any definite conclusions about the control of enzyme synthesis during the cycle. The two theories that we have considered are intriguing, but they are much more in the nature of working hypotheses than proven mechanisms. It is possible that linear reading applies to eukaryotes and oscillatory repression to prokaryotes, but there are cases that argue the other way (e.g. Masters & Pardee, 1965; Molloy & Schmidt, 1970). It is also possible that the controls in higher eukaryotes differ from those in lower eukaryotes. I think, however, that the most likely outcome from the present position is that it will be found that both types of control exist in part and in varying proportions in different groups of cells. Oscillatory repression may be the dominant mechanism in pro- karyotes and with biosynthetic enzymes but it may not apply to enzymes which are free of end-product repression. Linear reading might be the normal control in lower eukaryotes even though other mechanisms exist to transcribe a gene at any point of the cycle under an inductive stimulus. It may be, as Tauro *et al.* (1968) have suggested, that linear reading takes place over restricted regions of the genome rather than running sequen- tially along the whole chromosome. In any case, these two methods of regulation are almost certainly not the only ones. Continuous enzymes appear to be under a different control which, in the case of the linear pattern, involves a gene dosage effect. Some enzymes also are subject to a translational or post-transcriptional control (Martin, Tomkins & Bresler, 1969; Martin & Tomkins, 1970). Nor is it likely that either of the two theories can explain the single, widely spaced synthetic events that are the chemical basis of development in higher organisms or the periods of synthesis on stimulation that are characteristic of specialised cells that are outside the cell cycle.

There are many gaps in our present knowledge not only about the control mechanism but also about the facts of the cellular situation upon which control models depend. Throughout this chapter, I have been discussing enzyme synthesis, but what is actually measured in nearly all cases is enzyme activity. It is an assumption to equate rise in activity with synthesis of enzyme protein and it would be wrong if, for example, it turned out that activity were controlled by low molecular weight activators. Some of the possible errors can be eliminated by suitable assay techniques and the use of various controls, but it is obviously better, where possible, to measure the production of enzyme protein. Two recent papers on mammalian cells include such measurements, made with immunological techniques (Martin, Tomkins & Granner, 1969; Klevecz, 1969a).

Another assumption is that step patterns are produced by stable enzymes and peak patterns by unstable enzymes. But there is some evidence that step enzymes are, or at least can be, unstable (Donachie, 1965; Donachie & Masters, 1969). Steps could be produced by the continuous production of an unstable enzyme with a doubling of the rate of synthesis at the time of the step. This doubling would change the enzyme concentration from one steady state to another, higher one. As in the case of the linear pattern, this doubling in rate might occur at the time of replication of the structural gene. A prediction (so far untested) would be that this type of step, unlike that of a stable enzyme, would be stopped by an inhibition of DNA synthesis.

A desirable check, which is now being carried out more frequently, is to follow enzyme activity after adding a protein synthesis inhibitor such as cycloheximide or puromycin (e.g. Molloy & Schmidt, 1970; Klevecz, 1969a; Yagil & Feldman, 1969). This tests whether the enzyme is stable, and also whether rising activity depends on protein synthesis. It has revealed an interesting situation in *S. pombe* (Mitchison & Creanor, 1969). With alkaline phosphatase (but not acid phosphatase), there seems to be a delay of about half an hour between the synthesis of enzyme protein and its final activation. This delay is important in interpreting experimental results since it means that the doubling point of the rate of synthesis of this linear enzyme is half an hour before it is apparent in activity measurements.

Another large gap in our knowledge about the cellular background is that we do not know the importance of periodic patterns in the curve of increase of total cell protein. Is the periodic pattern for enzymes and other specific proteins the dominant one, with the periods of synthesis spread through the cycle to give a continuous increase of total protein? Or is it quite a minor component which is overshadowed by continuous synthesis not only of other enzymes but also of ribosomal and structural proteins? We need a balance sheet of the major protein components of the cell as well as knowledge of their patterns of synthesis.

Conclusions

The majority of enzymes are synthesised discontinuously during periods of the cell cycle which are characteristic for each enzyme. With a stable enzyme, this produces a 'step' pattern which is similar to that for DNA in eukaryotes. With an unstable enzyme, this produces a 'peak' pattern.

Two theories have been suggested for the control of periodic synthesis. 'Oscillatory repression', primarily developed for prokaryotes, attributes the periods of synthesis to oscillations set up by the negative feed-back of end-product repression, and entrained to the same frequency as the cell cycle. 'Linear reading', primarily developed for lower eukaryotes, suggests that the genes are transcribed during the cycle in a sequence which corresponds to their order on the chromosomes. Both theories fit some but not all of the facts.

Some enzymes are synthesised continuously through the cycle, and this pattern is commoner in mammalian cells than in lower eukaryotes and prokaryotes. In a few cases, continuous synthesis can be resolved as a linear pattern in which the rate of synthesis doubles at a characteristic part of the cycle. In the majority of cases in prokaryotes and in some eukaryotes, enzymes are inducible throughout the cycle and the rate of inducibility ('potential') doubles at a characteristic point in the cycle. The doubling point of the rate of synthesis and of inducibility appears to correspond with the functional replication of the appropriate gene. In prokaryotes, the functional replication occurs at the same time as DNA replication: in *Schizosaccharomyces*, it occurs later.

TABLE 8.1. Patterns of enzyme synthesis in synchronous cultures
of growing cells (prokaryotes).

S, step enzymes; P, peak enzymes; C(E), continuous exponential enzymes; C(L), continuous linear enzymes; (FR), fully repressed; (FD), fully derepressed or induced. Methods of synchronisation: *Sel*, selection; *Sta*, starvation and growth in fresh medium.

Organism and enzyme	Strain	Pattern	Method of syn-chrony	Reference
Escherichia coli				
β-Galactosidase (FR)	K12 Hfr CS-101-G-1	C(L)	*Sel*	Kuempel *et al.* (1965)
Alkaline phosphatase (FR)	K12 Hfr CS-101-G-1	C(L)	,,	,, ,, ,,
Aspartate transcarbamylase	K12 Hfr CS-101-G-1	S	,,	,, ,, ,,
Dihydroorotase	K12 Hfr CS-101-G-1	S	,,	,, ,, ,,
Histidinol dehydrogenase	K12 Hfr CS-101-G-1	S	,,	,, ,, ,,
Glycyl-glycine dipeptidase	K12 Hfr C(met⁻)	P	,,	Nishi & Hirose (1966)
Glycyl-glycine dipeptidase	K12 Hfr H	S	,,	,, ,, ,,
Glycyl-glycine dipeptidase	K12 E64 (F⁻, B₁⁻)	P	,,	,, ,, ,,
Glycyl-glycine dipeptidase	ML 308	S (or P?)	*Sta*	Kogoma & Nishi (1965)
Leucine aminopeptidase	ML 308	S (or P?)	,,	,, ,, ,,
Protease	ML 308	S (or P?)	,,	,, ,, ,,
β-Galactosidase (FD)	B/r	C(E)	*Sel*	Cummings (1965)
β-Galactosidase (FD)	B/r	C(L)	,,	Donachie & Masters (1969)
β-Galactosidase (FD)	B	C(E)	,,	Abbo & Pardee (1960)
Bacillus subtilis				
Aspartate transcarbamylase	W23	S	*Sta*	Masters & Pardee (1965), Donachie (1965), Kuempel *et al.* (1965)
Ornithine transcarbamylase	W23	S	,,	Masters & Pardee (1965), Donachie (1965)
Dehydroquinase	W23	S	,,	Masters & Pardee (1965)
Histidase	W23	S	,,	Masters & Pardee (1965), Kuempel *et al.* (1965)
Alkaline phosphatase (FR)	W23	C(L)	,,	Donachie (1965)
Sucrase	W23	S	,,	Masters & Donachie (1966)
Sucrase (FR)	W23	C(E)	,,	,, ,, ,,
Rhodopseudomonas spheroides				
Succinyl CoA thiokinase		S	*Sta*	Ferretti & Gray (1968)
Aminolevulinic acid synthetase		S	,,	,, ,, ,,
Aminolevulinic acid dehydrase		S	,,	,, ,, ,,
Alkaline phosphatase		S	,,	,, ,, ,,
Ornithine transcarbamylase		S?	,,	,, ,, ,,
Ornithine transcarbamylase (FR)		C?	,,	,, ,, ,,

TABLE 8.2. Patterns of enzyme synthesis in synchronous cultures
of growing cells (eukaryotes).

S, step enzymes (numerals after S indicate more than one step per cycle); P, peak enzymes (numerals after P indicate more than one peak per cycle); C(L), continuous linear enzymes. Methods of synchronisation: *Sel*, selection; *Sta*, starvation and growth; *Cyc*, cyclic illumination; *Fus*, fusion of microplasmodia; *DNA*, inhibition of DNA synthesis; *Col*, colcemid accumulation.

Organism and enzyme	Pattern	Method of synchrony	Reference
Saccharomyces cerevisiae			
Protease	P	*Sta*	Sylvén *et al.* (1959)
Peptidase	P	,,	,, ,,
α-Glucosidase	S	*Sta, Sel*	Gorman *et al.* (1964), Tauro & Halvorson (1966), Tauro *et al.* (1968)
α-Glucosidase	S (2 and 3)	,,	Tauro & Halvorson (1966)
Sucrase	S(2)	*Sta*	Gorman *et al.* (1964)
Alkaline phosphatase	S(2)	,,	Gorman *et al.* (1964), Cottrell & Avers (1970)
Histidinol dehydrogenase	S	*Sel*	Tauro *et al.* (1968)
Orotidine-5′-phosphate decarboxylase	S	,,	,, ,,
Aspartokinase	S	,,	,, ,,
Phosphoribosyl-ATP-pyrophosphorylase	S	,,	,, ,,
Threonine deaminase	S	,,	,, ,,
Argininosuccinase	S	,,	,, ,,
Saccharopine dehydrogenase	S	,,	,, ,,
Saccharopine reductase	S	,,	,, ,,
Alcohol dehydrogenase	S	*Sta*	Eckstein *et al.* (1966)
Hexokinase	S	,,	,, ,,
Glyceraldehyde-3-phosphate dehydrogenase	S	,,	,, ,,
DNA polymerase	P	,,	Eckstein *et al.* (1967)
Trehalase	S?	,,	Küenzi & Fletcher (1969)
Cytochrome C oxidase	S	,,	Cottrell & Avers (1970)
Malate dehydrogenase	S	,,	,, ,,
Galactokinase	S	*Sel*	Cox & Gilbert (1970)
o-Aminoadipic acid reductase	S	,,	,, ,,
Saccharomyces dobzhanskii			
β-Glucosidase	S	*Sta*	Gorman *et al.* (1964)
Saccharomyces dobzhanskii x fragilis			
α-Glucosidase	S	*Sta, Sel*	Tauro & Halvorson (1966)
β-Glucosidase	S(2)	,,	,, ,,
Alkaline phosphatase	S(2)	,,	,, ,,
Schizosaccharomyces pombe			
Aspartate transcarbamylase	S	*Sel*	Bostock *et al.* (1966)
Ornithine transcarbamylase	S	,,	,, ,,
Tryptophane synthetase	S	,,	Robinson (1971)
Alcohol dehydrogenase	S	,,	,, ,,
Homoserine dehydrogenase	S	,,	,, ,,
Alkaline phosphatase	C(L)	,,	Mitchison & Creanor (1969)

Table 8.2. (cont.)

Organism and enzyme	Pattern	Method of synchrony	Reference
Acid phosphatase	C(L)	*Sel*	Mitchison & Creanor (1969)
Sucrase	C(L)	,,	,, ,,
Maltase	C(?)	,,	Bostock et al. (1966)

Chlorella pyrenoidosa

Aspartate transcarbamylase	C(?)	*Cyc*	Cole & Schmidt (1964)
Deoxythymidine monophosphate kinase	P(?)	,,	Johnson & Schmidt (1966)
Deoxycytidine monophosphate deaminase	S(?)	,,	Shen & Schmidt (1966)
Alkaline phosphatase	S	,,	Knutsen (1968)
Acid phosphatase	S	,,	,, ,,
Ribulose-1,5-diphosphate carboxylase	S, C	*Cyc, Sel*	Molloy & Schmidt (1970)
Isocitrate lyase	S	,, ,,	Baechtel et al. (1970)

Physarum polycephalum

Thymidine kinase	P	*Fus*	Sachsenmeier & Ives (1965)
Glucose-6-phosphate dehydrogenase	C(?)	,,	Sachsenmeier & Ives (1965), Rusch (1969)
Ribonuclease	S	,,	Braun & Behrens (1969)
Histidase	C	,,	Rusch (1969)
β-Glucosidase	C	,,	,,
Acid phosphatase	C	,,	,,
Phosphodiesterase	C	,,	,,
Glutamic dehydrogenase	C	,,	,,
Isocitric dehydrogenase	C	,,	,,

Mouse L cells

Thymidine kinase	S	*DNA*	Littlefield (1966)
,, ,,	P	*Sel*	Mittermayer et al. (1968)
,, ,,	P	*Sta*	Adams (1969b)
DNA polymerase	P	*DNA*	Gold & Helleiner (1964)
,, ,,	C?	,,	Turner et al. (1968)
,, ,,	P	*Sta*	Adams (1969b)
Ribonucleotide reductase	P	*DNA*	Turner et al. (1968)
Deoxycytidine monophosphate deaminase	P	*Sel*	Mittermayer et al. (1968)

Mouse L5178Y cells

Collagen-galactosyl transferase	P	*DNA, Col*	Bosmann (1970a)
Collagen-glucosyl transferase	P	,, ,,	,, ,,
Uridine diphosphatase	P	,, ,,	,, (1970b)
Esterase	P	,, ,,	,, ,,
5′-Nucleotidase	P	,, ,,	,, ,,

Mouse P815Y cells

NADPH-cytochrome C reductase	C	*Sel*	Warmsley et al. (1970)
Uridine diphosphatase	C	,,	,, ,,
Cytochrome C oxidase	C	,,	,, ,,
Succinate-cytochrome C reductase	C	,,	,, ,,
Lactate dehydrogenase	C, P?	,,	,, ,,
Glucose-6-phosphate dehydrogenase	C, P?	,,	,, ,,
Glutamate dehydrogenase	S	,,	,,

TABLE 8.2. (cont.)

Organism and enzyme	Pattern	Method of synchrony	Reference
	Rat HTC cells		
Tyrosine aminotransferase	C?	*Col*	Martin, Tomkins & Granner (1969)
Lactate dehydrogenase	C?	,,	,, ,,
Alcohol dehydrogenase	C?	,,	,, ,,
Glucose-6-phosphate dehydrogenase	C?	,,	,, ,,
	Chinese Hamster Don C cells		
Thymidine kinase	P	*Col*	Stubblefield & Murphree (1967)
,, ,,	P(2)	,,	Klevecz (1969*b*)
,, ,,	P (G3 cells)	,,	,,
Ribonucleotide reductase	P?	,,	Murphree *et al.* (1969)
Glucose-6-phosphate dehydrogenase	P(3)	,,	Klevecz & Ruddle (1968)
Lactate dehydrogenase	P(3)	*Col, DNA*	Klevecz & Ruddle (1968), Klevecz (1969*a*)
,, ,,	C (G3 cells)	*Col*	Klevecz (1969*b*)
	Human HeLa cells		
Thymidine kinase	P	*Sel*	Brent *et al.* (1965)
Thymidylate kinase	P	,,	,, ,,
DNA polymerase	P(?)	*DNA*	Friedman & Mueller (1968), Friedman (1970)
Deoxycytidine monophosphate deaminase	P	*Sel*	Gelbard *et al.* (1969)
Alkaline deoxyribonuclease	S?	,,	Churchill & Studzinski (1970)
Acid phosphatase	S?	,,	,, ,,
Alkaline phosphatase	S?	,,	,, ,,
,, ,,	C	,,	Griffin & Ber (1969)
,, ,,	P	*DNA*	Melnykovich *et al.* (1967)
Ornithine transaminase	P?	?	Volpe (1969)
	Human KB cells		
Lactate dehydrogenase	C	*DNA*	Bello (1969)
Fumarase	C	,,	,,
	Human Henle cells		
Alkaline phosphatase	P	*DNA*	Melnykovych *et al.* (1967)

TABLE 8.3. Patterns of change in potential (inducibility) of enzymes
in synchronous cultures.

S, step pattern; C, continuous increase; R, restricted. Methods of synchronisation: *Sel*, selection; *Sta*,
starvation and growth; *Cyc*, cyclic illumination; *Col*, colcemid accumulation.

Organism and enzyme	Strain	Pattern	Method of syn-chrony	Reference
	Prokaryotes			
	Escherichia coli			
Aspartate transcarbamylase	K12 Hfr CS-101-G-1	S	*Sel*	Kuempel et al. (1965)
Alkaline phosphatase	K12 Hfr CS-101-G-1	S	,,	,, ,,
Tryptophanase	K12 Hfr CS-101-G-1	S	,,	,, ,,
Tryptophanase	K12 58–161(F⁻)	S	,,	Donachie & Masters (1966)
-Galactosidase	K12 58–161(F⁻)	S	,,	,, ,,
Tryptophanase	K12 58–161(F′Lac⁺/ Lac⁺)	S	,,	,, ,,
β-Galactosidase	K12 58–161(F′Lac⁺/ Lac⁺)	S(2)	,,	,, ,,
β-Galactosidase	K12 HfrH	S	,,	Nishi & Horiuchi (1966)
D-Serine dehydratase	K12 HfrH	S	,,	,, ,,
βGalactosidase	K12 E52 + F₁₃	S	,,	,, ,,
D-Serine dehydratase	K12 E52 + F₁₃	C	,,	,, ,,
β-Galactosidase	K12 E64(F⁻)	C	,,	,, ,,
D-Serine dehydratase	K12 E64(F⁻)	C	,,	,, ,,
D-Serine deaminase	15T⁻	S	,,	Donachie & Masters (1966)
β-Galactosidase	B/r	S	,,	Donachie & Masters (1966, 1969), Helmstetter (1968), Pato & Glaser (1968)
Tryptophanase	B/r	S	,,	,, ,,
D-Serine deaminase	B/r	S	,,	,, ,,
Aspartate transcarbamylase	B/r	S	,,	Donachie & Masters (1969)
Dihydroorotase	B/r	S	,,	,, ,,
Orotidine monophosphate pyrophosphorylase	B/r	S	,,	,, ,,
	Bacillus subtilis			
Sucrase	W23	S	*Sta*	Masters & Pardee (1965)
	Eukaryotes			
	Schizosaccharomyces pombe			
Sucrase		S	*Sel*	Mitchison & Creanor (1969)
Maltase		S	,,	Mitchison & Creanor (1971a)
	Chlorella pyrenoidosa			
Nitrite reductase		R	*Cyc*	Knutsen (1965)
Acid phosphatase		R	,,	Knutsen (1968)
Alkaline phosphatase		R	,,	,,
Isocitrate lyase		S	,,	Baechtel et al. (1970)
	Rat HTC cells			
Tyrosine aminotransferase		R	*Col*	Martin, Tomkins, Granner (1969), Martin, Tomkins & Bresler (1969), Martin & Tomkins (1970)

9 Organelles, respiration and pools

Cell organelles

There are major changes in cell morphology which take place during mitosis at the end of the eukaryotic cycle, but they are beyond the scope of this book. The most striking fact about the structure of the cell during the rest of the cycle is how little it changes at the level both of the light microscope and of the electron microscope (for mammalian cells see Blondell & Tolmach, 1965; Robbins & Scharff, 1966). Něsković (1968) finds changes during interphase in the shape and staining properties of mammalian cells but the differences are fine ones and may only apply to particular strains. In general, it is difficult, if not impossible, to distinguish a G1 cell from a G2 cell except on the basis of size, and sizing is only easy with those micro-organisms which grow unidirectionally.

Nucleus

The most conspicuous structures in the interphase nucleus are heterochromatin (chromocentres) and nucleoli. There are some reports of changes in these structures. González & Nardone (1968) found an increase in the size of nucleoli and a reduction in number (due to fusion) during the cycle in mouse L cells. The nucleoli were oval in shape and lay in the centre of the nucleus during the S period, but were irregular in shape, showed amoeboid movement, and were located near the nuclear membrane during G1 and G2. In female human fibroblasts in culture, the area of the sex chromatin increases significantly during the S period (while still heterochromatic) but there is no parallel increase in the total nuclear area during this period (Comings, 1967). In unsynchronised *Tetrahymena*, there is no change in macronuclear structure but the micronucleus shows a reduction in the number and density of the chromatin aggregates (Flickinger, 1967). In heat-synchronised *Tetrahymena*, there are several descriptions of changes in macronuclear and micronuclear structure (reviewed by Zeuthen & Rasmussen, 1971). On the other hand, Schwarzacher (1963) found a constant pattern of arrangement of the heterochromatic sex chromatin in cultured human cells, and, with mouse embryo fibroblasts, Abercrombie & Stephenson (1969) say that their 'predominant impression is of a high stability of the chromotype during interphase'. As with whole cells, the overall picture at present is that the changes in nuclei are only in fine detail although a general pattern might emerge from future observations. The critical point here is at what level the observations can be made, since light microscopy and electron microscopy on thin sections may not have sufficient resolving power. There are good reasons for supposing that there is an ordered arrangement in the fine structure of chromatin in the interphase nucleus (reviewed by Comings, 1968). We know that there is a restricted period in interphase in which the chromosomes are replicated and we can speculate that the periods of G1 and G2 which precede and follow it also involve changes in chromosome structure (p. 89). Whether these changes occur is uncertain, and whether they can be detected will depend on our success in isolating and examining interphase chromosomes.

More is known about the growth of the nucleus and the patterns of synthesis of its chemical constituents than for any other organelle. The evidence, however, is given elsewhere in this book – in Chapter 7 for volume, dry mass and protein, in Chapter 4 for DNA and in Chapter 6 for RNA.

Cytoplasmic organelles

Two types of question can be asked about cytoplasmic organelles such as mitochondria, plastids, centrioles or ribosomes which are clearly defined unitary structures with, in some cases, boundary membranes. One is the same question we have addressed to the cell as a whole – what are the patterns of growth and of synthesis of the chemical components? The other is concerned with the mode of division or reproduction. More specifically, we can ask whether new organelles arise *de novo* from rudiments or precursor structures, or whether they are formed by division of pre-existing organelles. In either case, we can also ask whether they originate synchronously and, if so, where this origin is in the cell cycle. The answers are mostly unknown, but it is safe to predict that this field of sub-cellular embryology will be a major growing point in the future.

There are two lines of evidence that indicate division is the normal mode of reproduction of mitochondria. One of them is microscopical. It has been known for some time that the mitochondria that can be seen in cultured vertebrate cells appear to divide (e.g. Lewis & Lewis, 1915). But they are very mobile organelles which are continuously moving and changing shape, and it could be argued that there are cycles of fusion and splitting which are not directly concerned with the process of growth. In recent years, however, the electron microscope has produced pictures of what appear to be division stages without any evidence for fusion. There are examples in mammalian liver (Bahr & Zeitler, 1962; Tandler *et al.*, 1969), in liverwort (Diers, 1968), in *Physarum* (Guttes *et al.*, 1969) and in a Foraminiferan *Boderia* (Hedley & Wakefield, 1968). In some small Protozoa, there is only one mitochondrion and this divides in two about the time of cell division (Manton, 1959; Manton & Parke, 1960; Steinert & Van Assel, 1967). The cleavage furrow cuts through the dividing mitochondrion, which occupies a relatively large part of the cell, so the position is rather different from that in larger cells with many mitochondria.

The second line of evidence for mitochondrial division comes from labelling experiments. The lipid components of the mitochondria of a choline-requiring strain of *Neurospora* can be labelled with radioactive choline (Luck, 1963*a*; 1963*b*; 1965). The distribution of label is then determined by biochemical separation and autoradiography after growth following a pulse of choline. There is random distribution of label throughout the mitochondrial population, which is consistent with division and with the continuous addition of new mitochondrial material in all the mitochondria. There are no signs of the development of an unlabelled fraction of the population, as would be expected with *de novo* synthesis from rudiments. Similar results are found after labelling mitochondrial DNA in *Tetrahymena* and examining autoradiographs (Parsons & Rustad, 1968).

The position with chloroplasts of lower plants is similar to that with mitochondria except that there are no labelling experiments. The microscopical evidence for division is rather stronger and the general assumption is that this is the normal mode of reproduction. A particularly convincing demonstration of chloroplast division in *Nitella* comes from the time-lapse photographs of Green (1964). As with the mitochondrion, the single plastids of some small Protozoa and Algae divide at the time of cell division (e.g. Manton, 1959; Manton & Parke, 1960; Soeder, 1965). The situation in higher plants is somewhat different since chloroplasts develop from proplastids and usually do not divide in fully differentiated cells – nor of course do these cells themselves divide (reviewed by Granick, 1961). The proplastids, however, appear to divide in the rapidly growing cells of shoot and root tips, so division seems to be a basic process in plastid reproduction.

Accepting that many mitochondria and plastids normally reproduce by division, the question then arises as to whether the divisions are synchronous. The answer seems to be no in most cases since the observations which have been mentioned above show a small but roughly constant proportion of dividing figures in growing cells. With synchrony, the proportion would be high in some cells but low in others. There is, however, some evidence of synchronous mitochondrial division in the growing hyphae of *Neurospora* (Hawley & Wagner, 1967) and in synchronised cultures of the fission yeast *Schizosaccharomyces pombe* (Osumi & Sando, 1969). This question merits further observations on synchronous cultures, preferably with several techniques of synchronisation since the results with induction synchrony might differ from those with selection synchrony.

There is only a little that can be said about the synthesis of the chemical components of mitochondria and chloroplasts. The replication of their DNA has been discussed earlier (p. 71). Warmsley *et al.* (1970) found that both mitochondrial and microsomal protein are synthesised continuously through the cycle of mouse P815Y cells, and that enzyme markers for the membranes of mitochondria (cytochrome C oxidase and succinate-cytochrome C reductase) and of microsomes (NADPH-cytochrome C reductase and uridine diphosphatase) also show continuous synthesis. On the other hand, glutamate dehydrogenase, a soluble enzyme of the mitochondrial matrix, shows periodic synthesis in these cells, as do some of the other mitochondrial enzymes in Table 8.2 (p. 177), e.g. cytochrome oxidase in *Saccharomyces* (Cottrell & Avers, 1970). The same conclusions about periodic synthesis that apply to whole cells may also apply to the mitochondrial population and perhaps to the individual mitochondrion.

The centriole has a mode of reproduction that is in sharp contrast to mitochondria and plastids (reviewed by Pitelka, 1969; see also Stubblefield & Brinkley, 1967; Brinkley & Stubblefield, 1970). It does not divide: instead, a new centriole forms near (but not touching) one end of a mature centriole and then grows outwards at right angles. The chronology of centriole duplication and splitting was first worked out in early sea urchin embryos by Mazia *et al.* (1960) who exploited the fact that mercaptoethanol inhibits the initial duplication but does not prevent the later separation. Duplication (the initiation of new centrioles) starts early in the cycle in late telophase or early interphase – the time of the S period in these embryos (p. 63). There are two adjacent mature centrioles, each of which starts duplication by the outgrowth of a young one. The two mature centrioles, each accompanied by its young, separate in early prophase and move to opposite sides of the nucleus, there to organise the mitotic apparatus. The mature and the young centriole have not, however, separated at this stage. This is delayed until the end of mitosis, when separation is completed and a new cycle of duplication starts. There is, therefore, a full cycle between the initiation of a new centriole and its separation from its mature neighbour. This plan of reproduction has been confirmed by electron microscope observations in synchronised HeLa cells (Robbins *et al.*, 1968). The production of a new centriole starts at the beginning of the S period and is complete by G2, but separation of the orthogonal pair of young and old does not occur until the following G1.

It is satisfying to be able to give a fairly clear picture of centriole reproduction in most animal cells – though not in all, since some ciliated and flagellated cells have centrioles with atypical structure and replication (Pitelka, 1969). It is also important to have a good example of the separation of an organelle which is out of phase with nuclear and cell separation. There remains, however, a great deal to be found out. Presumably the new centriole starts from a rudiment, but the source and nature of this rudiment is unknown. Nor have we any clue about an intriguing aspect of the outgrowth – that it is at right angles to the neighbouring mature centriole. The chemical composition also remains unknown, though the close similarities between centrioles and the basal bodies of cilia and flagella coupled with recent demonstrations of DNA in basal bodies (e.g. Randall & Disbrey, 1965) suggests the presence of DNA in centrioles.

The mode of reproduction has little meaning when considering ribosomes but there is a question that can be asked about their increase in numbers during the cycle. Does this exactly parallel the increase in ribosomal RNA (or ribosomal protein)? The answer need not be yes if the RNA has been extracted from whole cells without a separation of the ribosomal pellet since, in principle at any rate, there could be a pool of ribosomal RNA which had not been assembled into complete ribosomes. The RNA from pelleted ribosomes gives a better estimate of the number of ribosomes but there are not many cases where such a pellet contains all the ribosomal RNA of the cells and no other kind of RNA. The same argument also applies to ribosomal protein. For any detailed study of the control of ribosome production, it would be worth having figures for the increase in numbers in growing cells. So far, this has only been done in *Schizosaccharomyces* by Maclean (1965), but the method (counting ribosomes in electron microscope thin sections) is not very precise.

There is little or no information about changes through interphase in the membrane systems of the cytoplasm – the endoplasmic reticulum (but see p. 157), the Golgi apparatus and membrane bounded vacuoles such as lysosomes. All that can be done now is to emphasise that there are problems in their growth and development. Do they, for example, have particular growing regions or is there general growth by intercalation over all the membrane surface? Do lysosomes or the Golgi apparatus divide or do they grow from rudiments? Are ribosomes attached to the rough-surfaced reticulum as soon as it is formed?

The cytoplasmic components that we have been considering are those of the 'general cell' and have a widespread distribution among eukaryotic cells. There are, of course, more specialised cells which have other structures that do show changes in interphase. A striking example is the oral apparatus of Ciliates which will be considered in the next chapter.

Cell membrane

The surface area of a growing cell doubles during an ideal cell cycle, as do all other cell properties. Increasing volume implies increasing area, but the relationship between these two parameters is not proportional except in the case of an infinitely thin rod-shaped cell growing only in length (Fig. 9.1(A)). With a spherical cell, simple arithmetic shows that 60 per cent of the area increase during a cycle happens as the cell grows through the cycle, while the remaining 40 per cent takes place abruptly at division (Fig. 9.1(B)). In practice, there are many cells with shapes intermediate between a sphere and a very thin rod but they will all show some degree of sharp area increase at division.

These simple models suggest that membrane components might show a sharp increase at the time of division, and there is some recent evidence which supports this. Mammalian cells show an increased rate of incorporation into the surface membrane fraction at or just after division (Gerner *et al.*, 1970) and into lipids and glycolipids at or just before division (Bosmann & Winston, 1970; but see also Bergeron *et al.*, 1970). There is also a large but transient increase of glycerol incorporation into lipids at about the time of division in *Bacillus megaterium* and *Escherichia coli* (Daniels, 1969). There are, of course, many difficulties in equating area increase with membrane increase. Membranes are elastic and the area increase at division could be due to stretching the existing membrane followed only later by the insertion of new material. The stretching could also involve the opening out of folds in the surface. In many cells, the surface is highly mobile and there is almost certainly a reduction of its area in situations like fibroblasts at the time of cell division when processes are retracted and the cell rounds up. We do not know whether this reduction is accomplished by a simple thickening and folding of the membrane or whether there is also a flow of membrane material inwards to the cytoplasm. Membrane material certainly does move into the cytoplasm during pinocytosis and it has been estimated from experiments with labelled antibody that the membrane material in *Amoeba* has a turnover of 0.2 per cent/min (Wolpert & O'Neill, 1962). This degree of turnover may be unusually high

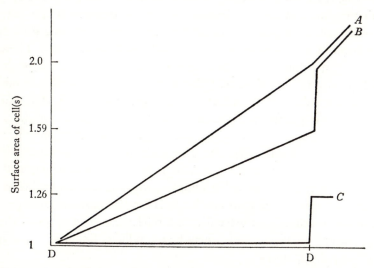

Fig. 9.1. Calculated curves for the increase in surface area of a cell during the cell cycle. *A*, growing rod-shaped cell. *B*, growing spherical cell. *C*, non-growing spherical cell. Area growth through the cycle is assumed to be linear for *A* and *B*. D is cell division.

since mouse L cells appear to have little turnover when growing, though it rises when they are not growing (Warren & Glick, 1968). But it raises a problem which applies both to membranes and to other cell structures, for example, ribosomes. There may be a difference between the time course of synthesis and that of assembly. The components of membranes could be synthesised continuously and accumulate in pools, but only come into position at particular times in the cycle. There is some evidence of the independent synthesis of membrane lipids and membrane proteins (e.g. Kahane & Razin, 1969), so they might accumulate as separate pools. A pool or pools of membrane components is a convenient postulate for the changing membrane of amoeboid cells. It might also be present in early embryos where the only major growth requirements are an increase in nuclei and an increase in cell membranes. A 'non-growing' spherical egg has an increase of 26 per cent in surface area at each division (Fig. 9.1(C)). This could be accompanied either by *de novo* synthesis of new membrane or by assembly from a pre-existing pool.

Another problem about membrane growth is whether there are only a few regions where new material is added or whether growth takes place by widespread intercalation over the whole area. One would guess that the latter type of growth is the common one, especially in mobile animal cells, but there is no real evidence one way or the other and it would be interesting to test this by, for instance, the immunofluorescent techniques that have worked with cell walls (p. 190). Localised regions of membrane growth may well occur in cells showing tip growth of the wall (as in Fungi). There is evidence from palmitic acid labelling that new membrane in *B. megaterium* is formed at the end of the cell (Morrison & Morowitz, 1970) – a mode of growth which is different from that shown by the wall (p. 191). Donachie & Begg (1970) have put forward a 'unit cell model' for the growth of *E. coli* in which there are either one or two localised sites for the formation of new membrane (but *not* of cell wall). The distribution of pigment granules in big Amphibian eggs also suggests that there is localised membrane growth in the cleavage furrow (Selman & Waddington, 1955). One of the problems, however, about eggs and perhaps other cells is how far movements of pigment granules anchored in a gelated cortex reflect the movements of the overlying plasma membrane which may be more like a liquid in physical properties (e.g. Dan, 1954).

A question that can be asked about the cell surface is whether or not its physical properties change during the cycle. Most of the work on this subject has been done on the early cleavage cycles of sea urchin eggs and has been well reviewed by Hiramoto (1970). There are certainly increases in the stiffness of the surface at fertilisation and at division, and, with some techniques of measurement, one or two points in interphase where the stiffness is minimal. One difficulty here is an uncertainty about the relative contributions made to surface stiffness by the plasma membrane, by the cortex, or by the interior of the egg when it contains structures like the sperm aster or the mitotic asters. A second difficulty is to know whether or when there is a period in the early cleavage cycles that corresponds to interphase in normal growing cells. It would be worth following surface changes during the cycle in growing cells with a technique such as the cell elastimeter which was originally developed for sea urchin eggs (Mitchison &

Swann, 1954) but can be used successfully with mammalian cells (Weiss, 1966).

It appears that the electrical properties of the cell surface also change during the cycle. Two kinds of mammalian cells show a minimum electrophoretic mobility (surface charge) in mid-cycle during the S period (Mayhew & O'Grady, 1965; Mayhew, 1966; Brent & Forrester, 1967).

It is not easy to assess the importance of these alterations in physical properties. There are, without doubt, changes in the surface at the time of division – indeed, it can be put more strongly that the immediate agent in dividing an animal cell is very probably the cell membrane together with its associated cortex. The changes in interphase may be a slow preparation of the membrane for the activity of cleavage, followed by a slow reversal. Alternatively, the changes may be related to a sequence of events in interphase about which we know little.

Cell wall

Far more is known about the plant cell wall than about the cell membrane because the wall is a relatively thick and rigid structure which is easy to observe and to separate. Much of the evidence, however, comes from the final growth phase of differentiating cells which are not strictly in a cell cycle. The two major techniques that have been used to measure and locate the growing regions are autoradiography after pulse-labelling with wall precursors and direct observation of wall expansion using surface markers such as the primary pit fields (reviewed by Roelofsen, 1959; 1965). In many higher plant cells, growth takes place throughout the whole length of the cell and autoradiographs show that this growth involves the deposition of new material and is not a simple stretching of the wall. There seem to be differences in the way this new material is added both in oat coleoptiles and in pea stems (Ray, 1967). New cellulose is added to the inner side of the wall by apposition, whereas non-cellulose material (e.g. hemicelluloses) is deposited within the structure of the pre-existing wall. In neither case, however, are there localised regions of deposition along the length of the cells.

This pattern of growth in all regions of the cell wall is not the universal rule and there are a number of higher plant cells, especially long thin ones, which show localised growth points. Tip growth occurs in pollen tubes, root hairs, tracheids and some phloem and xylem fibres (Roelofsen, 1959).

Tip growth also occurs in a number of lower plants, for instance fungal hyphae. It has recently been analysed quantitatively in three species of yeast by grain counting on autoradiographs of cell wall 'ghosts' prepared from cells which have been pulse-labelled with tritiated glucose (Johnson, 1965; Johnson & Gibson, 1966 a, b). The fission yeast *Schizosaccharomyces pombe* is a cylindrical cell with rounded ends and shows tip growth (at one end only, in the majority of cells). This confirms earlier results from microscopic observation (Mitchison, 1957) and immunofluorescence (May, 1962). The addition of new wall material at the tip, as judged by the amount of incorporation, follows a curve of exponential increase with time through the cycle which is very similar to the curve for cell volume shown in Fig. 7.6. There is no tip growth during the last quarter of the cycle but new wall material is being laid down in the cell plate. As well as this labelling of growing regions, there is also a substantial amount of incorporation into non-extensile parts of the wall. This incorporation, which stays at a constant rate through the cycle, may be due to reorganisation and turnover in the wall. But, whatever the explanation, it emphasises the danger of assuming that growth always parallels incorporation. Tip growth also seems to occur in the fusiform budding yeast *Pichia farinosa* and in the ellipsoid budding yeast *Saccharomyces cerevisiae*. Both species show some incorporation into the non-growing wall of adult cells but on a much smaller scale than in *S. pombe*.

In bacteria, the most widely used technique for following wall growth is immunofluorescence (reviewed by Cole, 1965). The essence of this method is to prepare an antibody against the wall, label it with a fluorochrome (usually fluorescein), apply it for a short period to growing cells, and then wash off and observe further growth with fluorescence microscopy. The regions of the wall present at the time of labelling will be fluorescent whereas any new wall formed after the labelling will be dark. It is best to combine this direct method with the reverse one in which cells are first exposed to unlabelled antibody then grown for a period and finally exposed to labelled antibody. In this case, old wall will be dark and new wall will be bright.

The results with immunofluorescence show that there are different modes of growth in different bacteria (Fig. 9.2). *Streptococcus pyogenes* has an equatorial ring in which new wall is formed both at the surface and in the septum between the daughter cells (Cole & Hahn, 1962). The secondary regions of growth appear on either side of the equatorial rings under conditions of fast growth leaving a narrow ring of old wall between them and the equator. The result of having limited regions of wall growth is that parts of the wall (the outer black hemispheres in Fig. 9.2) are conserved through successive generations. *Bacillus cereus* and *B. megaterium* also have limited regions of growth and show alternating regions of fluorescence and darkness when they have been grown for some time after fluorescent labelling (Chung *et al.*, 1964). In contrast, *Salmonella typhimurium* and *S. typhosa* have a large number of growth regions and show a gradual and uniform fading of fluorescence when grown after labelling (May, 1963; Cole, 1965). The number of regions has been kept small in Fig. 9.2 for ease of illustration. There would be many more of them in real cells and the gaps between them are not resolvable under normal growth conditions.

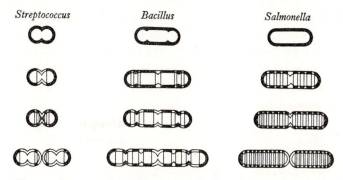

Fig. 9.2. Diagram of possible modes of cell wall growth in bacteria as seen by immunofluorescence. From Cole (1965).

Gaps, however, do appear in *S. typhosa* grown in the presence of chloramphenicol. A similar fading of fluorescent label occurs with *Spirillum volutans* and can be interpreted in the same way, as a diffuse intercalation of new wall material (McElroy *et al.*, 1967). There is some dispute about *Escherichia coli*, since Chung *et al.* (1964) found limited growth regions, whereas Beachey & Cole (1966) found the same pattern as in *Salmonella* both with fluorescent antibodies and with ferritin labelled antibody examined in the electron microscope. Considering the results from the closely related *Salmonella*, the weight of the evidence is in favour of growth by diffuse intercalation being the normal pattern in *E. coli*.

The pattern that emerges from this work is that the gram-positive bacterial cells which have been examined show only a few regions of growth and the gram-negative ones show growth over the whole surface. The limited resolution of the light microscope makes it impossible to be precise about the fine structure of these growth regions. The gaps between the points of intercalation in the gram-negative walls might be only a few molecules in size or they might not exist at all. It is only the appearance of the wall after chloramphenicol treatment that suggests the gaps are relatively large. In the same way, the limited regions of growth in the gram-positive walls might contain many points of intercalation. We should also remember that the pattern of wall growth may be different from that of membrane growth, and there is in fact evidence of this in *B. megaterium* and *E. coli* (p. 188).

Immunofluorescence has limitations which have been discussed by Cole (1965) and Rogers (1965). The effective surface antigens are probably not the basic mucopeptide of the bacterial wall, and they may be obscured in certain regions or they may move relative to the wall during growth and division. Turnover may also add new antigens in a non-growing region (Mauck & Glaser, 1970). Nevertheless this method gives us the best evidence that we have about wall growth and it only conflicts with other results at one point. A bacterial rod growing by either of the modes shown in Fig. 9.2 should elongate at both ends with respect to a fixed point in the background. Adler & Hardigree (1964) found this to be so for a number of strains of *E. coli* but not for one strain of B/r which elongated only at one end.

One other point about wall growth is that some of it may be due to passive stretching rather than active synthesis of new wall material. Most of the incorporation of diaminopimelic acid (a wall component) takes place at division in growing cells of *B. subtilis*, suggesting that the wall growth in interphase is largely stretching (Dadd & Paulton, 1968).

Respiration

There has been a long history of measurements of the rate of respiration in dividing eggs where the aim has been to throw light on the energetics of cell division. The early experiments used Warburg manometers and large quantities of eggs. There was then a great improvement in technique about thirty years ago with the introduction of the Cartesian diver respirometer which can be made sufficiently sensitive to measure the oxygen consumption of a single small egg. This method was used, primarily by Zeuthen, on a variety of Vertebrate and Invertebrate eggs (Zeuthen, 1946; 1955; Holter & Zeuthen, 1957; Frydenberg & Zeuthen, 1960). The general pattern, of which Fig. 9.3 is an example, is one in which there is respiratory rhythm with a minimum oxygen consumption during mitosis and with cleavage occurring during the early part of the subsequent rise. Scholander *et al.* (1952, 1958), using another form of diver respirometer, found this rhythm in some eggs but no consistent rhythm in others. Some possible

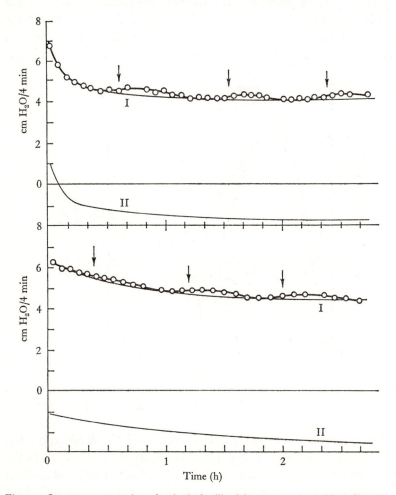

Fig. 9.3. Oxygen consumption of a single fertilised frog egg measured in a Cartesian diver respirometer. The arrows mark the appearance of the first, second and third cleavage furrows. The ordinate is proportional to oxygen uptake. From Zeuthen (1946).

explanations of this discrepancy are discussed by Zeuthen (1960). The most striking thing about the egg rhythms is the small size of the fluctuations. If the mechanical work of splitting the chromosomes and the cell requires appreciable amounts of energy, this is certainly not apparent as a major increase in respiration. But energy could of course be drawn from a pool which is steadily built up during the whole cycle (Epel, 1963).

Eggs do not grow, and we might well expect to find a different pattern through the cycle of growing cells. In fact, the pattern in several cell systems is similar though there are much greater differences between the troughs at mitosis and the peaks. Fig. 9.4 shows a halving in the oxygen consumption between the peak value in the S period and the minimum

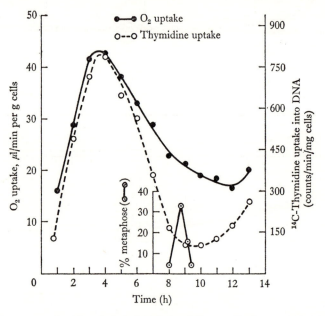

Fig. 9.4. Oxygen uptake (closed circles) and thymidine uptake (open circles) in a HeLa cell culture synchronised by a thymidine block. From Robbins & Morrill (1969).

value during mitosis and early G_1 in HeLa cells (Robbins & Morrill, 1969; see also Gerschenson *et al.*, 1965). There is a sharp drop at mitosis in the synchronous microspore divisions of *Lilium* (Erickson, 1947) and *Trillium* (Stern & Kirk, 1948). *Chlorella* shows a trough, though it is a little later than the cell divisions and is preceded by a minor trough (Curnutt & Schmidt, 1964*b*). Temperature-cycled cultures of *Astasia longa* also have a drop at mitosis, though, as always in these systems, it is difficult to separate the changes caused by the cell cycle from those caused by the alterations of the environment (James, 1965).

One interpretation of this pattern, which has been suggested by Robbins & Morrill (1969), is that the fall at mitosis is caused by the stoppage in macromolecular synthesis. Most cells cease to make RNA at that time, and in some of them there is also a sharp reduction in protein synthesis. This theory has some attractive features. It would explain why the fluctuations are so small in eggs, which have much lower synthetic rates than growing cells. It would fit in with the results from some preliminary experiments with inhibitors which suggest that the demands of macromolecular synthesis account for most of the oxygen consumption of growing cells (Robbins & Morrill, 1969). It would also explain why different patterns are found in yeast and *Tetrahymena*, as we shall see below. Against it is the fact that protein synthesis does not cease, or even decline, in some mammalian cells (p. 132). In *Chlorella* also, RNA synthesis shows a fall in rate at division but protein synthesis does not (Hermann & Schmidt, 1965; Hare & Schmidt, 1965; 1970).

Ciliates and yeast are unusual among cells in showing no decline in the rate of synthesis of RNA or protein at division, and they are also unusual in their respiratory patterns. Zeuthen (1953) showed that the rate of oxygen consumption in *Tetrahymena* increased linearly through nearly all of the cycle but there was a short period just before division when the rate stopped increasing or even decreased slightly. In later experiments, Lövlie (1963) found similar patterns though there was more variability between individual cells (Fig. 7.4). The main point, however, is that the respiratory changes at division are very much smaller than those shown by mammalian cells. The results with yeast agree in showing no fall in respiration at division, but the patterns through the cycle differ (perhaps because of differences in the synchronising procedure). Cottrell & Avers (1970) found a continuous exponential increase in oxygen consumption in *Saccharomyces cerevisiae*. With the same species, however, Scopes & Williamson (1964) discovered a quite different pattern in which there are a series of plateaux and steps (Fig. 9.5, see also Eckstein *et al.*, 1966). There are three points of interest here. One is that the steps are less than doublings, which is probably due to the method of synchronising by feeding after starvation. Volume and mass also fail to double for the first few cycles. The second point is that the interval between the steps tends to be less than a cycle (seen best in Fig. 9.5(*a*)). This is an unusual state of affairs and suggests perhaps that the timing clock for the growth cycle is running in these experiments at a different speed from the clock for the division cycle (p. 246). The third point is that this pattern is quite different from that of dry mass which follows a smooth upwards curve. So it is not legitimate to equate respiratory rate with total dry mass in yeast, even though it is, at any rate approximately, in *Tetrahymena* (Zeuthen, 1953; Lövlie, 1963). The reason for the steps is not clear. Although mitochondrial number may double abruptly at one stage of the cycle, it is improbable that mitochondrial mass will do so; and respiration is more likely to be proportional to mass than to number. A step might possibly be caused by the synthesis of one or

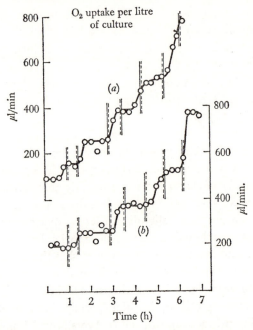

Fig. 9.5. Rate of oxygen consumption in two synchronous cultures of *Saccharomyces cerevisiae*. The vertical dashed lines indicate the limits of the periods of budding and cell division. From Scopes & Williamson (1964).

more key enzymes at that stage in the cycle. Similar respiratory steps have been reported by Osumi & Sando (1969) in synchronous cultures of *Schizosaccharomyces pombe*.

We can conclude that a good working hypothesis for most growing cells is that the rate of macromolecular synthesis is a major factor in controlling the respiratory rate through the cycle. But we cannot be more precise until there is a clearer picture of the energy balance sheet in a cell – how much is used for synthesis, for maintenance, for transport and for movement.

Pools and uptake

All cells contain acid-soluble pools of small molecules which are clearly important in the economics of growth. Food materials pass through the pools and whatever regulates entry into the pools also regulates overall cell growth. The molecules in the pools are not only the raw materials for growth but also regulators in their own right through enzyme repression and inhibition. Pools also play a practical role in experiments in the cell cycle because of their effect on pulse-labelling (p. 11). In spite of all this, the evidence on pool changes through the cycle is fragmentary. There is no doubt that the reason lies in the appalling complexity of the problems. Pools both contain a large number of molecular species and may also be partitioned unevenly within the cell (Miller *et al.*, 1964; Merriam, 1969). But we shall not be able to progress far in understanding the regulation of cell growth without more attention being paid to pools.

The behaviour of the total acid-soluble pool in *Schizosaccharomyces pombe* has already been described (p. 138). In a broth medium, it fluctuates during the cycle with a maximum in mid-cycle. But in a minimal medium where it lacks an expandable component containing amino-acids, its size is smaller and there is no fluctuation.

Robbins & Scharff (1966) compared the amino-acid pool of metaphase-arrested HeLa cells with that of interphase cells and found no significant difference in size or composition apart possibly from glycine. Klevecz (1969*a*) suggests that there is a fall in the amino-acid pool during the S period of Chinese hamster cells, but the evidence is indirect. The fullest analyses of amino-acid pools have been made on *Chlorella* (Hare & Schmidt, 1970) and on *S. pombe* (Stebbing, 1969). In *Chlorella*, but not in *S. pombe*, there are definite changes in the proportions of a number of the amino-acids (Fig. 9.6). In view of the oscillatory repression theory of enzyme synthesis (p. 169), it would be interesting to see whether these changes bore any relation to the steps in the enzymes of amino-acid biosynthesis.

Both in *Tetrahymena* (Stone *et al.*, 1965) and mouse L cells (Adams, 1969*a*, *b*), the uptake of thymidine into a pool only occurs in the S phase (and perhaps early G2). This pool is restricted to the nucleus and in the L cells at least is primarily thymidine triphosphate. Although it cannot be labelled with exogenous thymidine outside the S period, it exists at all times in the cycle with presumably a low turnover in G1 and G2. Adams (1969*a*) suggests that the pool is larger outside the S period and that this inhibits thymidine kinase and so prevents the phosphorylation and entry of exogenous thymidine.

Direct measurements have been made in several systems of the pool of nucleotide triphosphates (NTP) or adenosine triphosphate (ATP). The main interest here has been in whether the energetics of cell division cause changes in the ATP pool. Swann (1953; 1957) developed the concept of

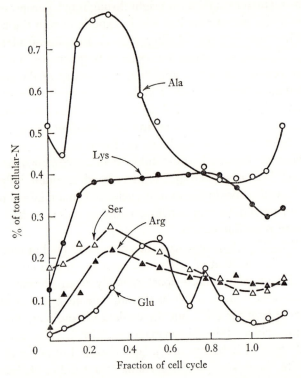

Fig. 9.6. Changes in proportions of free cellular amino-acids during the cycle of *Chlorella pyrenoidosa*. Alanine, lysine, serine, arginine and glutamate are shown here. From Hare & Schmidt (1970).

an 'energy reservoir' which was filled during the cycle and discharged at the start of mitosis. The experimental support, however, for this concept is in erious doubt because of the experiments of Epel (1963, p. 230). He showed that ATP was required throughout mitosis in sea urchin eggs and that there was normally no change in its level in the pool during the cycle. In the *Chlorella* cycle there is also no change in the relative level of ATP (as a proportion of the dry mass) and in the ATP/ADP ratio (Curnutt & Schmidt, 1964a). But other systems show alterations in the nucleotide pool. During nuclear division there is a rise followed by a fall in the ATP pool in *Physarum* (Chin & Bernstein, 1968; Sachsenmaier *et al.*, 1968) and in the NTP pool in heat-synchronised *Tetrahymena* (Plesner, 1964) and HeLa cells (Gerschenson *et al.*, 1965). No clear pattern emerges from this evidence and it should be remembered that ATP levels may be altered in growing cells not only by the energy requirements at division but also by the changes in respiration and macromolecular synthesis.

In one of the earliest biochemical studies of the cell cycle, Rapkine (1931) found fluctuations through the cycle in the sulphydryl content of an acid-soluble material extracted from sea urchin eggs. He interpreted this material as glutathione, but later work showed that it was an unusual protein soluble in 25 per cent trichloracetic acid (Sakai & Dan, 1959). This protein, which may be involved in the mitotic apparatus and the control of division, will be discussed later (p. 213). Variations have recently been found in the non-protein sulphydryl content of mammalian cells (Ohara & Terasima, 1969; Harris & Patt, 1969). The pattern of change has a minimum at the beginning of the S period and a maximum at the end. It is approximately the inverse of the pattern for protein sulphydryl and it also shows a marked similarity to the curve for survival after X-irradiation. There are no changes in the level of glutathione during the HeLa cell cycle (Klein & Robbins, 1970).

There is little information about changes in inorganic ions. Jung and Rothstein (1967) found variations in the sodium and potassium content of mouse cells, and the phosphate and sulphate pools have been followed in *Chlorella* by Baker & Schmidt (1963; 1964) and Johnson & Schmidt (1963).

The uptake of materials from the medium has a close relation to pools. Food molecules pass through the uptake mechanisms on their way to the pools, and the rate of uptake is one of the factors that govern pool size. It may be more profitable to think of this relationship the other way round – that pool size governs the rate of uptake. A neat way of regulating the size of a pool would be to have negative feedback which would cut off the uptake when the pool reached a given size. An example of this mechanism may be the uptake of adenine by *Schizosaccharomyces pombe* (Cummins & Mitchison, 1967). But whatever the machinery of regulation, it is clearly of considerable importance in the life of the cell since it may be the limiting factor in determining the rate of growth. It is also important experimentally since the rate of uptake of a labelled precursor is one of the variables which affect incorporation. Having said this, however, we are faced with the fact that little is known about changes in uptake during the cycle. The restriction of thymidine uptake to the S period in *Tetrahymena* and L cells has been mentioned above. Amino-acids are taken into the pool at the same rate in metaphase-arrested HeLa cells as in interphase cells (Robbins & Scharff, 1966). A more extensive survey has recently been made on the uptake rates and pool sizes of RNA bases during the cycle in *S. pombe* (Mitchison *et al.*, 1969). With adenine and uracil, the following properties increase steadily through the cycle in proportion to the dry mass: initial rate of uptake, rate of incorporation, size of the pre-existing precursor pool and, with adenine, the size of the expanded pool. The rising rate of incorporation indicates an approximately exponential curve for RNA synthesis and there is no evidence that this interpretation is distorted by restrictions on uptake or by fluctuations in the pools. These are only scattered examples, and we need to know much more about uptake. One intriguing possibility should be investigated. Uptake appears to happen in most cases with the mediation of a carrier which is probably a protein with some of the properties of an enzyme. Since many enzymes are synthesised in steps, the same may be true of transport carriers and the effect of this might be manifest in sharp increases in uptake rate (cf. Kubitschek, 1970).

10   The control of division[1]

There are two main ways of looking at the problems of the control of cell division. One of them is concerned with the stimulus which starts growth and division in a quiescent cell or tissue. This is a subject of major importance but beyond the scope of this book. The second way, which will concern us in this chapter, is to look at the narrower field of the cell cycle and to ask about the mechanisms which control division, the major event at the end of the cycle in growing cells. It should be made clear at the outset that there is no general answer to this question. There are only partial and incomplete answers which apply to some cell systems. I shall concentrate on a few of these because they seem to me to be the most illuminating at the present time, but this means that many interesting points will be left out, for instance the question of control by the attainment of a critical cell size or critical ratio of cytoplasm to nucleus.

These and other problems of division control have been reviewed many times in the last twenty years. Some of the more thoughtful of these general reviews are by Mazia (1961), Prescott (1961; 1964a, b; 1968b), Stern (1966) and Swann (1957; 1958).

Heat-shock synchrony in *Tetrahymena*

One of the first systems in which synchrony was induced in a random cell population was the heat-shock synchronisation of *Tetrahymena* developed by Zeuthen and Scherbaum nearly twenty years ago (Scherbaum & Zeuthen, 1954). It has been exploited by many people since then, and has generated a large literature, of which the key reviews are by Zeuthen & Rasmussen (1971) and Zeuthen (1964).[2] The importance of this system lies in the fact that it is the only case of induction synchrony where there is

---

[1] The following abbreviations are used at intervals in this chapter: DD-cycle = DNA-division cycle; DP = division protein; ED = excess delay; EH = end of heat shocks; OA = oral apparatus; TP = transition point.

[2] Other reviews are Zeuthen (1958; 1971) and from different standpoints Frankel & Williams (1971), Scherbaum (1960; 1963b, 1964), Scherbaum & Loefer (1964).

a well-developed theory, with experimental backing, on the mechanisms
that control division in individual cells and cause division synchrony in
cell populations.

Tetrahymena pyriformis is a Ciliate Protozoan, about 60 μm × 40 μm
in size, which grows relatively fast in sterile medium (generation time of
2.5 hours in rich medium). There are a number of strains but the one that
has been used most frequently for synchrony work is GL, which lacks

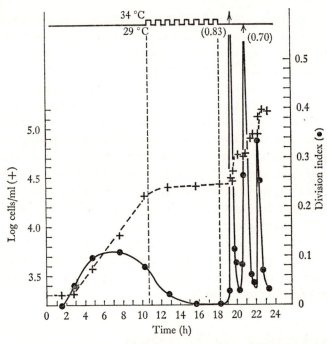

Fig. 10.1. Synchrony of *Tetrahymena pyriformis* induced by repetitive heat shocks (29 °C
to 34 °C). Continuous line is the division index. Dashed line is cell number. From Zeuthen
& Scherbaum (1954).

a micronucleus. The basic technique of heat shock synchrony is shown in
Fig. 10.1. An asynchronous culture growing in complex medium at the
optimum temperature of 29 °C is raised to the sub-lethal temperature of
34 °C, kept there for 30 minutes and then brought down to 29 °C again.
This is repeated at 30 minute intervals until eight heat shocks have been
given. After the end of the heat shocks (EH) the culture is brought down
to 29 °C. The cells stop dividing during the heat shocks and the division
index (proportion of dividing cells) falls to zero. At about 80 minutes after
EH there is a good synchronous division with the index rising to 0.83.

This is followed by other divisions which gradually become less synchronous. The time between them also lengthens until it eventually becomes the same as the normal generation time in asynchronous cultures.

Although division is stopped by the heat shocks, growth continues. The rate of growth is reduced and by EH the cells have roughly doubled in size, dry mass, RNA and DNA. There is no synchronisation of growth so the cells at EH are not only larger than exponential phase cells but also as variable in size. This lack of synchronisation also affects DNA synthesis and it is not until after the first synchronous division that DNA synthesis becomes synchronised with the division cycle – a point we shall return to later.

One important point to realise is that the cycles defined by the synchronous divisions are not the same as the cycles in an exponential culture. Although growth does take place in the complex medium after EH, it is much less than in exponential growth. The average increase in dry mass between EH and the first division or between the first and second divisions is only 15–20 per cent, instead of the 100 per cent in a normal cycle. The cycles are also shorter – about two-thirds of the normal cycle time. Broadly speaking, the effect of fast division and slow growth is to reduce the oversize cells at EH to normal proportions. But the details of this effect must be more complicated since there is a change from a population after EH which is synchronous yet has a large variation in size at division, to a population later which is asynchronous yet has a small variation in size at any one stage of the cycle. It suggests that rates of growth and perhaps cycle times may vary with the size of the cell during the desynchronisation.

There are various modifications that can be made to the temperature regime without seriously affecting the final degree of synchrony. Other timings are effective, for example 20 minutes at 34 °C followed by 40 minutes at 29 °C. The number of shocks can also be changed, though if their number is less than five, synchrony is reduced; and if it is greater than ten, the cells start to divide during the shocks as if they were becoming adapted to the high temperature. The temperature of the heat shocks depends to some extent on the strain of *Tetrahymena* so that WH-14 which has a higher optimum temperature (34 °C) than strain GL needs a temperature shift from 34 °C to 43 °C. Cold shocks can be as effective as hot shocks, for instance five periods of two hours at 10 °C alternating with 40 minutes at 29 °C.

Cultures transferred to an inorganic medium (salts only) before the heat shocks cannot be synchronised. But if this transfer is done after EH or one or two shock periods before, the cells will go through two synchronous divisions without any net growth. The dissociation of division from growth is the nearest thing in this system to 'laying an egg' (Zeuthen, 1964). An interesting intermediate situation occurs with cultures which are heat shocked in a medium which contains glucose and vitamins as well as salts but lacks amino-acids as a nitrogen source. There are no divisions but there is both synchronous rounding of the cells and replacement of the oral apparatus (discussed later) with the same timing after EH as the synchronous divisions in other media (Watanabe, 1963; Frankel, 1970). Some of the normal events of the cycle are therefore recurring in the absence of cell division itself.

The key to understanding *why* heat shocks produce synchrony lies in the fact that the response to these shocks changes during the cycle. A shock causes a delay in division and this delay follows a characteristic pattern. Fig. 10.2 shows this pattern for single cells from a normal random phase culture (Thormar, 1959). The ordinate is 'excess delay' or 'set-back' which is the time by which division is delayed over and above the time occupied by the temperature shock.[1] The total division delay is therefore the shock time plus ED. Young cells at the start of the cycle suffer little ED. It then increases for older cells later in the cycle up to the 'transition point' which is a short time before division (25 minutes). After the transition point there is little or no ED and the coming division is comparatively insensitive to the shock. Taking the top curve in Fig. 10.2, the effect of a 15 minute shock at 34.1 °C is to give a 35 minute ED to a cell 30 minutes 'old' (i.e. from the previous division). This cell would normally have begun division 90 minutes later so after the heat shock it will begin division at $90 + 35 = 125$ minutes. On the other hand, a cell 100 minutes old will be given a much greater ED of 95 minutes. It would have begun division 20 minutes later, so after the shock it will be delayed until $20 + 95 = 115$ minutes. The result of this is to achieve a fair degree of synchrony since the interval between the division times of these two cells has been reduced from 70 to 10 minutes (Fig. 10.3). A third cell, older than the transition point, will not be delayed and will divide much earlier than the previous two cells. But it will be caught by later shocks in a repetitive sequence and

---

[1] 'Set-back' is an attractive word used for *Tetrahymena*, but it carries with it the implication that the cell is set back in the division cycle and has to repeat a series of events rather than being halted in the cycle until some damage has been repaired. There are not enough marker events identified in the division cycle to allow a decision between these alternatives, so I have used the more neutral phrase 'excess delay' (ED) except in some cases where morphogenetic events are repeated.

it is for this reason that multiple shocks are needed to produce good synchrony.

There are some inconsistencies in this scheme, for instance the flatness of the early part of the ED curves in Fig. 10.2 suggests that cells in the

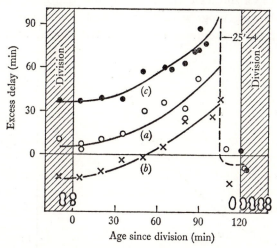

Fig. 10.2. Excess delay (total delay minus time of temperature shock) plotted against stage in the cycle for single cells from an exponential culture subjected to the following temperature shocks: (a) 9.3 °C for 30 min; (b) 31.1 °C for 20 min; (c) 34.1 °C for 15 min. Cells were grown and observed at 28.5 °C. From Thormar (1959) redrawn by Zeuthen (1964).

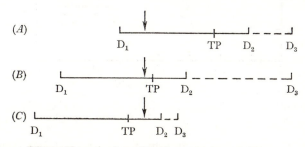

Fig. 10.3. Effect of heat shock on *Tetrahymena* in an asynchronous culture.

(A) Cell near beginning of cycle. (B) Cell towards end of cycle but before transition point. (C) Cell very near end of cycle and after transition point. (D_1) Previous division. (D_2) Forthcoming division if there was no heat shock. (D_3) Division after heat shock. $D_3 - D_2$ = total delay = excess delay + time of heat shock.

first 30 minutes of the cycle will all have the same ED so their division times will not be brought closer together by a single shock. This may not, however, be important in practice because there is evidence that the ED pattern changes during the heat shocks. Fig. 10.4 shows the ED response to a single heat shock given to a synchronised population *after* a preliminary series of heat shocks. The amount of ED is about 30 minutes less at all stages up to the transition point than it is with the cells from an asynchronous culture in Fig. 10.2(*c*). The ED also starts to rise immediately after EH, though in a shorter cycle. It seems then that the cells adapt to heat shocks by showing progressively less ED and eventually escape the effect and divide when there are more than ten shocks, as mentioned

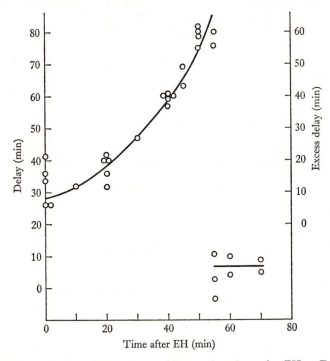

Fig. 10.4. Effect of heat shock (20 min at 34 °C) at various times after EH on *Tetrahymena* previously synchronised by repetitive heat shocks. From Frankel (1962).

earlier. This does not happen with cold shocks, where the ED pattern with asynchronous cells (Fig. 10.2(*a*)) is very similar to that with a synchronised population after EH (Frankel, 1962).

This explanation of heat-shock synchrony in terms of the physiological response of single cells has led to a molecular model, but before going on to this it is worth pausing a moment to consider the three main features of this response – excess delay, the transition point, and the continued growth during the synchronisation regime. ED or set-back is an essential and Zeuthen (1964) has said 'without set-backs there can be no temperature-induced synchrony in *Tetrahymena*'. This can be extended even further to say that in all systems of induction synchrony where there is substantial synchronisation by one pulse treatment, the cellular response *must* follow the ED pattern with increasing delay (or, theoretically, decreasing advance) through the cycle. In contrast, the transition point and the continued growth are not essential for synchronisation, and they may even be disadvantageous since the continued asynchronous synthesis of DNA seems to be responsible for the lack of perfect synchrony (p. 215).

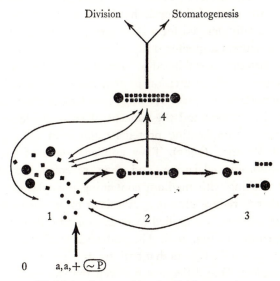

Fig. 10.5. Model for the assembly of a division protein structure.
From Zeuthen & Williams (1969).

Zeuthen and his colleagues have put forward several related molecular models of what may be happening in heat-shock synchrony and the latest and most sophisticated is shown in Fig. 10.5. It is suggested that cell division requires the assembly of some structure (Fig. 10.5(4)) from a number of components. At least two of these components are proteins (Fig. 10.5, small circles and squares) and one of them (small circles) has to be synthesised continuously if the assembly is to go forward. The protein components, which can be called 'division proteins', are linked to form a strand which connects other larger subunits to make an intermediate structure (Fig. 10.5(2)). This structure is highly labile, can exchange components by turnover, and will disintegrate if the supply of its components is interrupted. When this structure is complete, it develops into the final structure, shown with two strands, which is stable and not subject to breakdown, though it may show turnover. This structure then performs some essential function in cell division.

The effect of temperature shocks would be to interrupt the assembly of the division proteins into the intermediate structure either by slowing the assembly process or by accelerating the breakdown. The important point is that once this happens, even for a short time, the whole of the intermediate structure breaks down and it has to be built up again from the beginning. It is this aspect of its behaviour which is responsible for variable ED. A cell at the start of the cycle has little of this division protein assembled so it is only delayed for a short time, but an older cell has more assembled and suffers a greater delay. The transition point is where the intermediate structure is stabilised to form the final structure.

This presentation of the division protein model has anticipated a good deal of its supporting evidence. We must now look at the reasons why the model brings in proteins and the concept of an unstable structure. The initial evidence for implicating proteins came from the experiments of Rasmussen & Zeuthen (1962). They applied pulses of the amino-acid analogue, *p*-fluorophenylalanine, and then reversed the effect of the analogue by washing with medium containing phenylalanine. The effect should be either to reduce protein synthesis or to produce altered protein. The results of these pulses applied to synchronised cells at different times after EH are shown in Fig. 10.6. The pattern is similar to that with heat shocks in Fig. 10.4 and suggests that both treatments are affecting the same control mechanism. One difference between the two patterns is that the transition point with the analogue is about 10 minutes earlier than it was with heat. An explanation of this could be a delay in the penetration of the analogue. If it took 10 minutes to reach a critical concentration within the cells, the apparent transition point would be 10 minutes before the real transition point. Another difference between the patterns is that the ED produced immediately after EH is somewhat greater with the analogue than it is with heat. It may be that heat does not completely discharge the division protein so it can be rebuilt in a shorter time. This would explain why the first cycle after EH is shorter than normal, though not why the second cycle is also shorter. An important point that appears in Fig. 10.6 is that a pulse applied between the first transition point and the first division does not affect the first division, but does produce some ED to the second division. As we shall see, this phenomenon of 'carry-over' occurs in other systems. In terms of the model, it implies that as soon as the division protein structure for the first division has been stabilised, the intermediate structure for the second division starts to be assembled.

Cycloheximide, an efficient inhibitor of protein synthesis, has a bio-chemical effect which is better defined than amino-acid analogues and Frankel (1969) has recently shown that when it is applied in what are effectively pulses, it has a very similar effect on division, both with variable

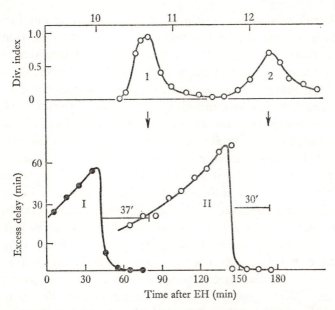

Fig. 10.6. Effect of pulse (20 min) of fluorophenylalanine at various times after EH on *Tetrahymena* previously synchronised by repetitive heat shocks. Upper curve shows division index in control culture. Lower curve shows the excess delay produced by pulses. From Rasmussen & Zeuthen (1962).

ED, and with a transition point at the same time as heat shocks. Puromycin, chloramphenicol and several amino-acid analogues show transition points but they have not been tested in the pulse experiments which are necessary to establish ED (Rasmussen & Zeuthen, 1962; Frankel, 1967c).

Inhibitors of protein synthesis are not the only agents that mimic temperature shocks. Variable ED and transition points occur with meta-bolic inhibitors (fluoride, fluoroacetate and azide – Hamburger, 1962), with anaerobiosis (Rasmussen, 1963), with the sulphydryl reagent mercapto-ethanol (Mazia & Zeuthen, 1966), and with high hydrostatic pressure (Zimmerman, 1969). These can all be fitted into the rather wide terms of the model since interference with metabolism might reduce the supply of division protein, mercaptoethanol would affect disulphide bonds in the protein or its intermediate structure, and high pressure dissolves protein gels. If these agents mimic temperature in variable ED, they should also be able to mimic temperature in causing synchrony. This has been shown as a partial effect with fluorophenylalanine (Zeuthen, 1964), with anaerobiosis (Rasmussen, 1963) and with multiple hypoxic shocks (Rooney & Eiler, 1967).

The production of division protein should be stopped by interference at levels which precede the final translation on the ribosomes. Genetic damage, inhibition of messenger RNA transcription or accelerated messenger breakdown could all reduce the synthesis of division protein, although the extent of this reduction and the time lag before it was apparent would depend on a number of factors, including in particular the stability of the messages. These possibilities have stimulated a good deal of work in recent years on RNA synthesis in synchronised *Tetrahymena*, which can only be touched on briefly here (for a fuller discussion see Zeuthen & Rasmussen, 1971). It is clear that RNA must be synthesised after EH if the first synchronous division is to occur with cells in nutrient medium. Actinomycin D, at rather high concentrations, will block this division up to a transition point (48 minutes after EH) which is near that for heat shocks and for cycloheximide (Nachtwey & Dickinson, 1967; see also Cleffman, 1965 for the effects on cells which have not been heat-shocked). Enucleation by microsurgery has a similar effect, though the transition point is not so sharply defined (Nachtwey, 1965). This correspondence between the transition points for RNA and protein synthesis suggests that the message for a division protein is relatively unstable. If it were not, the RNA transition point would be earlier than the protein one – as it is in mammalian cells (p. 222). But this argument fails if a sequence of different messages has to be produced up to the protein transition point in order that a succession of proteins be synthesised and either used separately in division or assembled into a complex structure.

Cells transferred to starvation medium after EH show a somewhat different response to inhibitors which can affect RNA synthesis. Their division is not inhibited by fluorinated pyrimidines whereas cells in nutrient medium are blocked (Holz *et al.*, 1963; Frankel, 1965). From this and other results, Zeuthen & Rasmussen (1971) argue that it is still uncertain whether or not RNA transcription is needed for division in the period after EH. I think that the balance of the evidence is in favour of transcription being required. The positive effects of inhibitors and enucleation in nutrient medium are more persuasive than the negative effects in starvation medium which may be due to other factors, for example, lack of penetration.

Byfield & Scherbaum (1967; see also Byfield & Lee, 1970) found that a 34 °C heat shock in the presence of actinomycin D caused the breakdown of RNA labelled after EH in starved cells. They suggest that heat shocks synchronise cells by promoting the breakdown of what are otherwise stable RNA messages which are accumulated up to the transition point and then translated into division proteins. This model fits some of the synchrony experiments but it fails to explain why pulses of protein inhibitors cause variable ED – a fact which is a major buttress for the division protein model. Another criticism is that the RNA which is labelled may well be a species other than messenger. The RNA metabolism of synchronised cells is well worth investigating but this should now be done with modern RNA technology with which the various types of RNA can be identified.

One of the most unusual features of the division protein model is that it involves an intermediate structure which breaks down if its synthesis is interrupted. If this synthesis can be compared to building a house, it is as though the whole house falls down when there is a temporary halt in construction. For our peace of mind, therefore, it is providential that there is a visible structure in *Tetrahymena* which behaves in just this way. This is the oral apparatus (OA), a complex arrangement of membranes, cilia and basal bodies which develops at the equator of the cell and forms at division the new mouth of the posterior daughter cell. The stages and timing of the development of the OA in a synchronous culture are shown in Fig. 10.7. If the culture is treated with heat shocks, cold shocks or *p*-fluorophenyl-alanine, there is a marked transition point or 'stabilisation point' at about the beginning of stage 5 (Frankel, 1962). Cells treated after this point will go on to divide and there is no effect on the OA. But if they are treated before this point, most of the cells do not divide and there is a complete regression of the OA. There are differences between the treatments in the speed of regression and in the exact timing of the transition point, but the important points here are that the structure of the OA does break down and disappear, and that this regression continues for some time after the treatment has been concluded. This work has been amplified in later papers by Frankel (1967 *a*, *b*, *c*; 1969) using other agents such as metabolic inhibitors and cycloheximide, and in all cases there is the same basic pattern of regression before a transition point. Regression, incidentally, is not confined to *Tetrahymena* and it has been demonstrated in other Ciliates, especially *Stentor* (refs. in Frankel, 1962).

A general point of interest with the OA is that it is a new organelle whose normal development lasts for about half a cycle in synchronous cultures (and rather less in the cycle of asynchronous cultures). It is an example of morphological differentiation during interphase which parallels the chemical differentiation shown by the enzyme patterns in Chapter 8. This differentiation of the new OA can also be completely reversed by simply stopping protein synthesis for a short period – an effect which is local rather than general since the existing mouth is not affected. But the particular relevance

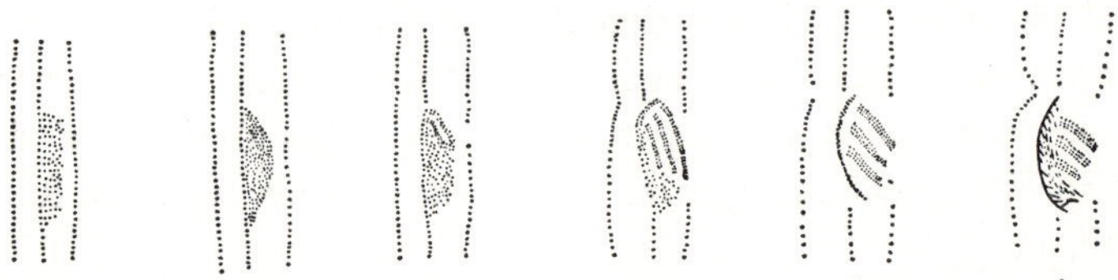

Stage:	1	2	3	4	5	6
Time after EH (min)						
PPL to 40		45–50	50–55	55–60	65–70	80
IM to 30		40	45	50	55–60	70
Approximate duration (min)						
		5–10	< 5	5–10	5–10	15

Fig. 10.7. Morphogenetic stages in the assembly of the oral apparatus in *Tetrahymena*. The timings are given after EH in cells previously synchronised by repetitive heat shocks, and then grown in organic medium (PPL) or inorganic medium (IM). From Frankel (1962).

to the control of cell division is that the OA behaves in a way very similar to the division protein in the Zeuthen model. This raises the question of whether the OA is in fact the protein structure in the model. Two lines of evidence argue against this. One is that the phenomena of variable ED and synchronisation which support the model also occur in bacteria and yeast which do not have an OA (p. 229). The second is that the presence of an OA is not a necessary prerequisite of division. In certain cases, both *Tetrahymena* (Nanney, 1967) and *Glaucoma*, a close relative (Frankel, 1961), can divide in the absence of an OA and produce a posterior daughter cell without a mouth. We can conclude then that the OA is an analogy to the division protein structure rather than the structure itself. But the resemblance may turn out to be closer than this if the division protein is found to be a component of the OA. The inner regions of the OA contain microtubules (Zeuthen & Williams, 1969) and Zeuthen & Rasmussen (1971) argue that there is a significant resemblance in heat-sensitivity between the division protein and the microtubules of the mitotic apparatus.

Another approach to the problem of division protein has been to try to isolate it by biochemical techniques. The story starts with the extraction from sea urchin eggs by 0.6 M KCl of protein which shows a fluctuation in

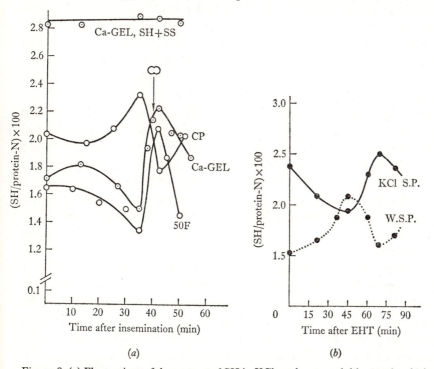

Fig. 10.8. (a) Fluctuations of the amount of SH in KCl- and water-soluble proteins during the first cycle of the fertilised eggs of the sea urchin *Anthocnidaris crassipina*. CP, KCl-soluble protein. Ca-GEL, water-soluble protein precipitated by CaCl₂. 50F, water-soluble protein precipitated by half-saturation with NH₄SO₄. From Dan (1966). (b) Fluctuations of the amount of SH in KCl- and water-soluble protein after EH in *Tetrahymena* synchronised by repetitive heat shocks. From Ikeda & Watanabe (1965).

SH groups during the cycle (Sakai & Dan, 1969; Sakai, 1960; reviewed by Dan, 1966). The SH-content rises during mitosis and falls again during cleavage, whereas reciprocal changes take place in the SH-content of water-soluble protein (Fig. 10.8(a)). The KCl-soluble protein can be drawn into threads and shows a contractility (when treated with polyvalent cations or mild oxidising agents) which is proportional to the SH-content. Although there is no direct evidence linking these proteins to division, it is tempting to suggest a connection since the KCl-soluble protein shows its greatest potential for contraction just before the start of the cleavage process (Dan, 1966). Returning now to *Tetrahymena*, Fig. 10.8(*b*) shows that there is a strikingly similar behaviour of the two protein fractions when extracted from synchronous cultures (Watanabe & Ikeda, 1965 *a*, *b*; Ikeda & Watanabe, 1965). There is the same peak in the SH-content of KCl-soluble protein at division, and a peak in the water-soluble protein at about the time of the transition point. These proteins have been fractionated chromatographically and one particular fraction (no. 7) of the water-soluble protein has been identified as the most likely candidate for the division protein. It has the largest number of the properties which might be expected from a division protein, for example high quantity at the transition point and sensitivity to heat and other division inhibitors. This approach is promising and the work should be extended, especially since Lowe-Jinde and Zimmerman (1971) have not been able to confirm all the results. There are potential snags since the division protein might turn out to be present in very small quantities and it is also uncertain what exactly happens to it after the transition point or on breakdown before the transition point. Nevertheless, predictions can be made from the physiological responses of synchronous cells which should provide good clues for biochemical identification.

Before leaving *Tetrahymena*, we should consider briefly the question of DNA synthesis in synchronised cells, a problem which has been of particular interest in recent years to the workers in Copenhagen (reviewed by Zeuthen, 1971). Heat shocks inhibit division but they do not inhibit DNA synthesis (Hjelm & Zeuthen, 1967). Replication continues asynchronously and the cells finish at EH not only larger than normal cells but also with a higher DNA content (approximately double). In the earlier part of the period between EH and the first synchronous division, DNA synthesis continues if the cells are in nutrient medium, but replication is still asynchronous and not all the cells are labelled with pulses of tritiated thymidine. Synchronous replication only starts after the first division, and in this cycle, which is shorter than the normal cycle in asynchronous growth, there is little or no G1.

This separation of the 'DNA-division cycle' (p. 246) in which division is inhibited but DNA synthesis continues, is not unusual and does occur in other systems (p. 249). But what is striking about *Tetrahymena* is that cellular events like the development of the OA occur synchronously in the cycle between EH and the first synchronous division, while DNA synthesis is still asynchronous. This asynchrony of DNA may be the limiting factor in achieving perfect synchrony after heat shocks. Zeuthen (1970) has pointed out that, in practice, only 80 per cent of the cells divide synchronously at the first division and the remainder are either delayed or do not divide. He suggests that these latter cells are those which do not finish their S period until late in the first cycle. As a result, there is not sufficient time for the completion of those processes which must occur after S if division is to succeed. In other words, there is minimum time for G2.

A more unusual separation of the DNA-division cycle takes place in *Tetrahymena* when DNA synthesis is blocked yet division continues. One example of this occurs when heat-shocked cells are suspended in a starvation medium after EH. They will then go through two synchronous divisions with no detectable synthesis of DNA, though they have of course accumulated an excess of DNA during the heat shocks (Lowy & Leick, 1969).[1] Another example is that DNA synthesis in exponential cultures can be blocked for a generation time by high concentrations of thymidine or uridine but cell division continues unaffected (Villadsen & Zeuthen, 1970). Nor is there any resulting synchrony of division as there would be with other systems such as mammalian cells. Another combination of inhibitors (methotrexate plus uridine) does, however, inhibit both division and DNA synthesis, and synchronous division follows reversal of this inhibition. A particularly interesting case of dissociation of DNA synthesis from division comes from microdissection experiments on another Ciliate *Urostyla* which has both macro- and micronuclei (Jerka-Dziadosz & Frankel, 1970). After transection of the cell, a macronuclear division can take place without prior DNA synthesis. This will be followed by two successive S periods before the next macronuclear division. Micronuclei, on the other hand, always have an S period before division. Cell division, micronuclear division, micronuclear S and the development of the OA appear to be closely linked. But this is not so with macronuclear division and macronuclear S. It is not altogether surprising that Ciliates can tolerate flexible relations in the macronucleus since it contains many copies of the genome and the exact gene-dosage may not be critical. One would also expect that this flexibility would not be permitted in the micronucleus, which is analogous in structure to the nucleus of other cells, and this appears to be so. We should bear in mind, therefore, that comparisons between Ciliates and

---

[1] Another recent set of experiments with heat shocks emphasises the lack of an obligatory temporal alternation of DNA synthesis and cell division in *Tetrahymena* (Jeffrey *et al.*, 1970).

most other eukaryotic cells may not be justified when they are concerned with the relations of the macronuclear S period to the cell cycle.

Conclusions

Heat shocks induce synchrony in *Tetrahymena* because they cause a division delay which varies through the cycle. There is a little or no delay in cells early in the cycle. The delay increases for cells later in the cycle up to a transition point before division after which there is no delay in the forthcoming division. A similar combination of variable delay and a transition point is produced by pulse treatments with other agents, in particular inhibitors of protein synthesis and amino-acid analogues. These effects can best be explained in terms of one or more division proteins which are synthesised through the cycle and which are necessary for division. Division protein is unstable and, until the transition point, the effect of pulse treatments with heat and other agents is to cause its complete breakdown. After the transition point, it is stabilised, perhaps in the form of a structure, and it is not affected by pulses. The oral apparatus, which develops in interphase as the precursor of the new mouth, behaves in the same way as the postulated division protein and shows both regression and a transition point. It is an important analogy for division protein and may contain the same components but it is probably not the protein itself. Some progress has been made in the biochemical isolation of division protein.

Although a series of heat shocks inhibit division, they do not inhibit growth or DNA synthesis in the macronucleus. After the heat shocks, synchronous division can take place both with asynchronous DNA synthesis and in the absence of DNA synthesis. The coupling between macronuclear S and division is relatively loose in *Tetrahymena* and other Ciliates.

Models for division control

Our next step is to proceed beyond *Tetrahymena* and see how far the concepts that have been developed in this system can be applied to a wider range of cells. But before doing this it is worth considering a number of the models that can be suggested for division control and can account for the experimental results which will be outlined in the succeeding two sections (Fig. 10.9).

The first model is the division protein (DP) scheme for *Tetrahymena* that has been outlined in the preceding section. A single DP is built up through the whole cycle between one transition point (TP) and the next, and the effect of a pulse (temperature or inhibitor) is to cause the break-down of the existing stock of DP. In an asynchronous culture, a pulse would not at first affect the increase in numbers or division index. These would continue until the cells after the TP had completed division. The division index would then drop to zero. It is drawn here and elsewhere in the diagram as a sharp drop, but it is not likely in practice to be so sharp since there would be differences between individual cells in the time taken to traverse the interval between the TP and division (y in the diagram). After a gap equal to $y +$ the cycle time, there would be a synchronous division followed by a second less synchronous division a cycle later.

The second model is similar to the first except that the synthesis of DP does not start until some way through the cycle. Excess delay, therefore, is only partial since it does not occur in cells in the early parts of the cycle. The effect of a pulse on an asynchronous culture would be to give a shorter period of inhibition of division than in model 1 followed by a less synchronous division produced only by that fraction of the original population which had been delayed. The difference between the two models would be easier to recognise in pulse treatments of synchronous cultures. Model 2 would give no excess delay either in cells early in the cycle or in cells after the TP which would not show a carry-over of delay into the next cycle.

The third model also involves DP, but in this case the DP is stable. The pulse stops the build-up of DP for the length of the pulse but it does not cause the existing stock to break down. There is delay up to the TP, but it is constant and equal to the pulse length. Thus there is no excess delay. In an asynchronous culture, there would be a period of division inhibition equal to the pulse length followed by a return to normal exponential growth and no burst of synchrony.

The next four models invoke the broader and rather vaguer concept of a sequence of events which take place during the cycle, all of which have to be completed before division can occur. These events might be morphological changes or the synthesis of proteins (structural or enzymatic) or the production of RNA messages. Model 4 assumes that the effect of a pulse is to block this sequence at any point in the cycle (before the TP). There is no

Model	Concept	Effect of pulse treatment	Diagram of concept D = Division P = Pulse TP = Transition point DP = Division protein	Effect of pulse on asynchronous culture	Effect of pulse on synchronous culture
1	Unstable division protein Full excess delay	Existing division protein breaks down			 Excess delay
2	Unstable division protein Partial excess delay	As 1			 Excess delay
3	Stable division protein No excess delay	Existing division protein conserved No increase during pulse			 Delay

4	Sequential events	Block at any stage of cycle. No cancellation of events	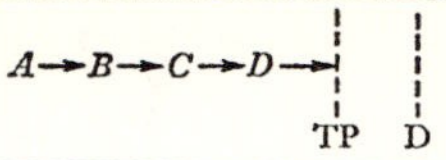 $A \to B \to C \to D$ TP D	As 3	As 3
5	Sequential events	Block at any stage of cycle with set-back	As 4	As 1 or 2	As 1 or 2
6	Sequential events	Block at any stage of cycle. With cancellation of event at that stage	As 4	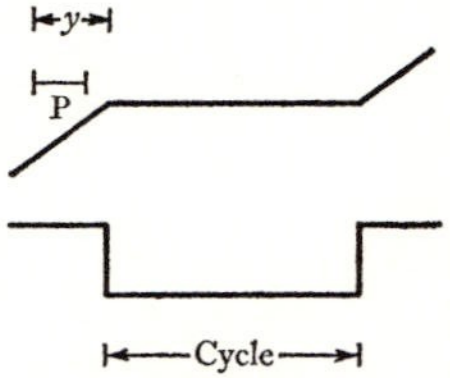 (Cycle, y, P)	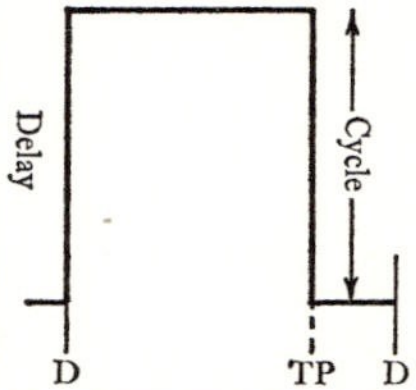 (Cycle, TP, D, Delay)
7	Sequential events	Block at one stage of cycle. Cells progress towards block during pulse	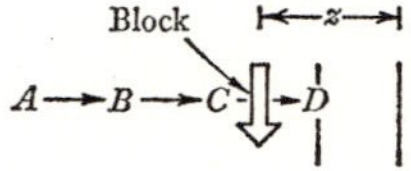 Block $A \to B \to C \to D$ z	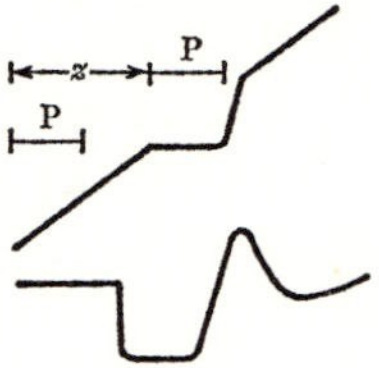 (z, P, P)	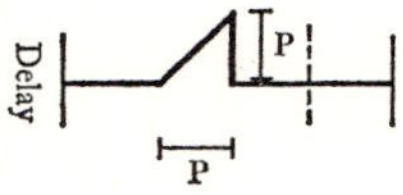 (Delay, P)
8	Constant repair time	Damage at any stage of cycle. Constant time for repair (R) which must be complete before passing TP	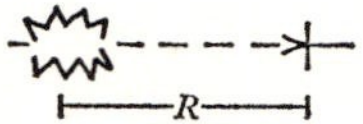 R	As 2 if R < cycle As 1 if R ⩾ cycle	As 2 if R < cycle As 1 if R ⩾ cycle

Fig. 10.9. Models for division control.

cancellation of an event and the cells are frozen in their progress through the sequence. The effect on both synchronous and asynchronous cultures would be the same as in Model 3 (with a stable DP). Model 5 includes set-back and assumes that if the sequence is interrupted by a pulse, a cell goes back and repeats all or some of the sequence. Here again the expected effect on cultures is the same as one of the DP models, in this case either Model 1 or Model 2. This emphasises the point that it is not easy to distinguish by this kind of analysis of cultures between a model which requires one species of DP to accumulate to a critical level and another model which requires the sequential synthesis of a set of different DPs.

Model 6 assumes that the pulse will affect cells at any stage of the cycle before the TP and that the effect is to cancel the event of this stage. The cells continue through the sequence after the pulse has finished but they do not divide because one of the necessary events has been cancelled. The sequence then starts again and it is only when this next round is completed that a full complement of events has been gone through and the cells can divide. Pulses in a synchronous culture would produce a large constant delay equal to the cycle time. In an asynchronous culture there would be inhibition of division lasting for one cycle.

In Model 7, the pulse is assumed to put a block (without cancellation) at one particular point in the sequence. During the pulse, cells at other stages of the cycle progress through the sequence until they meet the block. The maximum delay will be equal to the length of the pulse (or block) and will occur in cells at the point of the block. In an asynchronous culture, there would be no alteration in number increase or division index for a period equal to that between the block point and division (z). There would then be an interval without division equal to the pulse length, followed by a small burst of synchrony produced by cells which had accumulated at the block during the pulse. The effects here are similar to those with Model 2, though delay should not vary with pulse length in Model 2.

The last Model 8 invokes the concept of repair time, in this case a constant at all stages of the cycle before the TP. This concept has been largely developed for radiation-induced delay where a pulse of radiation may cause damage, for instance to the DNA, which has to be repaired before the cell can pass the TP and divide. During the repair time, the cell passes through the cycle up to the TP. If the repair time were less than the cycle time, the effects on synchronous and asynchronous culture would be similar to Model 2. If it were equal to the cycle time, the effects would be similar to Model 1. If it were greater than the cycle time, the effects would also resemble Model 1 but with an extended x. The reason for this similarity is that there is a basic resemblance between the concepts of repairing a structure and of rebuilding DP. Differences, of course, do exist and a conspicuous one is that repair time might be expected to vary with the extent of the damage. Division delay does in fact increase with radiation dose. On the other hand, times for rebuilding DP may also vary. There is some evidence (p. 208) that heat shocks do not completely discharge DP and it can be rebuilt in a shorter time than after an analogue pulse. This effect might be more marked in other cells and fit Model 2 better than Model 1.

Two general points should be made about these models. One is that they are only a simplified and selected collection, and that there are other possibilities and combinations. For instance, the partial delay in Model 2 and the block in Model 7 could occur at more than one place in the cycle, or the repair time in Model 8 could vary through the cycle. The second point is that the most powerful method of analysis at the level that we have been considering is to apply pulse treatments to cultures already synchronised. This can discriminate between models far better than pulses applied to asynchronous cultures, or continuous irreversible treatment which will only reveal the TP and not excess delay.

Transition points and delay in other systems

We can now see how far the two basic phenomena of the *Tetrahymena* system – variable excess delay and the transition point – appear in other cells. The main experimental technique to be considered here is the use of inhibitors of protein and RNA synthesis, and temperature shocks. The effects of radiation in inducing division delay will be left until the next section. Our main concern will be with work of the last decade, but the evidence for transition points as a general phenomenon, particularly during the process of mitosis, goes back before this. Earlier work has been discussed by Mazia (1961, p. 161) who uses the attractive phrase 'a point of no return'. It is better for present purposes to use the more neutral phrase 'transition point' with the operational definition as a point in the cycle before which an agent will delay or prevent the forthcoming division, and after which the agent will not. This definition is the same as that used by Mazia for the point of no return but he also links it to the concept of critical 'decisions' made by the cell in its progress through the cycle. Some transition points certainly involve such critical decisions but others may not. This important point is discussed later in this section.

It is clear that there are TPs both for protein synthesis and for RNA synthesis in cultured mammalian cells. They have been found in human KB cells with puromycin and chloramphenicol (E. W. Taylor, 1963), in kidney cells with puromycin and actinomycin (Kishimoto & Lieberman, 1964), and in HeLa cells with actinomycin, puromycin, and fluorophenylalanine (Mueller & Kajiwara, 1966*b*; Buck *et al.*, 1967). These points mostly lie in G2 with the RNA TP preceding the protein TP. A recent

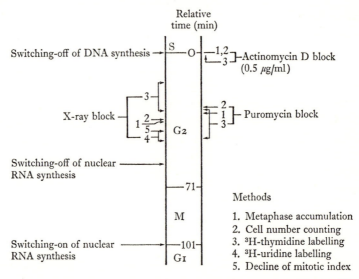

Fig. 10.10. Various molecular events in the G2 stage of mouse L5178Y cells. The blocks are transition points determined by a variety of methods. From Doida & Okada (1969).

and careful analysis by Doida & Okada (1969) has established these points in L5178Y cells by three different methods (Fig. 10.10). They lie rather nearer to mitosis than in other mammalian cells, but this cell line has an unusually short G2 and cycle time (9.5 hours).

An interesting set of experiments which throw light on both TPs and delay were carried out on hamster CHO cells by Tobey and his colleagues (Tobey, Petersen, Anderson & Puck, 1966; Tobey, Anderson & Petersen, 1966; Petersen *et al.*, 1969). Measurements both on asynchronous cultures and on cultures synchronised by thymidine block give an actinomycin TP at 1.9 hours before division and a cycloheximide TP at 1 hour. These are both in G2 which lasts from 2.8 hours before division to 0.8 hours before division in these cells (Puck *et al.*, 1964). There is also a puromycin TP which is 17 minutes later than the cycloheximide one. This apparent discrepancy may be due to the different modes of action of the two inhibitors (Petersen *et al.*, 1969). Although the actinomycin block is irreversible, the cycloheximide block is not and Fig. 10.11 shows the effect on cell numbers of cycloheximide pulses of varying length applied to an

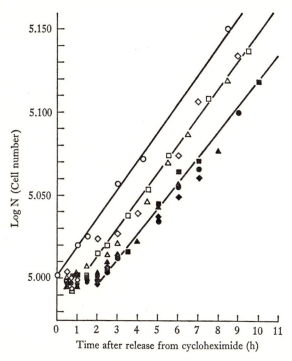

Fig. 10.11. Resumption of division of Chinese hamster ovary cells held in 2 μg/ml cycloheximide for varying periods of time prior to removal of the inhibitor at zero time. These periods of time are: ○, inhibitor-free control; □, 1.5 h; ◇, 2.0 h; △, 2.5 h; ■, 3.0 h; ●, 4.0 h: ▲, 5.0 hr; ◆, 8.0 h. From Tobey, Anderson & Petersen (1966).

asynchronous culture. With pulses between 1.5 and 2.5 hours, there is a delay of an hour after the end of the pulse in which there is no increase in numbers, and then the culture goes into steady exponential growth. With longer pulses, up to 8 hours, the delay increases to 2 hours. These results can be explained in terms of the two TPs, and of RNA messages which are synthesised up to the actinomycin TP but are not finally translated into necessary division protein until the cycloheximide TP an hour later. During the shorter pulses, there is no significant breakdown of the messages, and cells after the cycloheximide TP go through and divide, so emptying of cells a one hour gap in the cycle between this TP and division. At the end of the pulse, there is an hour's delay before division starts again because this gap has to be filled by cells moving into it from earlier stages of the cycle. With the longer pulses, however, the messages break down and have to be resynthesised. As a result, the gap and the delay increases to 2 hours – the time between the actinomycin TP and division. This effect appears with pulses of 3 hours or longer, so 3 hours is about the lifetime of the messages. More precise estimates of this lifetime can be got from double block experiments in which both inhibitors are applied to the same culture either in sequence or simultaneously. A second point in Fig. 10.11 is that there is no sign of a burst of synchronous division at the end of the one or two hour delay periods. This argues against the presence of variable ED, but the results are in full agreement with constant delay and the concept either of a stable DP as in Model 3 of Fig. 10.9 or of sequential events as in Model 4. Note that the delay time produced in an asynchronous culture in Model 3 is equal to the pulse length. This will happen if the pulse length is less than the gap between the TP and division, but if the pulse is greater than the gap then the delay time will equal the gap, as in Fig. 10.11.

Although variable ED and a burst of synchrony does not occur in these CHO cells, they do seem to happen in some other mammalian cells. Pulses of actinomycin at low concentration (0.04 μg/ml) are reversible after 3 hours in cultures of human amnion cells (Donnelly & Sisken, 1967). Their effect is to delay division in cells up to a TP in late S or G2. After the TP, the delay does not affect the coming division but is carried over to elongate the next cycle. The presence of variable ED (as in Model 1 or 5) is suggested both from some of the data on individual cells and from the presence of a burst of synchronous division at about a cycle time later (20 hours). Another example comes from an analysis of the mitotic index in rat intestinal crypt cells after a single injection *in vivo* of cycloheximide (Verbin & Farber, 1967). This is probably equivalent to a pulse though its total length is uncertain. There is an initial delay of about 1.5 hours before the mitotic index falls to zero which may be due to a protein TP in late G2. Later, at 7 hours after the injection, there is a marked peak in the mitotic index which suggests a synchronous burst of division. The authors interpret these results in terms of Model 7 with a G2 block, but Model 2 is equally applicable since the pulse time is unknown.

One of the features of the models of division proteins and of initiator proteins for DNA synthesis in bacteria (p. 108) is that once these proteins have been used for division, or initiation, they are not available for use again. There is, however, evidence in mammalian cells that this may not be true for all the proteins that are used in division. If human amnion cells are pulsed with fluorophenylalanine and other amino-acid analogues, those cells which were in G2 during the pulse show an abnormally prolonged metaphase not only in the division immediately following the pulse but also for the two succeeding divisions (Sisken & Wilkes, 1967; Sisken & Iwasaki, 1969). The suggestion here is that some protein is made in G2 which is defective because of incorporated analogue and that it is conserved and re-used in successive division. The defect manifests itself by metaphase delay.

Turning to lower eukaryotes, we might expect similarities between *Tetrahymena* and other Ciliates. These have been found in the effect of inhibitor pulses on *Paramecium aurelia*, grown in 'micro-cultures' (Rasmussen, 1967). Fluorophenylalanine, puromycin, chloramphenicol and fluorodeoxuridine all show TPs (about three-quarters through the cycle) and variable excess delay as in Model 1 of Fig. 10.9. Cells treated after the TPs are delayed to some extent in the next cycle. Actinomycin D (30 minute pulses) gives a similar TP but the delay before this point is very roughly constant through the cycle and varies with the concentration of the actinomycin – about 70 per cent of the cycle with 100 μg/ml and 50 per cent with 50 μg/ml. A TP and constant delay also appears in the experiments of Hanson and Kaneda (1968) with similar treatments by actinomycin on the same species, but the delay is longer and equal to a cycle time (Fig. 10.12). Their interpretation assumes that sequential gene action through the cycle up to the TP is needed for division and that the effect of an actinomycin pulse is to prevent the formation of a gene product at that time in the cycle. This is Model 6 in Fig. 10.9. The effect of one or two pulses before the transition point is that the gene products for the coming division are not available so this division does not occur (Ia and IIa of Fig. 10.12). The cell then progresses through the next cycle, accumulates all the required gene products and finally divides at the same time as the second division of the control. A pulse after the first transition point cancels the second division rather than the first (Ib of Fig. 10.12). The effects of two pulses in successive cycles are more complicated and more variable. When they are given at the same points in the two cycles, another complete cycle is usually needed before division (IIb1 and IIb3 of Fig. 10.12). When they are given at different points, division can happen earlier since all the requisite gene products have then been accumulated (IIb2 of Fig. 10.12).

Apart from this last case, all the results can be described in the formal terms of a constant delay equal to one cycle time and determined by the last pulse if there is more than one. It is not easy to assess the significance of these experiments. The model is certainly an interesting one and is markedly different from those involving a division protein, stable or unstable. In a simple form, however, it does demand that the delay should

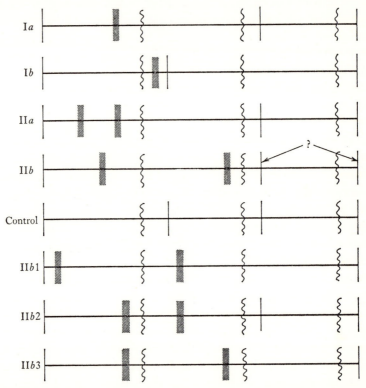

Fig. 10.12. Diagram of the effects of actinomycin D pulses on *Paramecium aurelia*. Single pulses are designated by I, and double pulses by II. Time is indicated by the horizontal lines from left to right. Vertical straight line represents division. Vertical wavy line represents the transition point. Hatched areas are exposures to actinomycin. From Hanson & Kaneda (1968).

equal a cycle and this is not so in the comparable experiments of Rasmussen (1967). It is clearly worth exploring the effect of actinomycin pulses in other systems where the effect of this inhibitor is reversible. More also needs to be known about the patterns of macromolecular synthesis during, and especially after such pulses. Is it possible, for instance, to identify the synthesis either of step enzymes or of DNA during the long delay? This might tell us how far a cell progresses between a pulse and the delayed division.

Two other cases of variable set-back in Protozoa should be mentioned, though they are not with respect to division. Burchill (1968) has examined the effect of actinomycin D, puromycin and cycloheximide on the regeneration of the oral apparatus in the Ciliate *Stentor*. There is a stage in development of this OA (similar to the stage in *Tetrahymena*) after which the inhibitors do not block the progress of morphogenesis. Cells in early regeneration, some time before this stage, show a set-back in morphogenesis with resorption of the OA. There is, however, an intermediate group of cells, absent in the *Tetrahymena* experiments, where development is blocked but the OA is not resorbed. The other case, also a morphogenetic one, occurs in the Amoeboflagellate *Naegleria* in which the development of flagella can be induced by washing (Wade & Satir, 1968). Pulses of mercaptoethanol during flagellation produce set-back to the start of the induction process.

Continuous treatment with cycloheximide in *Physarum* reveals two TPs late in the cycle (Cummins, Blomquist & Rusch, 1966). If the inhibitor is added prior to 13 minutes before metaphase, nuclear division does not occur. If it is added after 6 minutes before metaphase, division continues unaffected. But if the treatment starts between 13 and 6 minutes before metaphase, then division is delayed by about an hour. The suggestion made is that the first TP (13 minutes) represents the completion of essential structural proteins for mitosis and the second one (6 minutes) is for proteins, perhaps enzymes, concerned with energy requirements. Tests for variable delay with cycloheximide pulses have not been made. The position with actinomycin is not entirely clear (see summary by Rusch, 1970), but there does appear to be a TP at 35 minutes before telophase, earlier than the cycloheximide one, for actinomycin C which appears to be a better RNA inhibitor in this organism than actinomycin D (Sachsenmaier *et al.*, 1967).

Physarum is one of the few eukaryotes apart from *Tetrahymena* in which there have been studies of the effect of heat shocks through the cycle (Brewer & Rusch, 1968). Fig. 10.13 shows that there is both variable delay and a TP for heat shocks applied during interphase. There is, however, little or no delay in the first two hours and the maximum delay is only about a quarter of the cycle time. The pattern is much more like the partial ED of Model 2 in Fig. 10.9 than the complete ED of Model 1. Another point that should be noticed is that the heat TP is much earlier than the cycloheximide TPs. Heat has a different effect during mitosis. When applied in late prophase (the time of the cycloheximide TPs), nuclear division is delayed for a whole cycle but DNA synthesis occurs at the normal time so producing nuclei with double the normal DNA content.

Apart from Ciliates, the other clear example in eukaryotes of variable ED with pulses of protein synthesis inhibitors is in the fission yeast, *Schizosaccharomyces pombe* (Herring, 1971). Pulses of cycloheximide give a TP in late G2 just before nuclear division and a variable ED which is partial, like Model 2 in Fig. 10.9 (Fig. 10.14). The effect of a single pulse on an asynchronous culture is to induce partial synchrony. Heat shocks also induce a variable ED in interphase but the effect is complicated by the fact that they distort and elongate the process of cell plate formation and cleavage. There are possible similarities here with *Physarum*.

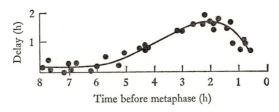

Fig. 10.13. Mitotic delay caused by a heat shock (30 min at 37 °C) applied at various times before mitosis in *Physarum*. From Brewer & Rusch (1968).

The alga *Ulva mutabilis* shows variable ED and a TP, when a dark period of 6 hours is given at various points of the cell cycle in the light (Lövlie, 1964). This result may explain why alternating periods of dark and light are effective in synchronising algae (p. 40).

An interesting and remarkably close analogy to heat shocks in *Tetrahymena* has been recently discovered in *Escherichia coli* (Smith & Pardee, 1970). A single heat shock of 16 minutes at 45 °C will synchronise a normal exponential culture. If the effect of such a shock is analysed in cultures synchronised by selection, the results show a variable and almost complete ED together with a TP just before cleavage (Fig. 10.15(*a*)). This, of course, is very similar to what occurs in *Tetrahymena* or in Model 1 of Fig. 10.9. Protein also seems to be involved since pulses of fluorophenylalanine will sensitise the cells to a shorter heat shock of 6 minutes applied after the pulse. The effect of these pulses may be to produce a protein structure which is more labile to heat because it contains the analogue in place of phenylalanine. Pulses by themselves, however, are not reported as being effective. Heat shocks (with or without fluorophenylalanine) also cause delay in total protein synthesis and in the initiation and synthesis of DNA, but in these cases the delay is nearly constant through the cycle (Fig. 10.15(*b*)). The fact that a shock at any point will delay DNA initiation is in accord with

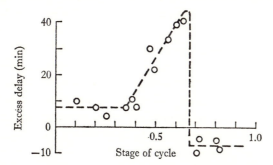

Fig. 10.14. Division delay caused by a pulse of cycloheximide (20 min at 100 μg/ml) applied at various times during the cell cycle of *Schizosaccharomyces pombe* synchronised by gradient selection. Cycle time = 150 min. From Herring (1971).

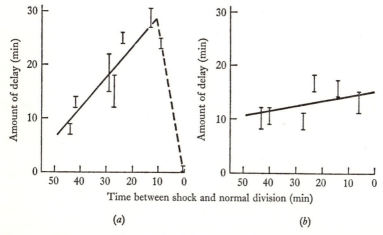

Fig. 10.15. Delays caused by heat shock (16 min at 45 °C) alone, or by heat shock (6 min at 45 °C) preceded by pulse of fluorophenylalanine (10 min at 9.1 × 10⁻⁴M) on cultures of *Escherichia coli* synchronised by gradient selection. (*a*) Delay of division caused by heat shock applied at varying points in the cycle. (*b*) Delay of initiation of DNA replication caused by heat shock plus fluorophenylalanine applied at varying points in the cycle. From Smith & Pardee (1970).

the concept of an initiator protein which is built up throughout the cycle (p. 107) but this protein would differ from the division protein in being stable during the heat shocks. Its behaviour with respect to initiation, though not to division, would be like Model 3 in Fig. 10.9. The result of having constant delay for the initiator protein and variable delay for the division protein, is that cultures which are division-synchronised by a heat shock are not synchronised for DNA replication, as also happens in *Tetrahymena*. The rule of the Cooper–Helmstetter model for normal cultures is that the interval between the end of replication and division is 20 minutes (p. 102). Heat-treated cultures break this rule since the interval varies from 12 to 30 minutes, but they do not break the more fundamental rule that replication must be completed before division. This partial separation between the DNA cycle and the division cycle may provide a useful method of identifying which other events are associated with each cycle (p. 249). For example, enzymes and other proteins involved in septum formation should be synthesised synchronously in heat-shocked cultures whereas those involved in replication should not.

We should finally consider the effects of inhibitor pulses on the early cleavage stages of fertilised eggs since it involves a more critical discussion of TPs than has been possible earlier. Swann (1953) followed the effects in sea urchin eggs of pulse treatments with carbon monoxide, a respiratory inhibitor, and found a TP in early prophase 35 minutes after fertilisation and 20 minutes before the first cleavage. Pulses applied before this point produce a constant delay to the first cleavage equal to the pulse length. After this point, they delay the second cleavage but not the first. These experiments led to the concept of an 'energy reservoir' for division which is filled continuously except when this process is interrupted by an inhibitor (Swann, 1953; 1954; 1957). At the TP, the contents of the reservoir are siphoned off (completely or partially) and used for the coming division but the flow into the reservoir still continues in preparation for the division after. The behaviour of the reservoir would be very like that of the stable division protein in Model 3 of Fig. 10.9. An essential feature of the reservoir hypothesis is that respiratory inhibition after the transition point should not block the coming division. This feature, however, was not confirmed in the later experiments of Epel (1963; for a good discussion, see Mazia, 1963). He also used carbon monoxide on sea urchin eggs but in a way which made it a more efficient inhibitor of ATP synthesis than in Swann's experiments. Under his conditions, an egg is blocked at any stage of mitosis provided the inhibition is applied a little time earlier. The egg behaves as if it had a small reserve, perhaps of ATP, which allows it to progress a short way through mitosis before it stops. As a specific example, an egg inhibited just late enough to allow cleavage will be blocked in telophase with condensed chromosomes (Mazia, 1963). Epel and Mazia interpreted Swann's results as due to partial inhibition which gave a larger reserve for progress through mitosis and thus an earlier TP.

These are interesting experiments which would repay further investigation since some of Swann's results do not exactly fit the later interpretation (for example, the lack of progress through interphase during inhibition and the sharp change in sensitivity to dinitrophenol at the TP – Swann, 1955) but they also have particular importance in underlining the fact that TPs may occur for two quite different reasons. One reason is that a TP marks a particular event in the cycle, a 'decision' as Mazia calls it. This could be the stabilisation of the oral apparatus or of a division protein structure, or it could be the production of the last protein or the last RNA message that is needed for division. If a cell is inhibited before the TP, it should not proceed further towards division and it might reverse. The second reason for a TP is illustrated by the sea urchin egg experiments. It occurs experimentally because there is a delay between the application of an inhibitor and the time when this inhibitor stops the progress of the cell towards division. If, for instance, an inhibitor takes 30 minutes to be effective, a cell treated 31 minutes before division will be blocked in late anaphase and will not divide. On the other hand, a cell treated 29 minutes before division will divide even though it will block immediately after this. The effect will be to give a TP at 30 minutes before division for this particular inhibitor, but this TP will not mark a special event in the cycle. We will be given useful information about inhibitor action but not about critical decisions in the cell.

The inhibitor delay could occur because of slow penetration and a lag before an effective concentration was reached within the cell. Alternatively, the inhibitor could be effective at once in stopping the synthesis of a key substance but the cell might have a reserve of this substance sufficient to allow a limited progress towards division. This substance might be an energy store such as ATP but it might equally well be a division protein. When the inhibitor is removed at the end of a pulse, there could well be a delay before progress starts again because the reserve has to be refilled.

Most of the TPs that have been mentioned in this section have been interpreted in the first way – as what can be called an 'event transition point'. Many of them, however, could equally well be 'delay transition points'. There may also be mixtures of the two, as in the case of fluorophenylalanine in *Tetrahymena* where a delay TP precedes what is probably an event TP by a period of 10 minutes (p. 208). How can these two types of TP be distinguished? Scoring divisions after a pulse in either a synchronous or asynchronous culture is not enough since the effects of both TPs are very similar. The best criterion for separating them is whether or not the cell progresses towards division during the inhibition. Progress is to be expected after a delay TP and is not to be expected after an event TP. The problem is how to measure this progress. What are needed are markers or signposts in the interval between the TP and division. There are good markers for the later portion of this interval in the form of the various stages of mitosis. These ought to be used in further studies of TPs in the way that they were by Epel on sea urchin eggs. For example, it would be illuminating to know whether or not any of CHO cells used by Tobey and his colleagues (p. 223) progressed into mitosis during the cycloheximide blocks. It will be more difficult to find markers before mitosis. Step enzymes might provide such markers but they would have to be closely associated with mitosis and division. Most step enzymes are probably associated with the 'growth cycle' (p. 246) and this cycle may continue even when progress towards division is blocked.

Before leaving eggs, we should consider the work of Geilenkirchen and his colleagues. Heat shocks on sea urchin eggs cause division delay with a TP and some indications of variable ED as in Model 1 of Fig. 10.9 (Geilenkirchen, 1964). Their effect on *Limnaea* eggs is more complicated (Geilenkirchen, 1966). There are signs of a TP and an overlap of delaying effect into the previous cycle, but maximum ED occurs not at the transition point but at the preceding cleavage. The delay curves are bimodal and it is possible that they are produced by a combination of the conventional variable ED with a high sensitivity at division, as in *Physarum*. Pulses of azide on *Limnaea* eggs also produce delay curves which are variable in the reverse way with the maximum delay at the preceding cleavage but with some traces of bimodality (Camey & Geilenkirchen, 1970). This is different from *Tetrahymena*, where azide pulses show maximum ED at the transition point (Hamburger, 1962). In all the three cases with the eggs, shocks or pulses at the time of mitosis cause morphogenetic abnormalities in later development.

The most important conclusion from this survey of reactions to heat shocks and inhibitor pulses is that nearly every case can be explained in terms of division protein or proteins provided that the *Tetrahymena* model can be extended in the following three ways. First, variable ED may be partial rather than complete, in the sense that cells before a particular point in the cycle do not show ED. This could be due to the synthesis of division protein starting at this point, though a somewhat similar effect would be produced if the shock or pulse caused the breakdown of only a portion of the protein. Secondly the division protein in some cases may be stable so that a pulse or shock does not cause the breakdown of pre-existing protein. This would produce constant delay, as in CHO cells. Thirdly, some cells (*Physarum*, *S. pombe*, perhaps *Limnaea* eggs) are particularly sensitive to heat shocks at the time of division suggesting that there is another heat-sensitive event at this time which is separate from the division protein.

Given these possible variations, the division protein hypothesis appears to be applicable to a wide range of cells. One of our next tasks is clearly to prove that these proteins exist and behave according to prediction. It is conceivable that a division protein structure could be identified with the electron microscope but far the most promising approach at the moment appears to be biochemical separation along the lines which have already been started in *Tetrahymena*. The problem, of course, is what sort of protein (or proteins) to look for. It should be associated with mitosis in eukaryotes (since mitosis comes before cleavage and both processes are delayed by inhibitors and heat) and with septum formation in prokaryotes. But we do not know whether it is present in large or small quantities, or whether it is enzymatic or structural. The suggestion that it is structural is attractive but the evidence is not compelling. It may be that this protein can be identified by the fact that it would disappear or be degraded by heat shocks and by temporary inhibition of protein synthesis before the TP, after which it would be unaffected. This seems to be a likely reaction, but it is possible for an enzyme to be inactivated or a protein structure to be disassembled without any significant change in the component molecules. So our task, though important, may not be easy.

Two important aspects of division control in *Tetrahymena* are the transition point and the variable delay. These also occur in other systems. Over a broad range from *E. coli* to mammals, all cells that have been tested show transition points with heat shocks and with inhibitors of protein and RNA synthesis. It is uncertain, however, how far these transition points are associated with a critical event in the cycle rather than with a delay before an inhibitor becomes effective.

Variable delay, either complete or partial, occurs with heat shocks in *E. coli* and *Physarum*, and possibly in sea urchin eggs. It also

occurs with inhibitors of protein synthesis in *Paramecium* and *Schizosaccharomyces*, and possibly in rat gut. Chinese hamster cells show a constant delay which could be interpreted in terms of a stable division protein.

The presence of variable delay in cells other than *Tetrahymena* suggests that the model of an unstable division protein may have wide application.

Division delays induced by radiation

The cellular effects of radiation is a large subject with a voluminous literature. All I shall do here is make a small and selective incursion into the field of radiation-induced division delay and try to show how it resembles and differs from delays caused by heat and inhibitors.

It has been known for many years that growing cells show delays in division when irradiated either with ionising radiation (X-rays and γ-rays) or ultraviolet light (u.v.) (Carlson, 1954; Lea, 1955; Rustad, 1964; Sinclair, 1968). As with heat shocks in *Tetrahymena*, growth continues during the period in which division is delayed so the cells become larger than normal.[1] Another similarity to heat shocks and inhibitor pulses is that there is a transition point after which radiation delays not the coming division but the one after. A significant point is that in three systems where the TPs for radiation and for protein synthesis inhibitors have been measured, the two TPs are identical. In CHO cells, the TP for X-ray doses of 25–800 rads is at 56 ± 4 minutes before division and for cycloheximide at 63 ± 5 minutes, though the puromycin TP is later. (Walters & Petersen, 1968*a*; Tobey, Anderson & Petersen, 1966). In mouse L5178Y cells, the TPs for X-ray doses (200 rads) and for puromycin are both in the middle of G2 about an hour before division (Fig. 10.10; Doida & Okada, 1969). In *Physarum*, the TP for X-rays and u.v., and the first TP for cycloheximide are both at 13 minutes before metaphase (pp. 227, 239).

An important point about radiation-induced delay is that it is dose-dependent.[2] An increase in the dose of ionising or u.v. radiation will cause longer delay, though the exact shape of the curve relating them may vary between different cells.

[1] E.g. Puck & Marcus (1956), Tolmach & Marcus (1960), Whitmore *et al.* (1961), Elkind *et al.* (1963), Rosenberg & Gregg (1969). Earlier references in Lea (1955).

[2] For X-rays on mammalian cells: Sinclair, (1968); Walters & Petersen (1968*a*); Doida & Okada (1969). For u.v. on sea urchin eggs: Rustad (1964). Earlier work in Lea (1955).

In many mammalian cells, the division delay produced by ionising radiation varies according to the stage in the cycle of the cells during the period of irradiation. Fig. 10.16 shows this effect for two types of mammalian cell, and there are some other examples quoted in the short review by Sinclair (1968). One way of interpreting these results, as suggested by Rustad (1970), is in terms of a model with two sensitive events during the

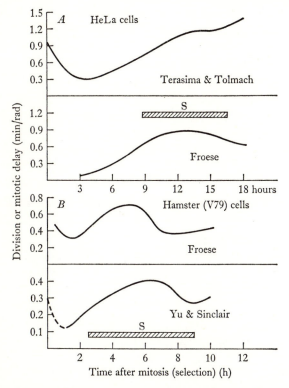

Fig. 10.16. Division delay in mammalian cells produced by X-rays applied at varying points in the cell cycle. *A*, for HeLa cells (Terasima & Tolmach, 1963*b*; Froese, 1966). *B*, for hamster cells (Froese, 1966; Yu & Sinclair, 1967). From Sinclair (1968).

cycle. One event is at or near division, and irradiation at this point causes a marked delay in the next division. The sensitivity falls sharply in early G1 and then rises slowly to a peak value which defines a second event towards the end of the cycle. It is not easy to establish the exact position of this peak, partly because cultures synchronised by wash-off are losing synchrony by the time they get to G2. In the hamster V79 cells in Fig. 10.16, the peak appears to be in mid- to late-S. But in the HeLa cells of Terasima &

Tolmach (1963 *b*) and in some other strains (Table II of Sinclair, 1968), the delays are greater for irradiation in G2 than in S. This would put the peak value in G2, presumably before the TP. A peak at this point, preceded by a long period of variable delay, is very like the pattern for a division protein in Models 1 and 2 of Fig. 10.9. There are, however, other interpretations, as we shall see later.

Variable division delay is not the only pattern found in mammalian cells. Elkind *et al.* (1963) found evidence of a constant delay through the cycle after irradiating asynchronous cultures of hamster V79 cells. The varied and careful experiments of Walters & Petersen (1968 *a*) showed the same pattern of constant delay (up to the TP) in synchronous and asynchronous cultures of hamster CHO cells. There is no very obvious explanation for the difference between the constant delay with these cells and the variable delay with the cells in Fig. 10.16, though Walters & Petersen suggest that it might be due to the fact that the constant delay occurs in cells which are in suspension culture whereas the variable delay occurs in cells grown as monolayers. One point to remember is that the constant delay with radiation is found in the same cells that show a constant delay with cycloheximide pulses (p. 223). Moreover, the TPs for both types of treatment are the same. The assembly of a stable division protein could be interrupted both by cycloheximide and by radiation.

Radiation effects on the neuroblasts of the grasshopper *Chortophaga* have been studied intensively by Carlson (review by Carlson, 1954). These cells are unusual in having a very long prophase which occupies the majority of the cell cycle. There is a TP in late prophase for X-rays, before which there is a maximum division delay – a situation similar to that for the second event in the model above. There is also some evidence for reversion of mitotic stages. After irradiation, prophase nuclei revert to a condition in which they resemble interphase nuclei. This does not seem to happen in most other cell systems after irradiation but there is a similarity to the resorption of the oral apparatus in *Tetrahymena*.

Fig. 10.17(*a*) shows the effect of X-rays in causing division delay in sea urchin eggs (Rustad, 1970). There is variable delay in the first part of the cycle followed by a fall to a TP about half-way through the cycle. Beyond the TP there is no delay to the first division but increasing delay to the second division. At first sight, the delay curves do not suggest two events. But with high doses, there is always a change in slope during the fall before the TP and there is sometimes a small plateau. This is even more striking in the earlier results of Henshaw & Cohen (1940) on the same species where there is a small subsidiary peak interrupting the fall. Rustad has therefore suggested that this biphasic response does indicate two events. These two events are far apart in the normal long cycles of growing cells, but they occur close together and their effects are nearly superimposed in the short cycles of the early cleavage stages of fertilised eggs.

For comparison, Fig. 10.17(b) gives the pattern of division delay produced by u.v. irradiation of the first cycle sea urchin eggs (Rustad, 1960). The TP occurs about half-way through the first cycle, as with X-rays, but otherwise the delay patterns are different with the two kinds of radiation.

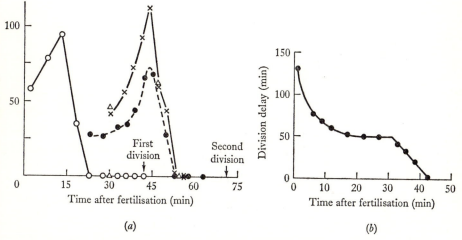

Fig. 10.17. Division delay produced by radiation applied at various stages of the cell cycle of fertilised sea urchin eggs. (a) X-rays applied during the first two cycles of *Arbacia punctulata*. Triangles and crosses in second cycle are with 10 kR; other data with 5 kR. From Rustad (1970). (b) u.v.-radiation applied during the first cycle of *Strongylocentrotus purpuratus*. First division is at 90 min. From Rustad (1960).

This is one of the few cases in which division delay has been followed with both ionising and u.v. radiation, even though two different species of sea urchin were used. The fact that the patterns are different in fertilised eggs, however, does not imply that they will necessarily be different in other cell systems. Eggs have unusually large quantities of cytoplasm which absorbs much of the u.v. radiation and experiments with enucleated half-eggs suggest that the u.v. target is in the cytoplasm, conceivably cytoplasmic DNA (Rustad, 1971). The target in thinner cells might well be nuclear.

There is a well-marked biphasic response to u.v. in *Schizosaccharomyces pombe* (Gill, 1965). Cells irradiated at the time of division show large delays (Fig. 10.18). There is then a sharp fall in sensitivity followed by a slow rise up to a TP in the later part of the cycle. The second event, marked by the TP and the preceding variable delay, gives a pattern similar to that from cycloheximide pulses in Fig. 10.14. There is a difference in the position of the two TPs which might be due to the fact that different strains of the yeast were used in the two sets of experiments and these strains may have had their S periods at different points in the cycle.

The response of heat-synchronised *Tetrahymena* to u.v. given in the first cycle after EH is not biphasic (Fig. 10.19, Nachtwey & Giese, 1968). There is a constant delay up to the TP with a constant dose, though the delay increases with higher doses. It should be remembered, however, that

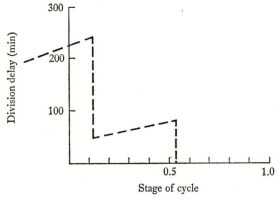

Fig. 10.18. Division delay produced by u.v.-radiation applied at various stages of the cell cycle of *Schizosaccharomyces pombe*. Cycle time = 140 min. From Gill (1965).

there is no division at the beginning of this cycle (i.e. at EH) so the first sensitive event may be missing. It would be interesting to know whether or not a biphasic response occurred in the next cycle. Fig. 10.19 shows that the TP for u.v. is nearly 30 minutes earlier than the TP for heat shocks. The u.v.-sensitive process which is essential for division is therefore completed well before the heat-sensitive process. These two processes could be independent, though there is some other evidence from this work indicating that the heat-sensitive process depends on the u.v.-sensitive process. They could also both be affecting a division protein structure which is stabilised against damage from u.v. before it becomes stabilised against heat, although in that case there should be a variable delay with u.v. before the TP.

Physarum has a biphasic response to X-rays (Sachsenmaier, Bohnert, Clausnizer & Nygaard, 1970). With a low dose (1000 R), the maximum delay is at mitosis and extends into the S period. With a higher dose (10000 R) however, a second sensitive period appears in late interphase. The delay rises for the last four hours of the cycle up to a transition point which exactly coincides with that for cycloheximide (p. 227). There is the same transition point for u.v. radiation (Devi *et al.*, 1968), but, as in sea urchin embryos, the response through the rest of the cycle is different. Devi *et al.* (1968) find a decreasing delay through the cycle, though some of their delay curves shows a plateau resembling that in the sea urchin

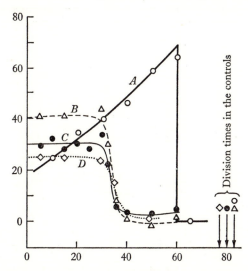

Fig. 10.19. Division delay produced by u.v.-radiation (and by heat shocks) applied at various times after EH in *Tetrahymena* previously synchronised by repetitive heat shocks. Abscissa, min after EH. Ordinate, division delay in min. Curve *A* is for heat shock (20 min at 34 °C). Curves *B*, *C* and *D* are for u.v. exposure of 230, 115 and 69 ergs/mm² respectively. From Nachtwey & Giese (1968).

embryo response in Fig. 10.17(*b*). This plateau is much more marked in similar experiments by Sachsenmaier, Dönges, Rupff & Czihak (1970).

Another system which has a biphasic reaction to X-rays is the flagellate *Astasia longa* synchronised by alternating temperatures (Padilla, Dreal & Anderson, 1966). There are two peaks of delay in Fig. 10.20. The first one, towards the start of the warm period, is preceded by a long stretch of variable delay. The second one comes towards the end of the warm period and refers to delay in the *next* division. The two peaks of delay for any one division are therefore separated by nearly one cycle time.

Before making any interpretation of these patterns of division delay, it is as well to emphasise how little we know about its fundamental causes.

Radiation has been shown to affect many cellular components and processes, but it is not clear which one is the chief target. Even the concept of a single target may be invalid since the delay patterns could be produced by the interactions of several processes. Moreover, ionising radiation may affect different systems from u.v. radiation. There are arguments against this, including the fact that delays induced by X-irradiation of sperm and by u.v.-irradiation of eggs are additive after fertilisation in sea urchins (for

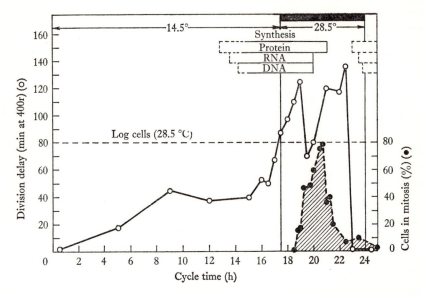

Fig. 10.20. Division delay produced by X-rays (400 R) applied at various times during the cell cycle of *Astasia longa* synchronised by alternating temperature cycles. Shaded curve shows the incidence of mitoses, and horizontal line the average division delay (80 min) for an asynchronous culture at 28.5 °C. From Padilla, Dreal and Anderson (1966).

this and other evidence, see Rustad, 1964). On the other hand, we have seen that the delay patterns through the cycle with the two types of radiation are not the same in sea urchin eggs and in *Physarum*.[1]

It is still, however, worth trying to define a general but tentative pattern for division delay in terms of the two sensitive events mentioned earlier; one at the beginning of the cycle and the other towards the end. There is not much that can be said about the first event except that it occurs near the time of mitosis and may therefore be nuclear or chromosomal since it is

[1] A recent paper on u.v.-induced delay in mouse L cells also gives a response which differs from that with X-rays (Thompson & Humphrey, 1970). The delay to the first division *decreases* through the cycle.

these organelles that are changing rapidly at that time in the cycle. Although most cells are not especially sensitive at this time to heat or inhibitors, there is a parallel in the effect of heat shocks on the eggs of *Limnaea* (p. 232).

Most of the systems show a fall in sensitivity to radiation after the time of the first event. Others do not, like the CHO cells which show constant delay between one transition point and the next (p. 236). A possible explanation for these latter cases is that the size and spread of the sensitivities for the two events is such that the diminishing sensitivity after the first event is counterbalanced by the increasing sensitivity before the second event.

The second event is particularly interesting because of its similarity to a TP for unstable division protein. Both happen towards the end of the cycle and both are preceded by a period of increasing delay. Is, therefore, the effect of radiation to cause the breakdown of division protein in the manner of heat shocks and inhibitor pulses? The most probable and the most widely accepted answer is no. Variable delay can be caused by an unstable division protein but it can also be caused by the constant repair time effect in Model 8 of Fig. 10.9. The likely target here is nuclear DNA with the damage being made good by the repair enzymes.

Various lines of evidence point to DNA as the site of the lesion. There is the general point that ionising radiation can break chromosomes, and that u.v. radiation produces molecular changes in DNA–thymine dimers which can be repaired by enzyme systems, one of which is photoreactivable. Bromodeoxyuridine, which is presumably incorporated into DNA, increases the delay caused by X-rays on mammalian cells (Schneider & Johns, 1966; Bootsma & Lohman, 1968) and by u.v. on sea urchin eggs (Cook, 1968). The action spectrum for mitotic delay induced by u.v.-irradiation of sea urchin sperm resembles that of a nucleoprotein, and this delay can be reduced by photoreactivation in the fertilised egg (Giese, 1946; Marshak, 1949; discussed by Rustad, 1971). It seems unlikely that division protein can be the target in sperm. A point, however, about sperm is that their chromatin is highly condensed and in this way is similar to mitotic cells. Their sensitivity, therefore, may be like growing cells at the first event in the cycle, which is near mitosis, and not at the second event which might be concerned with division protein. This criticism does not apply to the experiments with bromodeoxyuridine on mammalian cells where there is good evidence that cells late in the cycle have their sensitivity enhanced.

The dose-dependence with radiation argues against division proteins and for a damage plus repair mechanism, though the damage does not necessarily have to be in DNA. It is quite reasonable to expect that the repair time would increase with greater amounts of damage. If, on the other hand, a dose of radiation causes complete breakdown of the division protein, then an increase in the dose above this level should not cause any further delay. This is on the simplest hypothesis that radiation does not affect the production of new division protein. If, however, it did do this either at the level of transcription or of translation, then the delay could be dose-dependent.

We can now set out a model for the second event in more formal terms. Radiation causes damage to DNA in cells at all stages of the cycle. This damage can be repaired, and the repair time is constant for any given dose of radiation but increases as the dose is increased. The TP (sometimes called a 'block') is a stage in the cycle through which cells cannot pass until the damage is repaired. After this point, cells can go through mitosis irrespective of the amount of damage. The effect of this model is very similar to those with unstable division proteins (compare Model 8 in Fig. 10.9 with Models 1 and 2). A cell early in the cycle will have sufficient time to repair the damage before the TP is reached and so will not be delayed, whereas a cell late in the cycle (but before the TP) will be delayed at the TP until the damage is repaired. During the repair time, cells in an asynchronous culture will pass through the cycle and accumulate at the transition point. The rate of accumulation may be less than normal since there is evidence that radiation reduces the rate of DNA synthesis and the rate of progress through the S period (Sinclair, 1968). At the end of this time, they will pass through the transition point and then produce a burst of synchronous division (e.g. Whitmore *et al.*, 1961; Doida & Okada, 1969).

Although the evidence for the second event being concerned with DNA damage is persuasive, it is not compelling. There is some evidence pointing another way and suggesting that division proteins may be involved. The most striking point is the similarity of the timing of the TPs for radiation and for protein synthesis inhibitors in the three systems mentioned earlier (p. 234). Except with *Physarum*, the position of the TPs is well before the beginning of mitosis. There is no good reason why the time of synthesis of the last protein molecule needed for the coming division should be also the time after which DNA damage does not delay the division. But if radiation were acting on protein synthesis and either promoting the breakdown of an unstable division protein or delaying the completion of a stable one, we would expect the radiation TP to be the same as that for protein synthesis inhibition.

Another link with proteins is the finding that the recovery from X-radiation-induced delay in mammalian cells is stopped by inhibition of protein synthesis but not by inhibition of RNA and DNA synthesis (Walters & Petersen, 1968*b*; Doida & Okada, 1969). Similarly, the puromycin TP in sea urchin eggs occurs later in eggs that have been fertilised with X- and γ-irradiated sperm, and the radiation TP occurs later in eggs that have been treated with puromycin (Rustad & Burchill, 1966). Apart from the effect with the irradiated sperm, these results are in agreement with the concept of radiation interfering with division protein. But they are also consistent with another possible explanation – that continuing protein synthesis is needed for the repair of DNA damage. This type of experiment, however, with a combination of radiation and inhibitor pulses applied to synchronous cultures, is likely to be a powerful method for tracing the mechanisms of delay.

We cannot go much further at the moment in identifying the primary target in radiation-induced delay either as DNA or as division protein. Both indeed may be involved, with DNA damage being responsible for the first event and protein damage being responsible for the second. Another combination is to postulate that the division protein genes (or the division protein messages) are sensitive to radiation. This is a possibility but not a very attractive one, since the continued growth of cells after irradiation implies that translation at least has not been seriously affected.

The results of irradiating at different points in the cell cycle have been scored in other ways than measuring division delays (for example; the production of chromosome aberrations, lethality, or mutagenic effects) but in most cases they throw more light on the effects of radiation than on the normal events that occur in the cycle. One interesting exception, however, should be mentioned (review by Wolff, 1969*a*). If mammalian or plant cells are treated with ionising radiation early in the cycle, the aberrations in the chromosomes at the following metaphase indicate that they were single structures at the time of irradiation. If the cells are irradiated later in the cycle, the aberrations suggest that the chromosomes were double, i.e. two chromatids. One might expect the transition from singleness to doubleness to take place during the S period, but in fact it happens earlier, during G1. One interpretation of this is that the anaphase chromosome is double-stranded (this can be seen in some material) and that the two strands persist into the next interphase. In early G1, the strands are so closely wound together that radiation treats them as a single structure. Later in G1, they unravel themselves in preparation for replication and are then treated as two

targets for radiation. Two general points emerge: that here is an event in
G1, a period notably empty of markers; and that this unravelling fits into
the concept of a cycle of structural changes in the chromosome during
interphase (p. 89).

**The delays in division produced by ionising and u.v. radiation have
some similarity to those produced by heat and inhibitors. Transition
points are a common feature, and there is an interesting coincidence
between those for radiation and those for protein synthesis inhibitors. In some, but not all cases there is variable delay before the
transition point. The patterns of division delay through the cycle
can tentatively be described in terms of two peaks of sensitivity. The
first peak, which is usually absent with heat and inhibitors, is at the
beginning of the cycle. The second peak is near the transition point.
It could be caused by radiation acting on division protein in the
same way as heat does in *Tetrahymena*, but it is more likely to be
caused by radiation producing DNA damage which then takes a
constant time to be repaired.**

The 'DNA-division cycle' and the 'growth cycle'

The passage of a cell through the cycle can be described in terms of movement past a series of cycle 'markers', a concept that I have dealt with more
fully elsewhere (Mitchison, 1969 b). They can be defined as discrete events,
chemical, structural or physiological, which happen at a particular point
in the normal cycle. Some examples are the beginning and end of the S
period, the stages of mitosis, and the transition points.

A cycle with six markers (plus division) is shown in Fig. 10.21(a).
A marker map of this kind may be useful in itself as a measure of a cell's
progress through the cycle. If, for example, there were markers available
in early G1, it might be possible to tell whether or not a cell progresses into
G1 before it stops growing. But a more profitable use can be made of a
marker map if we are prepared to distort and change the cycle, since it may
then be possible to trace some of the causal connections between the
markers. One extreme case is where there is a direct causal connection
between one marker and the next (Fig. 10.21(b)). This can be called
a 'causal fixed sequence'. The spacing between the markers can be altered

but their order is fixed and a block placed on any one of them will halt
further progress through the cycle. Another type of fixed sequence is to
have markers without direct causal connection but dependent on signals
from some master timing mechanism. This 'non-causal fixed sequence'
differs from the causal one in allowing a marker to be blocked or omitted
without stopping the cycle. The other extreme case is a 'variable sequence'
(Fig. 10.21(*c*)). Each of the markers is the product of an independent chain
of events in the cell. In a normal cycle they will appear at particular points,
but their order as well as their spacing can be varied if the cycle is distorted.
Each of them has to happen at a point in the cycle but it may not matter to

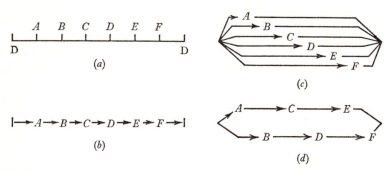

Fig. 10.21. (*a*) Hypothetical cell cycle map with six markers, *A* to *F*. These markers
as (*b*) a fixed sequence, (*c*) a variable sequence, and (*d*) a partially fixed sequence.

the cell when this point is. Between these two extremes, there are various
intermediate situations with 'partially fixed sequences'. One of these is
shown in Fig. 10.21(*d*) with two fixed sets of three markers each. The order
within the sets is fixed but the order between the sets can be varied, i.e. both
ABCDEF and *ACEBDF* are possible, but not *CDABFE*.

The average cell cycle is almost certainly a partially fixed sequence composed of sets of fixed sequences. Some of these are obvious, such as the
stages of mitosis or the beginning and end of the S period. But there is a way
in which this concept of sets of fixed sequences can be used with some
profit and novelty. This is to suggest that many of the existing cycle markers
can be assigned to one or other of two fixed sequences, or cycles since they
span the complete cell cycle. These can be called the 'DNA-division
cycle' or 'DD cycle', and the 'growth cycle' (Fig. 10.22). The main markers
in the DD cycle are the start and finish of the S period, the stages of mitosis,
nuclear division and cell division. Other markers may be the protein, RNA,
and radiation transition points for division which lie in G2; the RNA

transition point for DNA synthesis which may lie in G1 (p. 88); and the point in late G1 when the chromosome targets for radiation double (p. 243) As we have seen, the protein transition point for division may mark the completion and stabilisation of division protein. In addition, the DD cycle may include the synthesis of nuclear histones, of the enzymes associated with DNA synthesis and of proteins, structural and enzymatic, which are associated with mitosis, e.g. the proteins of the mitotic apparatus. The growth cycle, on the other hand, contains most of the macromolecular synthesis that takes place during the cycle. Its markers are the step and peak enzymes but there may well be portions of macromolecular synthesis,

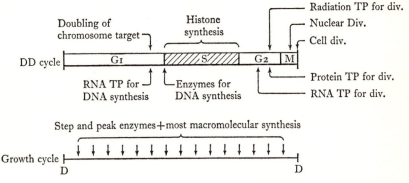

Fig. 10.22. The 'DNA-division cycle' and the 'growth cycle'.
TP, transition point; D, division.

for instance the production of ribosomal RNA, which have no markers because there are no sharp changes in the rate of synthesis. The existence of the growth cycle is much more hypothetical than that of the DD cycle, and, although it can be assumed to be a fixed sequence for the purpose of this argument, it may quite likely turn out to be only a partially fixed sequence when more is known about it.

The main evidence for the existence of these two cycles comes from the experiments where synchrony is induced by a temporary inhibition of DNA synthesis (p. 29). When DNA synthesis is blocked, for example by excess thymidine, the cells in the S period are arrested in their progress through the DD cycle. Division also stops as soon as the cells in G2 have divided. Meanwhile, however, the cells continue to grow at their normal rate in terms of macromolecular synthesis and volume increase. The two cycles have been dissociated since the growth cycle continues but the DD cycle is blocked. As a result, the DD cycles of all the cells are synchronised and there is a synchronous division when the DNA block is removed, but the growth cycles are *not* synchronised. Fig. 3.2 shows the difference between a synchronous division produced by this kind of induction and a synchronous division produced by selection. The use of both methods of synchronisation seems a promising way of finding out whether an event belongs to one or other cycle. For example, an enzyme step in the growth cycle should appear as a step with selection synchrony but not with induction synchrony whereas an enzyme step in the DD cycle should appear with both methods.

It is assumed here that a normal growth cycle does continue when the DD cycle is blocked. The justification for this comes from the evidence that enzyme steps and patterns continue normally when DNA synthesis is blocked in *Bacillus subtilis*, *Schizosaccharomyces pombe*, and mammalian cells (p. 168). This evidence, of course, is tenuous and we need to know a lot more about enzyme synthesis patterns after a DNA block. Two obvious questions are how long do the steps continue and do they appear in the same sequence as in the normal cycle.

The growth cycle might be expected to continue if the DD cycle was blocked at points other than the S period. There is some evidence that this happens after metaphase blocks induced by colchicine or colcemid (p. 32) though the position is complicated because condensed chromosomes are not transcribed and there may in some cases be an inhibition of translation during mitosis (p. 133). Another point of blocking is at the protein transition point when division protein is broken down by heat or inhibitor pulses. Growth certainly continues with *Tetrahymena* but it is not known whether it happens in the other systems that show variable delay. There are two possibilities with regard to the DD cycle in a system such as heat shocked *Tetrahymena*. One is that the cells are all set back by a heat shock to the start of the DD cycle and run through it again before dividing synchronously. The other is that cells which reach the transition point are held there until they have completed their division protein. The second possibility seems the most likely since the first one would produce synchronous synthesis of DNA and this does not happen. This point, however, needs further investigation, preferably in other systems because of the loose coupling of DNA synthesis to the cell cycle in Ciliates.

The DD cycle can also be blocked by radiation at a point in G2 or prophase. In some cases, it is clear that cells are held at this point until they have recovered (repaired DNA damage or rebuilt division protein). In sea urchins, for example, irradiated eggs pass through the first S period at the same time as controls and are delayed in early prophase (Rao & Hine-gardner, 1965; Zeitz *et al.*, 1968). Irradiated mammalian cells can go through the DD cycle until they are blocked in G2. There is no further DNA synthesis and no division (Whitfield & Rixon, 1959). Growth, how-ever, can continue and produce giant cells. If Ascites tumour cells are irradiated with 1100 R, division stops, DNA synthesis stops at the G2 value, but growth continues for five days until the protein per cell has increased several times over (Killander *et al.*, 1962). On the other hand, there are cases where DNA synthesis continues for some time after division has been blocked (Whitmore *et al.*, 1958; 1961). An interesting test of what was happening to growth after irradiation would be to see whether the enzyme markers of the growth cycle appeared in their normal sequence after the DD cycle had been blocked. Evidence from yeast suggests that this does happen (Eckstein *et al.*, 1966).

After the two cycles have been dissociated, there must presumably be a recovery process which allows them to lock together in their normal phase relation. If the DD cycle has been blocked while the growth cycle con-tinues, then the cells will become abnormally large. They can be brought back to normal size by speeding up the DD cycle relative to the growth cycle. Shortened cycles are in fact found after division blocks or delays, for instance in *Tetrahymena* after heat shocks, in *Schizosaccharomyces pombe* after a DNA block (Mitchison & Creanor, 1971c) and in *Physarum* after u.v. irradiation (Devi *et al.*, 1968; Sachsenmaier, Dönges, Rupff & Czihak, 1970). But the mechanisms behind these short cycles are obscure. If the blocking agent does not affect division protein (as with a DNA inhibitor), we can postulate the build-up of an excess of division protein during the block. After release from the block, less division protein will need to be made for the next division which can therefore occur sooner. This does not, however, easily explain why shortening occurs in several successive cycles or why short cycles follow a treatment which is supposed to act by destroy-ing division protein (though there might be a slow release of usable precursors from larger damaged pieces).

Physarum cultures are a case where the speed of the recovery process may affect the interpretation of results.[1] Plasmodia for experiments are started by amalgamating a number of micro-plasmodia which have been grown asynchronously in shake flasks. After amalgamation, the nuclear cycles of the micro-plasmodia are altered in such a way that the next mitosis is synchronised (Rusch *et al.*, 1966). The DD cycles are now synchronous

---

[1] This argument was developed in discussion with Dr R. C. Rustad.

but there could be a significant lag before the growth cycles also become synchronous. Any estimate of this lag would be guesswork, but we might remember that it takes a full cycle before DNA synthesis becomes synchronised with division after heat-shock treatment in *Tetrahymena* (p. 214). If the lag were as large as this, step enzymes would appear to be synthesized continuously in the first cycle of a *Physarum* plasmodium and would only emerge as steps in later cycles.

A question that can be asked of the DD cycle is whether it is a causal sequence in which every event has to occur if the cycle is to continue, or whether it is a non-causal sequence in which an event can be prevented without stopping progress through the cycle. This question can be answered for the two main events of this cycle, DNA synthesis and division. Although these two events are closely associated in normal cycles, there is not a universal causal connection between them because there are situations in which one or other of them is suppressed. It is relatively easy to find examples where DNA synthesis continues but division stops. This happens in *Tetrahymena* during heat shocks, in *Physarum* after a heat shock at mitosis (p. 228), in some cases in mammalian cells with radiation (Whitmore *et al.*, 1958; 1961), in mammalian cells after treatment with nitrogen mustard (Levis *et al.*, 1965) and in some differentiated cells where there is more than the diploid amount of DNA such as Dipteran salivary glands. Cases of the reverse situation – division without DNA synthesis – are much rarer, as might be expected since the biological penalty for this would normally be severe. They do, however, occur in special cases. Polyploid cells in crown gall tumours of bean stems can be reduced to the diploids by successive cell divisions without any DNA synthesis (Rasch *et al.*, 1959). Haploid Amphibian sperm injected into a mitotic egg can be induced to enter mitosis without any prior DNA synthesis (Graham, 1966*a*). There are also examples in the Ciliates which have been discussed earlier (p. 215). Little is known about the causal connections of other events in the DD cycle. The synthesis of histones and of the synthetic enzymes for DNA might have a close and causal connection with the start of the S period, or they might all be triggered by some other event. It would be interesting to know whether there was a causal connection between the completion of the division protein and division itself. In systems where division is inhibited, is it because a division protein is not completed or because it is complete but ineffective? These are problems for the future.

Postscript

This postscript mentions some of the papers which have come to my attention too late to be included in the main text. They have been arranged in the order of the relevant chapters.

Chapter 3. Synchronous cultures. There have been several developments in sedimentation techniques for selection synchrony. Shall & McClelland (1971) have used relatively short periods of gravity sedimentation in isotonic sucrose gradients to obtain synchronous cultures of mammalian cells. Surprisingly, the top layer of cells shows an initial burst of mitoses which indicates that there has been a partial selection of mitotic and pre-mitotic cells. Isolated mammalian nuclei can also be separated on sucrose gradients (McBride & Peterson, 1970). There is a sharper separation between G_1 and G_2 nuclei than there is with whole cells because of the major increase in total nuclear mass which happens towards the end of the cycle (p. 151). Counter-current (or elutriation) centrifugation seems to be a promising method of selecting cells of different size (Glick, Von Redlich, Juhos & McEwen, 1971). It needs a special centrifuge head, but there is no need for a gradient medium and the yield could be large. Hartwell (1970) has used equilibrium centrifugation in Renograffin gradients to produce synchronous cultures of budding yeast, exploiting the density changes through the cycle which were first described by Mitchison (1958, see p. 141).

Kovacs & Van't Hof (1970) have described a complex method of synchronising the meristem cells of pea roots by a combination of starvation and DNA inhibition, and Yeoman (1970) has reviewed some of the work on synchrony in plant callus cultures. Rooney *et al.* (1971) have developed a technique for synchronising *Chlamydomonas* with multiple cold shocks. The method developed by Cutler & Evans (1966) for synchronising bacteria (p. 36) also works with *Schizosaccharomyces pombe*, but not very reliably (Mitchison & Creanor, 1971*c*).

Chapter 4. DNA synthesis in eukaryotic cells. Johnson & Rao (1971) have published a thorough and timely review on nucleo-cytoplasmic interactions in the initiation of DNA synthesis and of mitosis. This covers in greater detail and with more examples the concepts discussed on pp. 83–6. It concludes *inter alia* that 'synchrony in multinucleate cells is achieved by the presence of specific inducers triggering DNA synthesis and mitosis'. Westerveld & Freeke (1971), however, find that the S periods of multinucleate human cells produced by virus-induced fusion are not perfectly synchronous in all cases.

Both heterochromatin and highly repetitive DNA appear to be replicated in the later parts of the S period (pp. 75–6). A more general statement about the DNA synthesis of normal mammalian cells is that the evidence so far shows that sequences rich in guanine and cytosine are preferentially replicated towards the start of the S period whereas sequences rich in adenine and thymine are replicated towards the end (Bostock & Prescott, 1971*a*, 1971*b*; Flamm *et al.*, 1971; Tobia *et al.*, 1970). These latter sequences may of course include types of DNA which differ from heterochromatin or highly repetitive DNA.

The DNA cycles of 7-day mouse embryos show definite differences between ectoderm and mesoderm (Solter *et al.*, 1971). As in Amphibia (Fig. 4.4), there seems to be a tendency for all phases of the DNA cycle (except mitosis) to lengthen, at any rate during some periods of development.

I have argued on p. 81 that there is a case for accurate determinations of the rate of DNA synthesis through the S period. This has been done in part by Schaer *et al.* (1971). who find a maximum rate of synthesis in mid-S in mouse P-815 cells. Possible distortions through pool changes have been minimised (but not excluded) by the use of four DNA precursors.

Nuclear DNA is synthesised periodically in cultures of *Chlorella pyrenoidosa* synchronised by light/dark cycles. A minor component of the total DNA which is probably chloroplast DNA is, however, synthesised continuously through the cycle (Wanka *et al.*, 1970). This is in contrast to the situation with chloroplast DNA in *Chlamydomonas* (p. 71). Another system where there is continuous synthesis of a minor DNA component is *Physarum*, though this component is nuclear rather than cytoplasmic (p. 83). There is evidence now that this component contains the genes for ribosomal RNA (Sonenshein *et al.*, 1970). This raises the possibility that there is continuous replication of some nuclear genes against a background of periodic replication for the rest.

Pederson & Robbins (1970*c*) have found significant changes in the capacity of chromatin to bind labelled actinomycin during the cycle of HeLa cells. This capacity decreases during G2 and before the start of mitosis. They suggest that chromosome condensation may occur some time in advance of prophase – which fits in well with the concept of a 'chromosome cycle' (p. 89).

In the Trypanosome *Crithidia luciliae*, the DNA both of the nucleus and of the kinetoplast (a single modified mitochondrion) are synthesised simultaneously during a restricted part of the cycle (Van Assel & Steinert, 1971). The S period is 220 min and G2 is about 60 min.

Some of the results on the molecular autoradiography of Amphibian DNA described on pp. 79–80 are mentioned by Priest & Callan (1970).

Chapter 5. DNA synthesis in prokaryotic cells. An important paper by Masters & Broda (1971) suggests that replication in *Escherichia coli* proceeds sequentially in *both* directions from the origin to a terminus elsewhere on the chromosome. This is in sharp contrast to the conclusion of unidirectional replication drawn from several independent methods (p. 99). If this evidence is confirmed, it will bring the mode of replication in *E. coli* in line with that in eukaryotic cells (p. 76). It will not alter most of the other conclusions in this Chapter, including the relations between replication and the cell cycle.

Chapter 6. RNA synthesis. Earlier evidence on the rate of RNA synthesis through the mammalian cell cycle shows two main patterns, a smooth increase in rate throughout the cycle and a relatively sharp increase restricted to the S period. The evidence, however, comes from different laboratories and different cell lines. Stambrook & Sisken (1970) have now shown that both patterns can be obtained from the same line of Chinese hamster cells by altering the culture medium. This indicates that the kinetics of RNA synthesis are not rigidly predetermined.

Acrylamide gel analysis of the RNA of *Physarum* after pulse labels through the cycle showed no qualitative changes, though some quantitative ones, in the gel profiles (Zellweger & Braun, 1971). The RNA labelled at different stages was also hybridised against the DNA replicated early and late in the S period, a method of examining the repetitive DNA sequences but not the unique ones. There was no preferential hybridisation of early-replicating DNA with early-transcribed RNA, and *vice versa*. This result is used as an argument against 'linear reading' (p. 170), though this conclusion is only valid if the order of transcription on the genome is the same as the order of replication – an assumption which is not an integral part of the hypothesis of linear reading. The hybridisation evidence also indicates that late-replicating DNA is less active in transcription than early-replicating DNA. This is in agreement with the fact that heterochromatin tends to be late-replicating and genetically inert (p. 75).

Pederson & Robbins (1970 *b*) mention that the synthesis of transfer RNA continues throughout mitosis in HeLa cells although the rate of synthesis is reduced at metaphase to 10–15 per cent of the interphase rate. The reason for this divergence from the general rule of a mitotic block to RNA synthesis is obscure. The characteristics of the RNA that returns from cytoplasm to nucleus in *Amoeba* (p. 121) have been described by Goldstein & Trescott (1970). The micronucleus of *Tetrahymena* resembles that of *Paramecium* (p. 123) in showing RNA labelling only during micronuclear S (Murti & Prescott, 1970). This may represent either synthesis *in situ* or accumulation of RNA synthesised in the macronucleus.

Chapter 7. Cell growth and protein synthesis. There is continuing interest in nuclear proteins and their relation to DNA synthesis in mammalian cells. A series of proteins can be isolated from whole mouse fibroblasts which bind to DNA but are not histones (Salas & Green, 1971). Like many other individual proteins or small groups of proteins, they show periodic synthesis during the cycle. One group is synthesised during the S period, and another group, which might be involved in the initiation of DNA synthesis, stops being synthesised at the end of G1. Auer *et al.* (1970) have shown that L cells continue to accumulate cytoplasmic protein for some time after DNA synthesis has been inhibited ('unbalanced growth', p. 29). There is, however, no accumulation of protein in the nucleus during this period and it only starts again when the DNA inhibition has been reversed. They suggest, therefore, either that the synthesis of nuclear proteins in the cytoplasm is linked to DNA synthesis or that the accumulation, perhaps binding, of protein in the nucleus is dependent on an increase in the DNA content. This suggestion, however, does not easily explain earlier results with other cell types (p. 151). Stein & Baserga (1970) have found that non-histone chromosomal proteins continue to be synthesised at an unchanged rate during mitosis although there is a fall of 70–90 per cent in the rate of synthesis of histones and of total cellular proteins at mitosis. There appear to be differences between the histone fractions in the time of acetylation in the S period and in the stability of this acetylation (Shepherd *et al.*, 1971).

The results of Stebbing (1969) on the growth of *Schizosaccharomyces pombe* (p. 140) have now been described in detail by Stebbing (1971). Total dry mass, protein and RNA, as well as pool dry mass, amino-acids and nucleotides all increase exponentially through the cycle of cells grown in minimal medium. In complex medium, there is an expandable pool of amino-acids whose fluctuation may account for linear growth in total dry mass.

Kubitschek (1971) has shown that a thymine-requiring strain of *Escherichia coli* will grow linearly in volume for two volume doublings when DNA synthesis and division are inhibited by thymine starvation. This is consistent with the concept of the control of linear growth being mediated through uptake sites for the binding, transport or accumulation of nutrients, which are constant in number through the cycle and double at the time of division (cf. p. 146).

The experiments of Kolodny & Gross (p. 157) on the rate of synthesis of soluble protein components during the cycle of HeLa cells have been repeated in more detail with *Schizosaccharomyces pombe* by Wain (1971). He did not, however, find marked changes in the rate of synthesis of these components through the cycle. There may be a difference here between higher and lower eukaryotes.

Chapter 8. Enzyme synthesis. Hütterman *et al.* (1970) gives a full account of the patterns of synthesis in *Physarum* of the enzymes referred to in Table 8.2 under Rusch (1969). Madreiter *et al.* (1971) describe the synthesis of DNA polymerase, triphosphatase and DNAase in L cells synchronised by wash-off. The first of these is probably a peak enzyme with maximum activity during the S period.

Chapter 9. Organelles respiration and pools. A detailed description of the changes in the ultrastructure during the HeLa cell cycle has been published by Erlandson & De Harven (1971). This does not differ greatly from the earlier accounts (p. 181) which showed little change during the cycle, but there appear to be some alterations in the nucleus which has more dense fibrillar bodies in G1 and more granular bodies in S and G2, and a change from polysomes to monosomes in late prophase.

Glick, Gerner & Warren (1971) have followed changes in the content of a number of carbohydrates during the cycle of mammalian KB cells. A substantial proportion of these carbohydrates are in membranes and their changing quantities suggest that there may be alterations in the membranes at the time of division. Fox *et al.* (1971). also find a membrane change at this time in mouse fibroblasts in that a specific binding site for an agglutinin is only exposed at the surface during mitosis. Interestingly, this site is exposed throughout the cycle when these cells have been transformed by polyoma virus.

Autoradiography with a labelled cell wall component has confirmed the immunofluorescence evidence (p. 190) in showing localised equatorial regions of wall growth in a gram-positive bacterium, *Diplococcus pneumoniae* (Briles & Tomasz, 1970).

The analysis by Stebbing (1969) of the changes in the amino-acid composition of the pool in *Schizosaccharomyces pombe* (p. 197) have now been described by Stebbing (1971). There is no significant deviation from exponential accumulation of these amino-acids through the cycle. The addition of arginine does not alter the step pattern of synthesis of ornithine transcarbamylase, which argues against the theory of 'oscillatory repression' (p. 169).

Chapter 10. The control of division. Two recent papers have been concerned with the relation between protein synthesis and division delay induced by ionising radiation. Van't Hof & Kovacs (1970) conclude that defective protein synthesis is the cause of delay in irradiated G1 cells of pea root meristem, though the time and cycle period when it occurs can vary. Bacchetti & Sinclair (1970) found the same transition point in Chinese hamster cells for X-rays and for high doses of cycloheximide (50 μg/ml), as has been found earlier (p. 234). With low doses (2–5 μg/ml), however, the transition point for cycloheximide was earlier. This might be due to delay in the inhibitor affecting the build-up of a specific division protein, though the effect of low and high doses are very similar in their depressive action on overall protein synthesis. Radiation also has the reverse effect of stimulating overall protein synthesis. They conclude that X-rays do not cause division delay by inhibiting the synthesis of specific proteins, though they admit the possibility that radiation may damage a structure (e.g. that formed by a division protein) which is not used after the end of G2 and which could be actively repaired during the delay period. This is interesting work which furthers our knowledge of division delay but it raises the problem of how to disentangle effects on general protein synthesis from those on specific division proteins or structures. These effects may not be identical.

References and Author Index

PREFACE TO BIBLIOGRAPHY

The following system has been used in arranging these references. Papers by a single author are listed in the order of dates of publication. These are followed by papers in which this author is the first author and there is one other collaborator. These two-author papers are arranged in chronological order for any one collaborator, and otherwise in alphabetical order of the collaborator. These are followed by papers with three or more authors, arranged in the alphabetical order of the second author.

Abbreviations of journals are from the *World List of Scientific Periodicals*. The figure(s) in square brackets at the end of each reference refer to the page in which this reference appears.

Abbo, F. E. & Pardee, A. B. (1960). Synthesis of macromolecules in synchronously dividing bacteria. *Biochim. biophys. Acta* **39**, 478–85. [4n, 54, 102, 125n, 144, 176]

Abe, M. & Tomizawa, J. (1967). Replication of the *Escherichia coli* K12 chromosome. *Proc. natn. Acad. Sci. U.S.A.* **58**, 1911–18. [98]

Abercrombie, M. & Stephenson, E. M. (1969). Observations on chromocentres in cultured mouse cells. *Nature, Lond.* **222**, 1250–3. [182]

Adams, R. L. P. (1969a). Phosphorylation of tritiated thymidine by L929 mouse fibroblasts. *Expl Cell Res.* **56**, 49–54. [16, 197]

Adams, R. L. P. (1969b). The effect of endogenous pools of thymidylate on the apparent rate of DNA synthesis. *Expl Cell Res.* **56**, 55–8. [81, 178, 197]

Adler, H. I. & Hardigree, A. A. (1964). Cell elongation in strains of *Escherichia coli*. *J. Bact.* **87**, 1240–2. [192]

Adolph, E. F. & Bayne-Jones, S. (1932). Growth in size of micro-organisms measured from motion pictures. II. *Bacillus megatherium*. *J. cell. comp. Physiol.* **1**, 409–27. [142]

Agrell, I. (1964). Natural division synchrony and mitotic gradients in metazoan tissues. In *Synchrony in Cell Division and Growth*, pp. 39–70. Ed. by E. Zeuthen. New York: Interscience. [55]

Alfert, M. & Das, N. K. (1969). Evidence for control of the rate of nuclear DNA synthesis by the nuclear membrane in eukaryotic cells. *Proc. natn. Acad. Sci. U.S.A.* **63**, 123–8. [83]

Alpen, E. L. & Johnston, M. E. (1967). DNA synthetic rate and DNA content of nucleated erythroid cells. *Expl Cell Res.* **47**, 177–92. [62, 81n]

Ammerman, D. (1970). The micronucleus of the Ciliate *Stylonichia mytilus*: its nucleic acid synthesis and its function. *Expl Cell Res.* **61**, 6–12. [69]

Anderson, E. C. (1970). Synchronous culture production by density selection. *Science, N.Y.* **170**, 97. [51]

Anderson, E. C., Bell, G. I., Petersen, D. F. & Tobey, R. A. (1969). Cell growth and division. IV. Determination of volume growth rate and division probability. *Biophys. J.* **9**, 246–63. [131, 132]

Anderson, E. C., Petersen, D. F. & Tobey, R. A. (1967). Biochemical balance and synchronized cell cultures. *Nature, Lond.* **215**, 1083–4. [4, 30]

Anderson, E. C., Petersen, D. F. & Tobey, R. A. (1970). Density invariance of cultured Chinese hamster cells with stage of the mitotic cycle. *Biophys. J.* **10**, 630–45. [132]

Anderson, P. A. & Pettijohn, D. E. (1960). Synchronization of division in *Escherichia coli*. *Science, N.Y.* **131**, 1098. [55]

Aoki, Y. & Moore, G. E. (1970). Comparative study of mitotic stages of cells derived from human peripheral blood. *Expl Cell Res.* **59**, 259–66. [60]

Auer, G., Zetterberg, A. & Foley, G. E. (1970). The relationship of DNA synthesis to protein accumulation in the cell nucleus. *J. Cell Physiol.* **76**, 357–64. [253]

Ayad, S. R., Fox, M. & Winstanley, D. (1969). The use of ficoll gradient centrifugation to produce synchronous mouse lymphoma cells. *Biochem. biophys. Res. Commun.* **37**, 551–8. [50]

Bacchetti, S. & Sinclair, W. K. (1970). The relation of protein synthesis to radiation-induced division delay in Chinese hamster cells. *Radiat. Res.* **44**, 788–806. [255]

Baechtel, F. S., Hopkins, H. A. & Schmidt, R. R. (1970). Continuous inducibility of isocitrate lyase during the cell cycle of the eucaryote *Chlorella*. *Biochim. biophys. Acta* **217**, 216–19. [164, 178, 180]

Bahr, G. F. & Zeitler, E. (1962). Study of mitochondria in rat liver. Quantitative electron microscopy. *J. Cell Biol.* **15**, 489–501. [183]

Baker, A. L. & Schmidt, R. R. (1963). Intracellular distribution of phosphorus during synchronous growth of *Chlorella pyrenoidosa*. *Biochim. biophys. Acta* **74**, 75–83. [44, 199]

Baker, A. L. & Schmidt, R. R. (1964). Further studies on the intracellular distribution of phosphorus during synchronous growth of *Chlorella pyrenoidosa*. *Biochim. biophys. Acta* **82**, 336–42. [199]

Barer, R. & Joseph, S. (1957). Phase contrast and interference microscopy in the study of cell structures. *Symp. Soc. exp. Biol.* **10**, 160–84. [10]

Barner, H. D. & Cohen, S. S. (1956). Synchronization of division of a thymineless mutant of *Escherichia coli*. *J. Bact.* **72**, 115–23. [36]

Barr, H. J. (1968). An effect of exogenous thymidine on the mitotic cycle. *J. cell. comp. Physiol.* **61**, 119–28. [26]

Baserga, R. (1962). A study of nucleic acid synthesis in ascites tumor cells by two-emulsion autoradiography. *J. Cell Biol.* **12**, 633–7. [18, 116n]

Baserga, R. (1965). The relationship of the cell cycle to tumor growth and control of cell division: a review. *Cancer Res.* **25**, 581–95. [62]

Baserga, R. & Nemeroff, K. (1962). Two-emulsion autoradiography. *J. Histochem. Cytochem.* **10**, 628–35. [22–3]

Baserga, R., Estensen, R. D. & Petersen, R. O. (1965). Inhibition of DNA synthesis in Ehrlich ascites cells by actinomycin D. II. The presynthetic block in the cell cycle. *Proc. natn. Acad. Sci. U.S.A.* **54**, 1141–8. [88]

Bassleer, R. (1968). Recherches sur les protéines nucléaires totales et les acides désoxyribonucléiques dans les fibroblastes cultivés in vitro et dans des cellules tumorales d'Ehrlich. *Archs Biol., Liège*. **79**, 181–325. [63, 151]

Bassleer, R. & Chèvremont-Comhaire, S. (1964). Nouvelles recherches concernant l'action de la thymidine tritiée sur des fibroblasts normaux cultivés *in vitro*. *C. r. Séanc. Soc. Biol.* **158**, 1405–8. [22]

Bayne-Jones, S. & Adolph, E. F. (1932). Growth in size of micro-organisms measured from motion pictures. I. Yeast *Saccharomyces cerevisiae*. *J. cell. comp. Physiol.* **1**, 387–407. [141]

Beachey, E. H. & Cole, R. M. (1966). Cell wall replication in *Escherichia coli*, studied by immunofluorescence and immunoelectron microscopy. *J. Bact.* **92**, 1245–51. [191]

Bello, L. J. (1968). Synthesis of DNA-like RNA in synchronized cultures of mammalian cells. *Biochim. biophys. Acta* **157**, 8–15. [121]

Bello, L. J. (1969). Studies on gene activity in synchronized cultures of mammalian cells *Biochim. biophys. Acta* **179**, 204–13. [121, 179]

Bergeron, J. J. M., Warmsley, A. M. H. & Pasternak, C. A. (1970). Phospholipid synthesis and degradation during the life cycle of P815Y mast cells synchronized with excess of thymidine. *Biochem. J.* **119**, 489–92. [186]

Bergter, R. (1965). Synchrone Zellteilung in Bakterienkulturen nach Überimpfung aus der stationären Phase. *Z. allg. Mikrobiol.* **5**, 92–5. [36]

Bernstein, F. (1968). Induction of synchrony in *Chlamydomonas moewusii* as a tool for the study of cell division. In *Methods in Cell Physiology*, Vol. 3, 119–45. Ed. by D. M. Prescott. New York and London: Academic Press. [45]

Bird, R. & Lark, K. G. (1968). Initiation and termination of DNA replication after amino-acid starvation of *E. coli* 15T⁻. *Cold Spring Harb. Symp. quant. Biol.* **33**, 799–808. [106]

Birky, C. W., Bignami, R. A. & Bentfeld, M. J. (1967). Nuclear and cytoplasmic DNA synthesis in adult and embryonic rotifers. *Biol. Bull. mar. biol. Lab., Woods Hole* **133**, 502–9. [84]

✓ Bleecken, S., Strohbach, G. & Sarfert, E. (1966). Autoradiography of bacterial chromosomes. *Z. allg. Mikrobiol.* **6**, 121–3. [93]

Blenkinsopp, W. K. (1968). Cell proliferation in stratified squamous epithelium in mice. *Expl Cell Res.* **50**, 265–76. [62]

Blenkinsopp, W. K. (1969). Cell proliferation in the epithelium of the oesophagus, trachea and ureter in mice. *J. Cell Sci.* **5**, 393–401. [62]

Bloch, D. P. & Goodman, G. C. (1955). 1. A microphotometric study of the synthesis of desoxyribonucleic acid and nuclear histone. 2. Evidence of differences in the deoxyribonucleoprotein complex of rapidly proliferating and non-dividing cells. *J. biophys. biochem. Cytol.* **1**, 17–28, 531–50. [153]

Bloch, D. P. & Teng, C. (1969). The synthesis of deoxyribonucleic acid and nuclear histone of the X chromosome of the *Rehnia spinosus* spermatocyte. *J. Cell Sci.* **5**, 321–32. [154]

Bloch, D. P., MacQuigg, R. A., Brack, S. D. & Wu, J-R. (1967). The synthesis of deoxyribonucleic acid and histone in the onion root meristem. *J. Cell. Biol.* **33**, 451–68. [22, 86n, 154]

Blondel, B. & Tolmach, L. J. (1965). Studies on nuclear fine structure. Three phases of HeLa cell cycle. *Expl Cell Res.* **37**, 497–501. [181]

Bode, H. R. & Morowitz, J. (1967). Size and structure of the *Mycoplasma hominis* H39 chromosome. *J. molec. Biol.* **23**, 191–9. [95]

Bonhoeffer, F. & Gierer, A. (1963). On the growth mechanism of the bacterial chromosome. *J. molec. Biol.* **7**, 534–40. [94]

Bonhoeffer, F. & Messer, W. (1969). Replication of the bacterial chromosome. *A. Rev. Genet.* **3**, 233–46. [91]

Bootsma, D. & Lohman, P. H. M. (1968). Mitotic delay after X-irradiation of synchronized mammalian cells cultivated in the presence of BUdR. *Int. J. Radiat. Biol.* **14**, 69. [241]

Bootsma, D., Budke, L. & Vos, O. (1964). Studies on synchronous division of tissue culture cells initiated by excess thymidine. *Expl Cell Res.* **33**, 301–9. [26]

Bosmann, H. B. (1970*a*). Glycoprotein synthesis; activity of collagen-galactosyl and collagen-glucosyl transferases in synchronized mouse leukemic cells L5178Y. *Expl Cell Res.* **61**, 230–3. [178]

Bosmann, H. B. (1970*b*). Cellular membranes: membrane marker enzyme activities in synchronized mouse leukemic cells L5178Y. *Biochim. biophys. Acta* **203**, 256–60. [178]

Bosmann, H. B. & Winston, R. A. (1970). Synthesis of glycoprotein, glycolipid, protein and lipid in synchronized L5178Y cells. *J. Cell Biol.* **45**, 23–33. [157, 186]

Bostock, C. J. (1970*a*). DNA synthesis in the fission yeast *Schizosaccharomyces pombe*. *Expl Cell Res.* **60**, 16–26. [19, 70]

Bostock, C. J. (1970*b*). The effects of 2-phenyl ethanol on the DNA synthesis cycle of *Schizosaccharomyces pombe*. *J. Cell Sci.* **7**, 523–30. [70]

Bostock, C. J. & Prescott, D. M. (1971*a*). Buoyant density of DNA synthesized at different stages of the S phase in mouse L cells. *Expl Cell Res.* **64**, 267–74. [76, 251]

Bostock, C. J. & Prescott, D. M. (1971*b*). Buoyant density of DNA synthesized at different stages of S phase in Chinese hamster cells. *Expl Cell Res.* **64**, 481–4. [251]

Bostock, C. J., Donachie, W. D., Masters, M. & Mitchison, J. M. (1966). Synthesis of enzymes and DNA in synchronous cultures of *Schizosaccharomyces pombe*. *Nature, Lond.* **210**, 808–10. [70, 177, 178]

Boyle, J. V., Goss, W. A. & Cook, T. M. (1967). Induction of excessive deoxyribonucleic acid synthesis in *Escherichia coli* by nalidixic acid. *J. Bact.* **94**, 1664–71. [108]

Braun, R. & Behrens, K. (1969). A ribonuclease from *Physarum*. Biochemical properties and synthesis in the mitotic cycle. *Biochim. biophys. Acta* **195**, 87–98. [178]

Braun, R. & Evans, T. E. (1969). Replication of nuclear satellite and mitochondrial DNA in the mitotic cycle of *Physarum*. *Biochim. biophys. Acta* **182**, 511–22. [71, 83]

Braun, R. & Wili, H. (1969). Time sequence of DNA replication in *Physarum*. *Biochim. biophys. Acta* **174**, 246–52. [80]

Braun, R., Mittermayer, C. & Rusch, H. P. (1965). Sequential temporal replication of DNA in *Physarum polycephalum*. *Proc. natn. Acad. Sci. U.S.A.* **53**, 924–31. [70, 81, 84]

Braun, R., Mittermayer, C. & Rusch, H. P. (1966*a*). Sedimentation patterns of pulse labelled RNA in the mitotic cycle of *Physarum polycephalum*. *Biochim. biophys. Acta* **114**, 27–35. [122]

Braun, R., Mittermayer, C. & Rusch, H. P. (1966*b*). Ribonucleic acid synthesis in vivo in the synchronously dividing *Physarum polycephalum* studied by cell fractionation. *Biochim. biophys. Acta* **114**, 527–35. [122]

Brent, T. P. & Forrester, J. A. (1967). Changes in the surface charge of HeLa cells during the cell cycle. *Nature, Lond.* **215**, 92–3. [189]

Brent, T. P., Butler, J. A. V. & Crathorn, A. R. (1965). Variations in phosphokinase activities during the cell cycle in synchronous populations of HeLa cells. *Nature, Lond.* **207**, 176–7. [179]

Bresciani, F. (1965). Effects of ovarian hormones on the duration of DNA synthesis in cells of the C_3H mouse mammary gland. *Expl Cell Res.* **38**, 13–32. [62]

Brewer, E. N. & Rusch, H. P. (1965). DNA synthesis by isolated nuclei of *Physarum polycephalum*. *Biochim. biophys. Res. Commun.* **21**, 235–41. [85]

Brewer, E. N. & Rusch, H. P. (1968). Effects of elevated temperature shocks on mitosis and on the initiation of DNA replication in *Physarum polycephalum*. *Expl Cell Res.* **49**, 79–86. [227, 228]

Briles, E. B. & Tomasz, A. (1970). Radioautographic evidence for equatorial wall growth in a gram-positive bacterium. Segregation of choline-³H-labelled teichoic acid. *J. Cell Biol.* **47**, 786–90. [255]

Brinkley, B. R. & Stubblefield, E. (1970). Ultrastructure and interactions of the kinetochore and centriole in mitosis and meiosis. *Adv. Cell Biol.* **1**, 119–85. [184]

Brostrom, M. A. & Binkley, S. B. (1969). Synchronous growth of *Escherichia coli* after treatment with fluorophenylalanine. *J. Bact.* **98**, 1271. [38]

Brown, S. W. (1966). Heterochromatin. *Science, N.Y.* **151**, 417–25. [75]

Bruce, V. G., Lark, K. G. & Maaløe, O. (1955). Turbidimetric measurement on synchronized *Salmonella typhimurium* cultures. *Nature, Lond.* **176**, 563–4. [40n]

Buck, C. A., Granger, G. A. & Holland, J. J. (1967). Initiation and completion of mitosis in HeLa cells in the absence of protein synthesis. *Currents in Modern Biology* **1**, 9–13. [133, 222]

Buell, D. N. & Fahey, J. L. (1969). Limited periods of gene expression in immunoglobulin-synthesizing cells. *Science, N.Y.* **164**, 1524–5. [157]

Burchill, B. R. (1968). Synthesis of RNA and protein in relation to oral regeneration in the ciliate *Stentor coeruleus*. *J. exp. Zool.* **167**, 427–38. [227]

Burnett-Hall, D. G. & Waugh, W. A. O'N. (1967). Indices of synchrony in cellular cultures. *Biometrics* **23**, 693–716. [3n, 56, 57]

Burns, V. W. (1959). Synchronized cell division and DNA synthesis in a *Lactobacillus acidophilus* mutant. *Science, N.Y.* **129**, 566–7. [36]

Burns, V. W. (1961). Relations among DNA and RNA synthesis and synchronized cell division in *Lactobacillus acidophilus*. *Expl Cell Res.* **23**, 582–94. [125n, 126, 144]

Byars, N. & Kidson, C. (1970). Programmed synthesis and export of immunoglobulin by synchronized myeloma cells. *Nature, Lond.* **226**, 648–50. [157]

Byfield, J. E. & Lee, Y. C. (1970). Do synchronizing temperature shifts inhibit RNA synthesis in *Tetrahymena pyriformis*? *J. Protozool.* **17**, 445–53. [211]

Byfield, J. E. & Scherbaum, O. H. (1967). Temperature dependent decay of RNA and of protein synthesis in a heat-synchronized protozoan. *Proc. natn. Acad. Sci. U.S.A.* **57**, 602–6. [211]

Cairns, J. (1963). The bacterial chromosome and its manner of replication as seen by autoradiography. *J. molec. Biol.* **6**, 208–13. [77, 93, 105]

Cairns, J. (1966). Autoradiography of HeLa cell DNA. *J. molec. Biol.* **15**, 372–3. [77]

Calkins, J. & Gunn, G. (1967). The time of DNA synthesis during the interdivision growth cycle of *Tetrahymena pyriformis* fed on living bacteria. *J. Protozool.* **14**, 210–13. [67n]

Callan, H. G. & Taylor, J. H. (1968). A radioautographic study of the time course of male meiosis in the newt *Triturus vulgaris*. *J. Cell Sci.* **3**, 615–26. [63]

Callan, H. G., Priest, J. & Lloyd, L. (1971). In preparation. [79, 81]

Cameron, I. L. (1964). Is the duration of DNA synthesis in somatic cells of mammals and birds a constant? *J. Cell Biol.* **20**, 185–8. [62, 63]

Cameron, I. L. (1966). A periodicity of tritiated-thymidine incorporation into cytoplasmic deoxyribonucleic acid during the cell cycle of *Tetrahymena pyriformis*. *Nature, London.* **209**, 630–1. [71]

Cameron, I. L. & Greulich, R. C. (1963). Evidence for an essentially constant duration of DNA synthesis in renewing epithelia of the adult mouse. *J. Cell Biol.* **18**, 31–40. [62]

Cameron, I. L. & Nachtwey, D. S. (1967). DNA synthesis in relation to cell division in *Tetrahymena pyriformis*. *Expl Cell Res.* **46**, 385–95. [67n, 68]

Cameron, I. L. & Prescott, D. M. (1961). Relations between cell growth and cell division. V. Cell and macronuclear volumes of *Tetrahymena pyriformis* HSM during the cell life cycle. *Expl Cell Res.* **23**, 354–60. [135, 151]

Cameron, I. L. & Stone, G. E. (1964). Relation between the amount of DNA per cell and the duration of DNA synthesis in three strains of *Tetrahymena pyriformis*. *Expl Cell Res.* **36**, 510–14. [67n, 68, 82]

Camey, T. & Geilenkirchen, W. L. M. (1970). Cleavage delay and abnormal morphogenesis in *Limnaea* eggs after pulse treatment with azide of successive stages in a cleavage cycle. *J. Embryol. exp. Morph.* **23**, 385–94. [64, 232]

Campbell. A. (1957a). Synchronization of cell division. *Bact. Rev.* **21**, 263–72. [4]

Campbell. A. (1957b). Division synchronization in a respiratory deficient yeast. *J. Bact.* **74**, 559–64. [35]

Carlson, J. G. (1954). Immediate effects on division, morphology, and viability of the cell. In *Radiation Biology* pp. 763–824. Ed. by A. Hollaender. New York: McGraw-Hill Book Co. Inc. [234, 236]

Caro, L. C. & Berg, C. M. (1968). Chromosome replication in some strains of *Escherichia coli* K12. *Cold Spring Harb. Symp. quant. Biol.* **33**, 559–73. [99, 112]

Caro, L. C. & Berg, C. M. (1969). Chromosome replication in *Escherichia coli*. II. Origin of replication in F⁻ and F⁺ strains. *J. molec. Biol.* **45**, 325–36. [98]

Carrière, R., Leblond, C. P. & Messier, B. (1961). Increase in the size of liver cell nuclei before mitosis. *Expl Cell Res.* **23**, 625–8. [151]

Caspersson, T. O. & Lomakka, G. M. (1962). Scanning microscopy technique for high resolution quantitative cytochemistry. *Ann. N.Y. Acad. Sci.* **97**, 449–63. [9]

Caspersson, T., Farber, S., Foley, G. E., Killander, D. & Zetterberg, A. (1965). Cytochemical evaluation of metabolic inhibitors in cell cultures. *Expl Cell Res.* **39**, 365–85. [86n, 88n]

Cave, M. (1966). Incorporation of tritium-labelled thymidine and lysine into chromosomes of cultured human leukocytes. *J. Cell Biol.* **29**, 209–22. [154]

Cave, M. D. (1967). Chromosomal ³H-lysine incorporation and patterns of deoxyribonucleic acid synthesis in human cells. *Expl Cell Res.* **45**, 631–7. [73n]

Cerdá-Olmeda, E. & Hanawalt, P. C. (1968). The replication of the *Escherichia coli* chromosome studied by sequential nitrosoguanidine mutagenesis. *Cold Spring Harb. Symp. quant. Biol.* **33**, 599–607. [99]

Cerdá-Olmeda, E., Hanawalt, P. C. & Guerola, N. (1968). Mutagenesis of the replication point by nitrosoguanidine: map and pattern of replication of the *Escherichia coli* chromosome. *J. molec. Biol.* **33**, 705–19. [99]

Chalkley, G. R. & Maurer, H. R. (1965). Turnover of template-bound histone. *Proc. natn. Acad. Sci. U.S.A.* **54**, 498–505. [156]

Chan, H. & Lark, K. G. (1969). Chromosome replication in *Salmonella typhimurium*. *J. Bact.* **97**, 848–60. [92]

Charret, R. (1969). L'ADN nucléolaire chez *Tetrahymena pyriformis*: chronologie de sa réplication. *Expl Cell Res.* **54**, 353–61. [67n, 76]

Chernick, B. (1968). Late synthesis of protein on the presumptive X chromosome of the human female lymphocyte. *Nature, Lond.* **220**, 195–6. [155]

Chèvremont-Comhaire, S. & Chèvremont, M. (1956). Action de températures subnormales suivies de réchauffement sur l'activité mitotique en culture de tissue. Contribution à l'étude de la préparation à la mitose. *C. r. Séanc. Soc. Biol.* **150**, 1046–9. [39]

Chiang, K-W. & Sueoka, N. (1967). Replication of chloroplast DNA in *Chlamydomonas reinhardii* during vegetative cell cycle: its mode and regulation. *Proc. natn. Acad. Sci. U.S.A.* **57**, 1506–13. [45, 71]

Chin, B. & Bernstein, I. A. (1968). Adenosine triphosphate and synchronous mitosis in *Physarum polycephalum*. *J. Bact.* **96**, 330–7. [199]

Christensson, E. G. (1967). Histones and basic proteins during cell growth and cell division in heat-synchronized mass cultures of *Tetrahymena pyrifomis* GL. *Ark. Zool.* **19**, 297–308. [155n]

Chung, K. I., Hawirko, R. Z. & Isaac, P. K. (1964). Cell Wall Replication. I. Cell wall growth of *Bacillus cereus* and *Bacillus megaterium*. II. Cell wall growth and cross wall formation of *Escherichia coli* and *Streptococcus faecalis*. *Can. J. Microbiol.* **10**, 43–8, 473–82. [191]

Church, K. (1967). Patterns of DNA replication in binucleate cells occurring in mouse embryo cell cultures. *Expl Cell Res.* **46**, 639–41. [84]

Churchill, J. R. & Studzinski, G. P. (1970). Thymidine as synchronizing agent. III. Persistence of cell cycle patterns of phosphatase activities and elevation of nuclease activity during inhibition of DNA synthesis. *J. Cell Physiol.* **75**, 297–304. [168, 179]

Clark, D. J. (1968*a*). Regulation of deoxyribonucleic acid replication and cell division in *Escherichia coli* B/r. *J. Bact.* **96**, 1214–24. [110]

Clark, D. J. (1968*b*). The regulation of DNA replication and cell division in *Escherichia coli* B/r. *Cold Spring Harb. Symp. quant. Biol.* **33**, 823–38. [110, 111]

Clark, D. J. & Maaløe, O. (1967). DNA replication and the division cycle in *Escherichia coli*. *J. molec. Biol.* **23**, 99–112. [40, 105]

Clason, A. E. & Burdon, R. H. (1969). Synthesis of small nuclear ribonucleic acids in mammalian cells in relation to the cell cycle. *Nature, Lond.* **223**, 1063–4. [121]

Cleaver, J. E. (1965). The relationship between the duration of the S-phase and the fraction of cells which incorporate 3-H-thymidine during exponential growth. *Expl Cell Res.* **39**, 697–700. [19]

Cleaver, J. E. (1967). *Thymidine Metabolism and Cell Kinetics*. Amsterdam: North-Holland Publishing Co. [11, 58, 60, 61, 62, 73n]

Cleaver, J. E. & Holford, R. M. (1965). Investigations into the incorporation of [³H] thymidine into DNA in L-strain cells and the formation of a pool of phosphorylated derivatives during pulse labelling. *Biochim. biophys. Acta* **103**, 654–71. [16, 81]

Cleffmann, G. (1965). Die Schwellen der Hemmung der Nucleinsäuresynthese und der Teilung durch Actinomycin bei *Tetrahymena pyriformis*. *Z. Zellforsch. mikrosk. Anat.* **67**, 343–50. [88, 210]

Cleffmann, G. (1967). Temperaturabhängigkeit der Phasen der Teilungszyklen von *Tetrahymena pyriformis* HSM. *Z. Zellforsch. mikrosk. Anat.* **79**, 599–602. [67n]

Clowes, F. A. L. (1965). The duration of the G1 phase of the mitotic cycle and its relation to radiosensitivity. *New Phytol.* **64**, 355–9. [66, 67]

Cohn, N. S. (1968). DNA synthetic rates among cells of populations *in vitro*. *Experientia* **24**, 822–4. [81n]

Cole, F. E. & Schmidt, R. R. (1964). Control of aspartate transcarbamylase activity during synchronous growth of *Chlorella pyrenoidosa*. *Biochim. biophys. Acta* **90**, 616–18. [178]

Cole, R. M. (1965). Symposium on the fine structure and replication of bacteria and their parts. III. Bacterial cell-wall replication followed by immunofluorescence. *Bact. Rev.* **29**, 326–44. [190, 191, 192]

Cole, R. M. & Hahn, J. J. (1962). Cell wall replication in *Streptococcus pyogenes*. *Science, N.Y.* **135**, 722–3. [190]

Collins, J. F. & Richmond, M. H. (1962). Rate of growth of *Bacillus cereus* between divisions. *J. gen. Microbiol.* **28**, 15–33. [21, 142]

Comings, D. E. (1967). Sex chromatin, nuclear size and the cell cycle. *Cytogenetics* **6**, 120–44. [181]

Comings, D. E. (1968). The rationale for an ordered arrangement of chromatin in the interphase nucleus. *Am. J. Hum. Genet.* **20**, 440–60. [112, 182]

Comings, D. E. (1969). Incorporation of tritium of ³H-arginine into DNA as the explanation of 'late synthesis of protein' on the human X chromosome. *Nature, Lond.* **221**, 570. [11, 155]

Comings, D. E. & Kakefuda, T. (1968). Initiation of deoxyribonucleic acid replication at the nuclear membrane in human cells. *J. molec. Biol.* **33**, 225–9. [112]

Comings, D. E. & Mattoccia, E. (1970). Replication of repetitious DNA and the S period. *Proc. natn. Acad. Sci. U.S.A.* **67**, 448–55. [76]

Cook, J. R. (1966). The synthesis of cytoplasmic DNA in synchronized *Euglena*. *J. Cell Biol.* **29**, 369–73. [71]

Cook, J. R. & James, T. W. (1964). Age distribution of cells in logarithmically growing cell populations. In *Synchrony in Cell Division and Growth*. pp. 485–95. Ed. by E. Zeuthen. New York: Interscience Publishers Inc. [3n, 18]

Cook, J. S. (1968). On the role of DNA in the ultraviolet-sensitivity of cell division in sand dollar zygotes. *Expl Cell Res.* **50**, 627–38. [241]

Coombs, J., Halicki, P. J., Holm-Hansen, O. & Volcani, B. E. (1967a). Studies on the biochemistry and fine structure of silica shell formation in diatoms. I. Changes in concentration of nucleoside triphosphates during synchronized division of *Cylindrotheca fusiformis* Reimann and Lewin. *Expl Cell Res.* **47**, 302–14. [39]

Coombs, J., Halicki, P. J., Holm-Hansen, O. & Volcani, B. E. (1967b). Studies on the biochemistry and fine structure of silica shell formation in diatoms. II. Changes in concentration of nucleoside triphosphates in silicon-starvation synchrony of *Navicula pelliculosa* (Bréb) Hilse. *Expl Cell Res.* **47**, 315–28. [36]

Cooper, S. & Helmstetter, C. E. (1968). Chromosome replication and the division cycle of *Escherichia coli* B/r. *J. molec. Biol.* **31**, 519–40. [102, 103, 105, 106]

Cota-Robles, E. H. (1963). Electron microscopy of plasmolysis in *Escherichia coli*. *J. Bact.* **85**, 499–503. [111]

Cottrell, S. F. & Avers, C. J. (1970). Evidence of mitochondrial synchrony in synchronous cell cultures of yeast. *Biochem, biophys. Res. Commun.* **38**, 973–80. [71, 177, 184, 195]

Cowie, D. B. & McLure, F. T. (1959). Metabolic pools and the synthesis of macromolecules. *Biochim. biophys. Acta* **31**, 236–45. [13, 15, 140]

Cowie, D. B. & Walton, B. P. (1956). Kinetics of formation and utilization of metabolic pools in the biosynthesis of protein and nucleic acid. *Biochim. biophys. Acta* **21**, 211–26. [16]

Cox, C. G. & Gilbert, J. B. (1970). Nonidentical times of gene expression in two strains of *Saccharomyces cerevisiae* with mapping differences. *Biochem. biophys. Res. Commun.* **38**, 750–7. [171, 177]

Crippa, M. (1966). The rate of ribonucleic acid synthesis during the cell cycle. *Expl Cell Res.* **42**, 371–5. [119]

Cummings, D. J. (1965). Macromolecular synthesis during synchronous growth of *Escherichia coli* B/r. *Biochim. biophys. Acta* **85**, 341–50. [51, 52, 125n, 162, 176]

Cummings, D. J. (1970). Synchronization of *E. coli* K12 by membrane selection. *Biochem. biophys. Res. Commun.* **41**, 471–6. [54]

Cummins, J. E. (1969). Nuclear DNA replication and transcription during the cell cycle of *Physarum*. In *The Cell cycle. Gene-enzyme interactions*. pp. 141–58. Ed. by G. M. Padilla, G. L. Whitson, and I. L. Cameron. New York and London: Academic Press. [122]

Cummins, J. E. & Mitchison, J. M. (1964). A method for making autoradiographs of yeast cells which retain pool components. *Expl Cell Res.* **34**, 406–9. [10]

Cummins, J. E. & Mitchsion. J. M. (1967). Adenine uptake and pool formation in the fission yeast *Schizosaccharomyces pombe*. *Biochim. biophys. Acta* **136**, 108–20. [13, 199]

Cummins, J. E. & Rusch, H. P. (1966). Limited DNA synthesis in the absence of protein synthesis in *Physarum polycephalum*. *J. Cell Biol.* **31**, 577–83. [87]

Cummins, J. E. & Rusch, H. P. (1967). Transcription of nuclear DNA isolated from plasmodia at different stages of the cell cycle of *Physarum polycephalum*. *Biochim. biophys. Acta* **138**, 124–32. [122]

Cummins, J. E., Blomquist, J. C. & Rusch, H. P. (1966). Anaphase delay after inhibition of protein synthesis between late prophase and prometaphase. *Science, N.Y.* **154**, 1343–4. [227]

Cummins, J. E., Weisfeld, G. E. & Rusch, H. P. (1966). Fluctuations of ^{32}P distribution in rapidly labelled RNA during the cell cycle of *Physarum polycephalum. Biochim. biophys. Acta* **129**, 240–8. [122, 123]

Curnutt, S. G. & Schmidt, R. R. (1964*a*). Possible mechanisms controlling the intracellular level of inorganic polyphosphate during synchronous growth of *Chlorella pyrenoidosa*. II. ATP/ADP ratio. *Biochim. biophys Acta* **86**, 201–3. [194, 198]

Curnutt, S. G. & Schmidt, R. R. (1964*b*). Possible mechanisms controlling the intracellular level of inorganic polyphosphate during synchronous growth of *Chlorella pyrenoidosa*. Endogenous respiration. *Expl Cell Res.* **36**, 102–10. [45]

Cutler, R. G. & Evans, J. E. (1966). Synchronization of bacteria by a stationary-phase method. *J. Bact.* **91**, 469–76. [36, 37, 38, 111, 125n, 126, 250]

Cutler, R. G. & Evans, J. E. (1967). Relative transcription activity of different segments of the genome through the cell division cycle of *Escherichia coli*. The mapping of ribosomal and transfer RNA and the determination of the direction of replication. *J. molec. Biol.* **26**, 91–105. [125n, 126]

Dadd, A. H. & Paulton, R. J. L. (1968). The cell wall of *Bacillus subtilis* during synchronous growth. *J. gen. Microbiol.* **54**, iii. [192]

Dan, K. (1954). The cortical movement in *Arbacia punctulata* eggs through cleavage cycles. *Embryologia* **2**, 115–22. [188]

Dan, K. (1966). Behaviour of sulphydryl groups in synchronous division. In *Cell Synchrony*, pp. 307–27. Ed. by I. L. Cameron and G. M. Padilla. New York and London: Academic Press. [213, 214]

Danielli, J. F. (1958). *General Cytochemical Methods*. Vol. I. New York and London: Academic Press. [10]

Danielli, J. F. (1961). *General Cytochemical Methods*. Vol. II. New York and London: Academic Press. [10]

Daniels, J. W. & Baldwin, H. H. (1964). Methods of culture for plasmodial myxomycetes. *Methods in Cell Physiology*. Vol. I. pp. 9–41. Ed. by D. M. Prescott. New York and London: Academic Press. [56]

Daniels, M. J. (1969). Lipid synthesis in relation to the cell cycle of *Bacillus megaterium* KM and *Escherichia coli. Biochem. J.* **115**, 697–701. [186]

Darnell, J. E. (1968). Ribonucleic acids from animal cells. *Bact. Rev.* **32**, 262–90. [115, 119]

Das, N. K. (1963). Chromosomal and nucleolar RNA synthesis in root tips during mitosis. *Science, N.Y.* **140**, 1231–3. [116n]

Das, N. K. & Alfert, M. (1968). Cytochemical studies on the concurrent synthesis of DNA and histone in primary spermatocytes of *Urechis caupo. Expl Cell Res.* **49**, 51–8. [154]

Das, N. K., Siegel, E. P. & Alfert, M. (1965). On the origin of labelled RNA in the cytoplasm of mitotic root tip cells of *Vicia faba. Expl Cell Res.* **40**, 178–81. [116]

Davidson, D. (1964). RNA synthesis in roots of *Vicia faba. Expl Cell Res.* **35**, 317–25. [116n, 121]

Davies, H. G. (1958). The determination of mass and concentration by microscope interferometry. In *General Cytochemical Methods* Vol. I, pp. 57–162. Ed. by J. F. Danielli. New York and London: Academic Press. [9]

Davies, H. G. & Deeley, E. M. (1956). An integrator for measuring the 'dry mass' of cells and isolated components. *Expl Cell Res.* **11**, 169–85. [9]

Davies, L. M., Priest, J. H. & Priest, R. E. (1968). Collagen synthesis by cells synchronously replicating DNA. *Science, N.Y.* **159**, 91. [157]

Dawson, P. S. S. (1965). Continuous phased growth with a modified chemostat. *Can. J. Microbiol.* **11**, 893–903. [38]

Defendi, V. & Manson, L. A. (1963). Analysis of the life cycle in mammalian cells. *Nature, Lond.* **198**, 359–61. [60]

Dendy, P. P. & Cleaver, J. E. (1964). An investigation of (*a*) variation in rate of DNA-synthesis during S-phase in mouse L-cells (*b*) effect of ultraviolet radiation on rate of DNA-synthesis. *Int. J. Radiat. Biol.* **8**, 301–15. [60, 81]

De Terra, N. (1967). Macronuclear DNA synthesis in *Stentor*: regulation by a cytoplasmic initiator. *Proc. natn. Acad. Sci. U.S.A.* **57**, 607–14. [69, 85]

De Terra, N. (1969). Cytoplasmic control over the nuclear events of cell reproduction. *Int. Rev. Cytol.* **25**, 1–29. [83]

Devi, V. R., Guttes, E. & Guttes, S. (1968). Effects of ultraviolet light on mitosis in *Physarum polycephalum. Expl Cell Res.* **50**, 589–98. [239, 248]

Dewey, D. L. & Howard, A. (1963). Cell dynamics in the bean root tip. *Radiat. Bot.* **3**, 259–63. [66]

Dick, C. & Johns, E. W. (1968). The effect of two acetic acid containing fixatives on the histone content of calf thymus deoxyribonucleoprotein and calf thymus tissue. *Expl Cell Res.* **51**, 626–32. [155]

Diers, L. (1968). On the plastids, mitochondria, and other cell constituents during oogenesis of a plant. *J. Cell Biol.* **28**, 527–43. [183]

Doida, Y. & Okada, S. (1967*a*). Synchronization of L5178Y cells by successive treatment with excess thymidine and colcemid. *Expl Cell Res.* **48**, 540–8. [28]

Doida, Y. & Okada, S. (1967*b*). Determination of switching-off and switching-on time of overall RNA synthesis. *Nature, Lond.* **216**, 272–3. [116]

Doida, Y. & Okada, S. (1969). Radiation-induced mitotic delay in cultured mammalian cells (L5178Y). *Radiat. Res.* **38**, 513–29. [222, 234, 242, 243]

Donachie, W. D. (1964). The regulation of pyrimidine biosynthesis in *Neurospora crassa*. II. Heterokaryons and the role of the 'regulatory mechanisms'. *Biochim. biophys. Acta* **82**, 293–302. [166]

Donachie, W. D. (1965). Control of enzyme steps during the bacterial cell cycle. *Nature, Lond.* **205**, 1084–6. [107, 111, 174, 176]

Donachie, W. D. (1968). Relationship between cell size and the time of initiation of DNA replication. *Nature, Lond.* **219**, 1077–9. [88, 108, 109]

Donachie, W. D. (1969). Control of cell division in *Escherichia coli*: experiments with thymine starvation. *J. Bact.* **100**, 260–8. [108, 110]

Donachie, W. D. & Begg, K. J. (1970). Growth of the bacterial cell. *Nature, Lond.* **227**, 1220–4. [188]

Donachie, W. D. & Masters, M. (1966). Evidence for polarity of chromosome replication in F⁻ strains of *Escherichia coli. Genet. Res.* **8**, 119–24. [49, 165, 180]

Donachie, W. D. & Masters, M. (1969). Temporal control of gene expression in bacteria. In *The cell cycle. Gene-enzyme interactions*, pp. 37–76. Ed. by G. M. Padilla, G. L. Whitson and I. L. Cameron. New York and London: Academic Press. [99, 108, 159, 165, 166, 169n, 174, 176, 180]

Donachie, W. D., Hobbs, D. G. & Masters, M. (1968). Chromosome replication and cell division in *Escherichia coli* 15T after growth in the absence of DNA synthesis. *Nature, Lond.* **219**, 1079–80. [108]

Donnelly, G. M. & Sisken, J. E. (1967). RNA and protein synthesis required for entry of cells into mitosis and during the mitotic cycle. *Expl Cell Res.* **46**, 93–105. [224]

Doudney, C. (1960). Inhibition of nucleic acid synthesis by chloramphenicol in synchronized cultures of *Escherichia coli. J. Bact.* **79**, 122–4. [40n]

Dubnau, D., Smith, I. & Marmur, J. (1965). Gene conservation in *Bacillus* species. II. The location of genes concerned with the synthesis of ribosomal components and soluble RNA. *Proc. natn. Acad. Sci. U.S.A.* **54**, 724–30. [126]

Duffus, J. H. (1971). The isolation and properties of nucleohistone from the fission yeast *Schizosaccharomyces pombe*. *Biochim. biophys. Acta* **228**, 627–35. [155n]

√ Duffus, J. H. & Mitchell, C. J. (1970). Effect of high osmotic pressure on DNA synthesis in the fission yeast, *Schizosaccharomyces pombe*. *Expl Cell Res.* **61**, 213–16. [70]

Duynstee, E. E. & Schmidt, R. R. (1967). Total starch and amylase levels during synchronous growth of *Chlorella pyrenoidosa*. *Archs Biochem. Biophys.* **119**, 382–6. [162n]

Eberle, H. & Lark, K. G. (1967). Chromosome replication in *Bacillus subtilis* cultures growing at different rates. *Proc. natn. Acad. Sci. U.S.A.* **57**, 95–101. [107]

Ecker, R. E. & Kokaisl, G. (1969). Synthesis of protein, ribonucleic acid, and ribosomes by individual bacterial cells in balanced growth. *J. Bact.* **98**, 1219–26. [125n, 144]

Eckstein, H., Paduch, V. & Hilz, H. (1966). Teilungssynchronisierte Hefezellen. II. Enzymsynthese nach Hemmung der Zellteilung durch Röntgenstrahlen. *Biochem. Z.* **344**, 435–45. [168, 177, 195, 248]

Eckstein, H., Paduch, V. & Hilz, H. (1967). Synchronized yeast cells. 3. DNA synthesis and DNA polymerase after inhibition of cell division by X-rays. *Europ. J. Biochem.* **3**, 224–31. [168, 177]

Edmunds, L. N. (1965). Studies on synchronously dividing cultures of *Euglena gracilis* Klebs (Strain 2). I. Attainment and characterization of rhythmic cell division. *J. cell. comp. Physiol.* **66**, 147–58. [45]

Edmunds, L. N. (1966). Studies on synchronously dividing cultures of *Euglena gracilis* Klebs (Strain 2). III. Circadian components of cell division. *J. Cell Physiol.* **67**, 35–44. [46]

Edmunds, L. N. & Funch, R. R. (1969a). Circadian rhythm of cell division in *Euglena*: effects of a random illumination regimen. *Science, N.Y.* **165**, 500–3. [46]

Edmunds, L. N. & Funch, R. (1969b). Effects of 'skeleton' photoperiods and high frequency light–dark cycles on the rhythm of cell division in synchronized cultures of *Euglena*. *Planta* **87**, 134–63. [46]

Edward, J. L., Koch, A. L., Youcis, P., Freese, H. L., Laite, M. B. & Donalson, J. T. (1960). Some characteristics of DNA synthesis and the mitotic cycle in Ehrlich ascites tumor cells. *J. biophys. biochem. Cytol.* **7**, 273–82. [81n]

Elkind, M. M., Han, A. & Volz, K. W. (1963). Radiation responses of mammalian cells grown in culture. IV. Dose dependence of division delay and postirradiation growth of surviving and nonsurviving Chinese hamster cells. *J. natl. Cancer Inst.* **30**, 705–21. [234n, 236]

Engelberg, J. (1964a). Measurement of the degree of synchrony in cell populations. In *Synchrony in Cell Division and Growth*, pp. 497–508. Ed. by E. Zeuthen. New York: Interscience. [56]

Engelberg, J. (1964b). The decay of synchronization of cell division. *Expl Cell Res.* **36**, 647–62. [3n, 57]

Engelberg, J. & Hirsch, H. R. (1966). On the theory of synchronous cultures. In *Cell Synchrony*, pp. 14–37. Ed. by I. L. Cameron and G. M. Padilla. New York and London: Academic Press. [56]

Enger, M. D. & Tobey, R. A. (1969). RNA synthesis in Chinese hamster cells. II. Increase in rate of RNA synthesis during G_1. *J. Cell Biol.* **42**, 308–15. [48, 119, 120]

Epel, D. (1963). The effects of carbon monoxide inhibition on ATP level and the rate of mitosis in the sea urchin egg. *J. Cell Biol.* **17**, 315–19. [194, 198, 230]

Erickson, R. O. (1947). Respiration of developing anthers. *Nature, Lond.* **159**, 275–6. [194]

Erickson, R. O. (1964). Synchronous cell and nuclear division in tissues of the higher plants. In *Synchrony in Cell Division and Growth*, pp. 11–37. Ed by E. Zeuthen. New York: Interscience. [55]

Eriksson, T. (1966). Partial synchronization of cell division in suspension cultures of *Haplopappus gracilis*. *Physiologia Pl.* **19**, 900–10. [28]

Erlandson, R. A. & De Harven, E. (1971). The ultrastructure of synchronized HeLa cells. *J. Cell Sci.* **8**, 353–97. [254]

Esposito, R. E. (1968). Genetic recombination in synchronized cultures of *Saccharomyces cerevisiae*. *Genetics, Princeton* **59**, 191–210. [28]

Evans, H. J. (1959). Nuclear behaviour in the cultivated mushroom. *Chromosoma* **10**, 115–35. [84]

Evans, H. J. (1964). Uptake of ^{3}H-thymidine and patterns of DNA replication in nuclei and chromosomes of *Vicia faba*. *Expl Cell Res.* **35**, 381–93. [16, 73n, 81]

Evans, G. M. & Rees, H. (1966). The pattern of DNA replication at mitosis in the chromosomes of *Scilla campanulata*. *Expl Cell Res.* **44**, 150–60. [73n]

Evans, H. J. & Scott, D. (1964). Influence of DNA synthesis on the production of chromatid aberrations by X-rays and maleic hydrazide in *Vicia faba*. *Genetics, Princeton* **49**, 17–38. [66]

Evenson, D. P. & Prescott, D. M. (1970). RNA metabolism in the macronucleus of *Euplotes eurystomus* during the cell cycle. *Expl Cell Res.* **61**, 71–8. [123]

Falcone, G. & Szybalski, W. (1965). Biochemical studies on the induction of synchronized cell division. *Expl Cell Res.* **11**, 486–9. [40n]

Fan, H. & Penman, S. (1970a). Mitochondrial RNA synthesis during mitosis. *Science, N.Y.* **168**, 135–8. [117]

Fan, H. & Penman, S. (1970b). Regulation of protein synthesis in mammalian cells. II. Inhibition of protein synthesis at the level of initiation during mitosis. *J. molec. Biol.* **50**, 655–70. [133]

Fangman, W. L., Gross, C. & Novick, A. (1967). Continuation of basal enzyme synthesis after cessation of DNA synthesis. *J. molec. Biol.* **29**, 317–20. [168]

Feinendegen, L. E. (1967). *Tritium-labelled molecules in biology and medicine*. New York and London: Academic Press. [10, 16]

Feinendegen, L. E. & Bond, V. P. (1963). Observations on nuclear RNA during mitosis in human cancer cells (HeLa-S$_3$) studied with tritiated cytidine. *Expl Cell Res.* **30**, 393–404. [116n]

Feinendegen, L. E., Bond, V. P., Shreeve, W. W. & Painter, R. B. (1960). RNA and DNA metabolism in human tissue culture cells studied with tritiated cytidine. *Expl Cell Res.* **19**, 443–59. [116n]

Ferretti, J. J. & Gray, E. D. (1968). Enzyme and nucleic acid formation during synchronous growth of *Rhodopseudomonas spheroides*. *J. Bact.* **95**, 1400–6. [111, 125n, 126, 144, 176]

Firket, H. & Mahieu, P. (1966). Synchronisme des divisions induit dans des cellules HeLa par un excès de thymidine. Etude des perturbations éventuelles du cycle cellulaire. *Expl Cell Res.* **45**, 11–22. [26]

Firket, H. & Verly, W. G. (1957). Autoradiographic visualization of synthesis of deoxyribonucleic acid in tissue culture with tritium labelled thymidine. *Nature, Lond.* **181**, 274–5. [63]

Flamm, W. G., Bernheim, N. J. & Brubacker, P. E. (1971). Density gradient analysis of newly replicated DNA from synchronized mouse lymphoma cells. *Expl Cell Res.* **64**, 97–104. [76, 251]

Fleming, L. W. & Duerksen, J. D. (1967). Evidence for multiple molecular forms of yeast β-glucosidase in a hybrid yeast. *J. Bact.* **93**, 142–50. [171]

Flickinger, C. J. (1967). The fine structure of the nuclei of *Tetrahymena pyriformis* throughout the cell cycle. *J. Cell Biol.* **27**, 519–29. [181]

Fox, T. O., Sheppard, J. R. & Burger, M. M. (1971). Cyclic membrane changes in animal cells: transformed cells permanently display a surface architecture detected in normal cells only in mitosis. *Proc. natn. Acad. Sci. U.S.A.* **68**, 244–7. [255]

Frankel, J. (1961). Spontaneous astomy: loss of oral areas in *Glaucoma chattoni. J. Protozool.* **8**, 250–6. [212]

Frankel, J. (1962). The effects of heat, cold and *p*-fluorophenylalanine on morphogenesis in synchronized *Tetrahymena pyriformis* GL. *C. r. Trav. Lab. Carlsberg* **33**, 1–52. [206, 211, 212]

Frankel, J. (1965). The effect of nucleic acid antagonists on cell division and oral organelle development in *Tetrahymena pyriformis. J. exp. Zool.* **159**, 113–48. [210]

Frankel, J. (1967a). Studies on the maintenance of development in *Tetrahymena pyriformis* GL-C. I. An analysis of the mechanism of resorption of developing oral structures. *J. exp. Zool.* **164**, 435–60. [211]

Frankel, J. (1967b). Critical phases of oral primordium development in *Tetrahymena pyriformis* GL-C: an analysis employing low temperature treatment. *J. Protozool.* **14**, 639–49. [211]

Frankel, J. (1967c). Studies on the maintenance of oral development in *Tetrahymena pyriformis* GL-C. II. The relationship of protein synthesis to cell division and oral organelle development. *J. Cell Biol.* **34**, 841–58. [209, 211]

Frankel, J. (1969). The relationship of protein synthesis to cell division and oral development in synchronized *Tetrahymena pyriformis* GL-C: an analysis employing cyclo-heximide. *J. Cell Physiol.* **74**, 135–48. [209]

Frankel, J. (1970). The synchronization of oral development without cell division in *Tetrahymena pyriformis* GL-C. *J. exp. Zool.* **173**, 79–100. [204]

Frankel, J. & Williams, N. E. (1971). Cortical development in *Tetrahymena.* In *The Biology of Tetrahymena.* Ed. by A. M. Elliott. New York: Appleton, Century and Crofton. [201n]

Frankfurt, O. S. (1968). Effect of hydrocortisone, adrenalin and actinomycin D on transition of cells to the DNA synthesis phase. *Expl Cell Res.* **52**, 222–32. [88]

Fraser, R. S. S., Loening, U. E. & Yeoman, M. M. (1967). Effect of light on cell division in plant tissue cultures. *Nature, Lond.* **215**, 873. [34]

Friedberg, S. H. & Davidson, D. (1970). Duration of S phase and cell cycles in diploid and tetraploid cells of mixoploid meristem. *Expl Cell Res.* **61**, 216–18. [82]

Friedman, D. L. (1970). DNA polymerase from HeLa cell nuclei: levels of activity during a synchronized cell cycle. *Biochem. biophys. Res. Commun.* **39**, 100–9. [179]

Friedman, D. L. & Mueller, G. C. (1968). A nuclear system for DNA replication from synchronized HeLa cells. *Biochim. biophys. Acta* **161**, 455–68. [85, 179]

Froese, G. (1966). Division delay in HeLa cells and Chinese hamster cells. A time lapse study. *Int. J. Radiat. Biol.* **10**, 353–67. [235]

Froland, A. (1967). Internal asynchrony in late replicating X chromosomes. *Nature, Lond.* **213**, 512–13. [73n]

Frydenberg, O. & Zeuthen, E. (1960). Oxygen uptake and carbon dioxide output related to the mitotic rhythm in the cleaving eggs of *Dendraster excentricus* and *Urechis caupo. C. r. Trav. Lab. Carlsberg* **31**, 423–55. [9, 192]

Fujiwara, Y. (1967). Role of RNA synthesis in DNA replication of synchronized populations of cultured mammalian cells. *J. Cell. comp. Physiol.* **70**, 291–300. [60, 61, 88, 119]

Fulton, C. & Guerrini, A. M. (1969). Mitotic synchrony in *Naegleria* amoebae. *Expl Cell Res.* **56**, 194–200. [39]

Fulwyler, M. J. (1965). Electronic separation of biological cells by volume. *Science, N.Y.* **150**, 910–11. [55]

Gaffney, E. V. & Nardone, R. M. (1968). Nucleolar RNA synthesis in synchronous cultures of strain L-929. *Expl Cell Res.* **53**, 410–16. [61, 119]

Galavazi, G. & Bootsma, D. (1966). Synchronization of mammalian cells *in vitro* by inhibition of the DNA synthesis. II. Population dynamics. *Expl Cell Res.* **41**, 438–51. [26]

Galavazi, G., Schenk, H. & Bootsma, D. (1966). Synchronization of mammalian cells *in vitro* by inhibition of DNA synthesis. I. Optimal condition. *Expl Cell Res.* **41**, 428–37. [26]

Gall, J. G. (1959). Macronuclear duplication in the ciliated Protozoan *Euplotes*. *J. biophys. biochem. Cytol.* **5**, 295–308. [153]

Gallwitz, D. & Mueller, G. C. (1969). Histone synthesis *in vitro* on HeLa cell microsomes. The nature of the coupling to deoxyribonucleic acid synthesis. *J. biol. Chem.* **244**, 5947–52. [155]

Gamow, E. I. & Prescott, D. M. (1970). The cell life cycle during embryogenesis of the mouse. *Expl Cell Res.* **59**, 117–23. [66]

Ganesan, A. T. & Lederberg, J. (1965). A cell-membrane bound fraction of bacterial DNA. *Biochem. biophys. Res. Commun.* **18**, 824–35. [100]

Gaulden, M. E. (1956). DNA synthesis and X-ray effects at different mitotic stages in grasshopper neuroblasts. *Genetics, Princeton* **41**, 645. [63]

Gavosto, F., Pegoraro, L., Masera, P. & Rovera, G. (1968). Late DNA replication pattern in human haemopoietic cells. A comparative investigation using a high resolution quantitative autoradiography. *Expl Cell Res.* **49**, 340–58. [73n, 74, 75]

Geilenkirchen, W. L. M. (1964). The cleavage schedule and the development of *Arbacia* eggs as separately influenced by heat shocks. *Biol. Bull. mar. biol. Lab., Woods Hole* **127**, 370. [232]

Geilenkirchen, W. L. M. (1966). Cell division and morphogenesis of *Limnaea* eggs after treatment with heat pulses at successive stages in early division cycle. *J. Embryol. exp. Morph.* **16**, 321–7. [232]

Gelbard, A. S., Kim, J. H. & Perez, A. G. (1969). Fluctuation in deoxycytidine monophosphate deaminase activity during the cell cycle in synchronous populations of HeLa cells. *Biochim. biophys. Acta* **182**, 564–6. [168, 179]

Gentry, G. A., Morse, P. A. & Potter, V. R. (1965). Pyrimidine metabolism in tissue cultures derived from rat hepatomas. III. Relationship of thymidine to the metabolism of other pyrimidine nucleosides in suspension cultures derived from the Novikoff hepatoma. *Cancer Res.* **25**, 517–24. [26]

German, J. (1964). The pattern of DNA synthesis in the chromosomes of human blood cells. *J. Cell Biol.* **20**, 37–55. [73n]

Gerner, E. W., Glick, M. C. & Warren, L. (1970). Membrane of animal cells. V. Biosynthesis of the surface membrane during the cell cycle. *J. Cell Physiol.* **75**, 275–80. [186]

Gerschenson, L. E., Strasser, F. F. & Rounds, D. E. (1965). Variations in the nucleoside triphosphate content and oxygen uptake of synchronized HeLa S3 cultures. *Life Sciences* **4**, 927–35. [194, 199]

Giese, A. C. (1946). Comparative sensitivity of sperm and eggs to ultraviolet radiations. *Biol. Bull. mar. biol. Lab., Woods Hole* **91**, 81–7. [241]

Gilbert, C. W., Muldal, S. & Lajtha, L. G. (1965). Rate of chromosome duplication at the end of the deoxyribonucleic acid synthetic period in human blood cells. *Nature, Lond.* **208**, 159–61. [73n, 75]

Gill, B. F. (1965). The effects of ultraviolet radiation during the cell cycle. *Ph.D. thesis*, Univ. of Edinburgh. [237, 238]

Giménez-Martín, G., González-Fernández, A. & Lopez-Sáez, J. F. (1966). Duration of the division cycle in diploid, binucleate and tetraploid cells. *Expl Cell Res.* **43**, 293–300. [24]

Glick, D. (1949). *Techniques of histo- and cytochemistry.* New York: Interscience. [10]

Glick, D. (1963). *Quantitative chemical techniques of histo- and cytochemistry.* Vol. 2. New York: Interscience. [10]

Glick, D., Von Redlich, D., Juhos, E. Th. & McEwen, C. R. (1971). Separation of mast cells by centrifugal elutriation. *Expl Cell Res.* **65**, 23–6. [250]

Glick, M. C., Gerner, E. W. & Warren, L. (1971). Changes in the carbohydrate content of the KB cell during the growth cycle. *J. Cell Physiol.* **77**, 1–6. [254]

Gold, M. & Helleiner, C. W. (1964). Deoxyribonucleic acid polymerase in L cells. I. Properties of the enzyme and its activity in synchronized cell cultures. *Biochim. biophys. Acta* **80**, 193–203. [178]

Goldstein, L. & Prescott, D. M. (1967). Proteins in nucleocytoplasmic interactions. I. The fundamental characteristics of the rapidly migrating proteins and the slow turnover proteins of the *Amoeba proteus* nucleus. *J. Cell Biol.* **33**, 637–44. [150, 156]

Goldstein, L. & Prescott, D. M. (1968). Proteins in nucleocytoplasmic interactions. II. Turnover and changes in nuclear protein distribution with time and growth. *J. Cell Biol.* **36**, 53–62. [150]

Goldstein, L. & Trescott, O. H. (1970). Characterization of RNAs that do and do not migrate between cytoplasm and nucleus. *Proc. natn. Acad. Sci. U.S.A.* **67**, 1367–74. [253]

Goldstein, L., Rao, M. V. N. & Prescott, D. M. (1969). The migration of RNA from cytoplasm to nucleus in *Amoebae proteus. Annales d'embryologie et d'morphogenèse* Suppl. 1. 189–97. [121]

González-Fernández, A., López-Sáez, J. F. & Giménez-Martín, G. (1966). Duration of the division cycle in binucleate and mononucleate cells. *Expl Cell Res.* **43**, 255–67. [24]

González, P. & Nardone, R. M. (1968). Cyclic nucleolar changes during the cell cycle. I. Variation in number, size, morphology and position. *Expl Cell Res.* **50**, 599–615. [181]

Goodwin, B. C. (1966). An entrainment model for timed enzyme synthesis in bacteria. *Nature, Lond.* **209**, 479–81. [169]

Goodwin, B. C. (1969a). Synchronization of *E. coli* B in chemostat by periodic phosphate feeding. *Europ. J. Biochem.* **10**, 511–14. [38, 164]

Goodwin, B. C. (1969b). Control dynamics of β-galactosidase in relation to the bacterial cell cycle. *Europ. J. Biochem.* **10**, 515–22. [38, 164]

Goodwin, B. C. (1969c). Growth dynamics and synchronization of cells. *Symp. Soc. gen. Microbiol.* **19**, 223–36. [169n]

Gorman, J., Tauro, P., La Berge, M. & Halvorson, H. (1964). Timing of enzyme synthesis during synchronous division in yeast. *Biochem. biophys. Res. Commun.* **15**, 43–9. [141, 170n, 171, 177]

Graham, C. F. (1966a). The regulation of DNA synthesis and mitosis in multinucleate frog eggs. *J. Cell Biol.* **1**, 363–74. [64, 249]

Graham, C. F. (1966b). The effect of cell size and DNA content on the cellular regulation of DNA synthesis in haploid and diploid embryos. *Expl Cell Res.* **43**, 13–19. [82, 83]

Graham, C. F. (1971). Nucleic acid metabolism during early mammalian development. In *The Regulation of Mammalian Reproduction.* National Institutes of Health; USA. In press. [66]

Graham, C. F. & Morgan, R. W. (1966). Changes in the cell cycle during early Amphibian development. *Devl Biol.* **14**, 439–60. [64, 65]

Graham, C. F., Arms, K. & Gurdon, J. B. (1966). The induction of DNA synthesis by frog egg cytoplasm. *Devl Biol.* **14**, 349–81. [82, 84]

Granick, S. (1961). The chloroplasts: inheritance, structure and function. In *The Cell*

Vol. 2, pp. 489–602. Ed. by J. Brachet and A. E. Mirsky. New York and London: Academic Press. [183]

Graves, J. A. M. (1967). DNA synthesis in chromosomes of cultured leucocytes from two marsupial species. *Expl Cell Res.* **46**, 37–57. [73n]

Green, P. B. (1964). Cinematic observations on the growth and division of chloroplasts in *Nitella*. *Am. J. Bot.* **51**, 334–42. [183]

Greulich, R. C., Cameron, I. L. & Thrasher, J. D. (1961). Stimulation of mitosis in adult mice by administration of thymidine. *Proc. natn. Acad. Sci. U.S.A.* **47**, 743–8. [22]

Griffin, M. J. & Ber, R. (1969). Cell cycle events in the hydrocortisone regulation of alkaline phosphatase in HeLa S3 cells. *J. Cell Biol.* **40**, 297–304. [179]

Griffith, J. S. (1968). Mathematics of cellular control processes. I. Negative feedback to one gene. *J. theor. Biol.* **20**, 202–8. [170]

Grivell, A. R. & Jackson, J. F. (1968). Thymidine kinase: evidence for its absence from *Neurospora crassa* and some other micro-organisms, and the relevance of this to the specific labelling of deoxyribonucleic acid. *J. gen. Microbiol.* **54**, 307–17. [11]

Gross, P. R. & Fry, B. J. (1966). Continuity of protein synthesis through cleavage metaphase. *Science, N.Y.* **153**, 749–51. [132]

Grossman, L. I., Goldring, E. S. & Marmur, J. (1969). Preferential synthesis of yeast mitochondrial DNA in the absence of protein synthesis. *J. molec. Biol.* **46**, 367–76. [86]

Grumbach, M. M., Morishima, A. & Taylor, J. H. (1963). Human sex chromosomal abnormalities in relation to DNA replication and heterochromatization. *Proc. natn. Acad. Sci. U.S.A.* **49**, 581–9. [73n]

Gurdon, J. B. & Woodland, H. R. (1968). The cytoplasmic control of nuclear activity in animal development. *Biol. Rev.* **43**, 233–67. [84, 87, 151]

Gurley, L. R. & Hardin, J. M. (1968). The metabolism of histone fractions. I. Synthesis of histone fractions during the life cycle of mammalian cells. *Archs Biochem. Biophys.* **128**, 285–92. [155n]

Gurley, L. R. & Hardin, J. M. (1969). The metabolism of histone fractions. II. Conservation and turnover of histone fractions in mammalian cells. *Archs Biochem. Biophys.* **130**, 1–6. [156]

Guttes, E. & Guttes, S. (1960). Incorporation of tritium-labelled thymidine into the macronucleus of *Stentor coeruleus*. *Expl Cell Res.* **19**, 626–8. [69]

Guttes, E. & Guttes, S. (1964). Mitotic synchrony in the plasmodia of *Physarum polycephalum* and mitotic synchronization by coalescence of microplasmodia. In *Methods in Cell Physiology*. Vol. 1, pp. 43–54. Ed. by D. M. Prescott. New York and London: Academic Press. [56]

Guttes, E. & Guttes, S. (1969). Replication of nucleolus-associated DNA during the 'G2 phase' in *Physarum polycephalum*. *J. Cell Biol.* **43**, 229–36. [83]

Guttes, E., Guttes, S. & Devi, R. V. (1969). Division stages of the mitochondria in normal and actinomycin-treated plasmodia of *Physarum polycephalum*. *Experientia* **25**, 66–8. [183]

Guttes, E. W., Hanawalt, P. C. & Guttes, S. (1967). Mitochondrial DNA synthesis and the mitotic cycle in *Physarum polycephalum*. *Biochim. biophys. Acta* **142**, 181–94. [71]

Guttes, S. & Guttes, E. (1968). Regulation of DNA replication in the nuclei of the slime mold *Physarum polycephalum*: transplantation of nuclei by plasmodial coalescence. *J. Cell Biol.* **37**, 761–72. [85]

Halvorson, H. O., Bock, R. M., Tauro, P., Epstein, R. & La Berge, M. (1966). Periodic enzyme synthesis in synchronous cultures of yeast. In *Cell Synchrony*, pp. 102–16. Ed. by I. L. Cameron, and G. M. Padilla. New York and London: Academic Press. [170, 171]

Halvorson, H. O., Carter, B. L. A. & Tauro, P. (1971). Synthesis of enzymes during the cell cycle. *Adv. microb. Physiol.* **6**. In press. [50, 159, 162, 163, 170n]

Halvorson, H., Gorman, J., Tauro, P., Epstein, R. & La Berge, M. (1964). Control of enzyme synthesis in synchronous cultures of yeasts. *Fedn Proc. Am. Socs. exp. Biol.* **23**, 1002–8. [170n]

Hamburger, K. (1962). Division delays induced by metabolic inhibitors in synchronized cells of *Tetrahymena pyriformis*. *C. r. Trav. Lab. Carlsberg* **32**, 359–70. [232]

Hamilton, A. I. (1969). Cell population kinetics: a modified interpretation of the graph of labelled mitosis. *Science, N.Y.* **164**, 952–4. [81n]

Hanawalt, P. C., Maaløe, O., Cummings, D. J., & Schaechter, M. (1961). The normal DNA replication cycle. II. *J. molec. Biol.* **3**, 156–65. [95]

Hancock, R. (1969). Conservation of histones in chromatin during growth and mitosis *in vitro*. *J. molec. Biol.* **40**, 457–66. [155n]

Hanson, E. D. & Kaneda, M. (1968). Evidence for sequential gene action with the cell cycle of *Paramecium*. *Genetics, Princeton* **60**, 793–805. [225, 226]

Hardin, J. A., Einem, G. E. & Lindsay, D. T. (1967). Simultaneous synthesis of histone and DNA in synchronously dividing *Tetrahymena pyriformis*. *J. Cell Biol.* **32**, 709–17. [155n]

Hardy, C. & Binkley, S. B. (1967). The effects of *p*-fluorophenylalanine on nucleic acid biosynthesis and cell division in *Escherichia coli*. *Biochemistry, N.Y.* **6**, 1892–8. [108]

Hare, T. A. & Schmidt, R. R. (1965). Nitrogen metabolism during synchronous growth of *Chlorella pyrenoidosa*. 1. Protein amino acid distribution. *J. cell. comp. Physiol.* **65**, 63–8. [195]

Hare, T. A. & Schmidt, R. R. (1968). Continuous-dilution method for the mass cultures of synchronized cells. *Appl. Microbiol.* **16**, 496–9. [44]

Hare, T. A. & Schmidt, R. R. (1970). Nitrogen metabolism during synchronous growth of *Chlorella*. II. Free-, peptide-, and protein-amino acid distribution. *J. Cell Physiol.* **75**, 73–82. [195, 197, 198]

Harris, H. (1966). Hybrid cells from mouse and man: a study in genetic regulation. *Proc. R. Soc. B* **166**, 358–68. [84]

Harris, H. (1967). The reactivation of the red cell nucleus. *J. Cell Sci.* **2**, 23–32. [84, 151]

Harris, H. (1970). *Cell fusion*. Oxford: University Press. [84]

Harris, H. & La Cour, L. F. (1963). Site of synthesis of cytoplasmic ribonucleic acid. *Nature, Lond.* **200**, 227–9. [116]

Harris, J. W. & Patt, H. M. (1969). Non-protein sulphydryl content and cell cycle dynamics of Ehrlich ascites tumor. *Expl Cell Res.* **56**, 134–41. [199]

Hartwell, L. H. (1970). Periodic density fluctuation during the yeast cell cycle and the selection of synchronous cultures. *J. Bact.* **104**, 1280–5. [250]

Harvey, E. B. (1956). *The American Arbacia and other sea urchins*. Princeton, N.J.: Princeton Univ. Press. [55]

Harvey, R. J. (1968). Measurement of cell volumes by electric sensing zone instruments. In *Methods in Cell Physiology*. Vol. 3, pp. 1–24. Ed. by D. M. Prescott. New York and London: Academic Press. [9]

Harvey, R. J., Marr, A. G. & Painter, P. R. (1967). Kinetics of growth of individual cells of *Escherichia coli* and *Azotobacter agilis*. *J. Bact.* **93**, 605–17. [142]

Hawley, E. S. & Wagner, R. P. (1967). Synchronous mitochondrial division in *Neurospora crassa*. *J. Cell Biol.* **35**, 489–99. [184]

Hedley, R. H. & Wakefield, J. St. J. (1968). Formation of mitochondria in *Boderia* (Protozoa: Foraminifera). *Z. Zellforsch. mikrosk. Anat.* **87**, 429–34. [183]

Hegarty, C. P. & Weeks, O. B. (1940). Sensitivity of *Escherichia coli* to cold-shock during the logarithmic growth phase. *J. Bact.* **39**, 475–84. [40]

Hegner, R. W. & Wu, H-F. (1921). An analysis of the relation between growth and nuclear division in a parasitic infusorian, *Opalina* sp. *Am. Nat.* **55**, 335–46. [84]

Helmstetter, C. E. (1967). Rates of DNA synthesis during the division cycle of *Escherichia coli* B/r. *J. molec. Biol.* **24**, 417–27. [40, 52, 53, 54, 104]

Helmstetter, C. E. (1968). Origin and sequence of chromosome replication in *Escherichia coli* B/r. *J. Bact.* **95**, 1634–41. [54, 99, 165, 180]

Helmstetter, C. E. (1969a). Regulation of chromosome replication and cell division in *Escherichia coli*. In *The Cell Cycle. Gene-enzyme interactions*, pp. 15–35. Ed. by G. M. Padilla, G. L. Whitson and I. L. Cameron. New York and London: Academic Press. [91, 102]

Helmstetter, C. E. (1969b). Sequence of bacterial reproduction. *A. Rev. Microbiol.* **23**, 223–38. [91]

Helmstetter, C. E. (1969c). Method for studying the microbial division cycle. In *Methods in Microbiology* Vol. 1. pp. 327–63. Ed. by J. R. Norris and D. W. Ribbons. New York and London: Academic Press. [25]

Helmstetter, C. E. & Cooper, S. (1968). DNA synthesis during the division cycle of rapidly growing *Escherichia coli* B/r. *J. molec. Biol.* **31**, 507–18. [104]

Helmstetter, C. E. & Cummings, D. J. (1963). Bacterial synchronization by selection of cells at division. *Proc. natn. Acad. Sci. U.S.A.* **50**, 767–74. [51]

Helmstetter, C. E. & Cummings, D. J. (1964). An improved method for the selection of bacterial cells at division. *Biochim. biophys. Acta* **82**, 608–10. [51]

Helmstetter, C. E. & Pierucci. O. (1968). Cell division during inhibition of deoxyribonucleic acid synthesis in *Escherichia coli*. *J. Bact.* **95**, 1627–33. [110]

Helmstetter, C., Cooper, S., Pierucci, O. & Revelas, E. (1968). On the bacterial life sequence. *Cold Spring Harb. Symp. quant. Biol.* **33**, 809–22. [102, 104, 106, 108, 113]

Henshaw, P. & Cohen, I. (1940). Further studies on the action of Roentgen rays on the gametes of *Arbacia punctulata*. IV. Changes in radiosensitivity during the first cleavage cycle. *Am. J. Roentgenol. Radium Ther.* **43**, 917–20. [236]

Herring, A. J. (1971). *Ph.D. thesis*. Univ. of Edinburgh. [228, 229]

Herrmann, E. C. & Schmidt, R. R. (1965). Synthesis of phosphorus-containing macromolecules during synchronous growth of *Chlorella pyrenoidosa*. *Biochim. biophys. Acta* **95**, 63–75. [125, 195]

Hewitt, R. & Billen, D. (1965). Reorientation of chromosome replication after exposure to ultraviolet light of *Escherichia coli*. *J. molec. Biol.* **13**, 40–53. [108]

Hilz, H. & Eckstein, H. (1964). Teilungssynchronisierte Hefezellen. I. Unterschiede Wirkungen von Röntgenstrahlen und cytostatischen Verbindungen auf Stoffwechsel und Zellteilung. *Biochem. Z.* **340**, 351–82. [35, 141]

Hinegardner, R. T., Rao, B. & Feldman, D. E. (1964). The DNA synthetic period during the early development of the sea urchin egg. *Expl Cell Res.* **36**, 53–61. [63, 64]

Hiramoto, Y. (1970). Rheological properties of sea urchin eggs. *Biorheology* **6**, 201–34. [188]

Hjelm, K. K. & Zeuthen, E. (1967). Synchronous DNA-synthesis induced by synchronous cell division in *Tetrahymena*. *C. r. Trav. Lab. Carlsberg* **36**, 127–60. [214]

Hodge, L. D., Borun, T. W., Robbins, E. & Scharff, M. D. (1969). Studies on the regulation of DNA and protein synthesis in synchronized HeLa cells. In *Biochemistry of Cell Division*. pp. 15–38. Ed. by R. Baserga. Springfield, Ill.: C. C. Thomas. [86n, 157]

Hodge, L. D., Robbins, E. & Scharff, M. D. (1969). Persistence of messenger RNA through mitosis in HeLa cells. *J. Cell Biol.* **40**, 497–507. [133]

Hoffman, J. G. (1949). Theory of the mitotic index and its application to tissue growth measurement. *Bull. math. Biophys.* **11**, 139–44. [19]

Hoffman, H. & Frank, M. E. (1965). Time-lapse photomicrography of cell growth and division in *Escherichia coli*. *J. Bact.* **89**, 212–16. [142]

Holt, C. E. & Gurney, E. G. (1969). Minor components of the DNA of *Physarum polycephalum*. *J. Cell Biol.* **40**, 484–96. [71, 83]

Holter, H. (1961). The cartesian diver. In *General Cytochemical Methods*. Vol. II, pp. 93–129. Ed. by J. F. Danielli. New York and London: Academic Press. [9]

Holter, H. & Zeuthen, E. (1957). Dynamics of early echinoderm development, as observed by phase contrast microscopy and correlated with respiration measurements. *Pubbl. Staz. zool. Napoli* **29**, 285–306. [192]

Holz, G. G., Rasmussen, L. & Zeuthen, E. (1963). Normal versus synchronized division in *Tetrahymena pyriformis*. A study with metabolic analogs. *C. r. Trav. Lab. Carlsberg* **33**, 289–300. [210]

Hoogenhout, H. (1963). Synchronous cultures of algae. *Phycologia* **2**, 135–47. [45]

Hopkins, H. A., Sitz, T. O. & Schmidt, R. R. (1970). Selection of synchronous *Chlorella* cells by centrifugation to equilibrium in Ficoll. *J. Cell Physiol.* **76**, 231–4. [51, 71]

Hotchkiss, R. D. (1954). Cyclical behaviour in Pneumococcal growth and transformability occasioned by environmental changes. *Proc. natn. Acad. Sci. U.S.A.* **40**, 49–55. [40n]

Hotta, Y., Ito, M. & Stern, H. (1966). Synthesis of DNA during meiosis. *Proc. natn. Acad. Sci. U.S.A.* **56**, 1184–91. [55]

Houtermans, T. Von. (1953). Ueber den Einfluss des Wachstumszustandes eines Bakterium auf seine Strahlenempfindlichkeit. *Z. Naturf.* **8b**, 767–71. [36]

Howard, A. & Dewey, D. L. (1960). Variations in the period preceding deoxyribonucleic acid synthesis in bean root cells. In *The Cell Nucleus*, pp. 156–62. Ed. by J. S. Mitchell. London: Butterworth. [66]

Howard, A. & Dewey, D. L. (1961). Non-uniformity of labelling rate during DNA synthesis. *Expl Cell Res.* **24**, 623–4. [81n]

Howard, A. & Pelc, S. R. (1953). Synthesis of desoxyribonucleic acid in normal and irradiated cells and its relation to chromosome breakage. *Heredity, Lond.* (*Suppl*) **6**, 261–73. [21, 58, 66]

Hsu, T. C. (1964). Mammalian chromosomes *in vitro* XVIII. DNA replication sequences in the Chinese hamster. *J. Cell Biol.* **23**, 53–62. [73n]

Hsu, T. C., Schmid, W. & Stubblefield, E. (1964). DNA replication sequences in higher animals. In *The role of chromosomes in development. 23rd Symp. Soc. for Study of Development and Growth*. pp. 83–112. Ed. by M. Locke. New York and London: Academic Press. [73n]

Huberman, J. A. (1967). Autoradiographic contributions to the study of eukaryotic chromosomal DNA replication. In *Recent developments in Biochemistry*. pp. 64–81. Ed. by H. U. Li, P. O. P. Ts'O., P. C. Huang and T. T. Kuo. Taipei: Academia Sinica, Republic of China. [73n]

Huberman, J. A. & Riggs, A. D. (1968). On the mechanism of DNA replication in mammalian chromosomes. *J. molec. Biol.* **32**, 327–41. [77, 78, 79, 81]

Hunter-Szybalska, M. E., Szybalski, W. & Delamater, E. D. (1956). Temperature synchronization of nuclear and cellular division in *Bacillus megaterium*. *J. Bact.* **71**, 17–24. [40n]

Hüttermann, A., Porter, M. T. & Rusch, H. P. (1970). Activity of some enzymes in *Physarum polycephalum* I. In the growing plasmodium. *Arch. Mikrobiol.* **74**, 90–100. [254]

Ikeda, M. & Watanabe, Y. (1965). Studies on protein-bound sulfhydryl groups in synchronous cell division of *Tetrahymena pyriformis*. *Expl Cell Res.* **39**, 584–90. [213, 214]

Inouye, M. & Pardee, A. B. (1970). Requirement of polyamines for bacterial division. *J. Bact.* **101**, 770–6. [38]

Iwamura, T. (1966). Nucleic acids in chloroplasts and metabolic DNA. *Prog. Nucleic Acid Res. & molec. Biol.* **5**, 133–55. New York and London: Academic Press. [41]

Jacob, F. & Wollman, E. (1961). *Sexuality and the genetics of bacteria.* New York and London: Academic Press. [94]

Jacob, F., Brenner, S. & Cuzin, F. (1963). On the regulation of DNA replication in bacteria. *Cold Spring Harb. Symp. quant. Biol.* **28**, 329–48. [99]

James, T. W. (1959). Synchronization of cell division in *Amoebae. Ann. N.Y. Acad. Sci.* **78**, 501–14. [46]

James, T. W. (1964). Induced division synchrony in the flagellates. In *Synchrony in Cell Division and Growth.* pp. 323–49. Ed. by E. Zeuthen. New York: Interscience. [45, 46]

James, T. W. (1965). Dynamic respirometry of division synchronized *Astasia longa. Expl Cell Res.* **38**, 439–53. [194]

James, T. W. (1966). Cell synchrony, a prologue to discovery. In *Cell Synchrony* pp. 1–13. Ed. by I. L. Cameron and G. M. Padilla. New York and London: Academic Press. [3, 25]

Jarrett, R. M. & Edmunds, L. N. (1970). Persisting circadian rhythms of cell division in a photosynthetic mutant of *Euglena. Science, N.Y.* **167**, 1730–3. [46]

Jeffery, W. R., Stuart, K. D. & Frankel, J. (1970). The relationship between deoxyribonucleic acid replication and cell division in heat-synchronized *Tetrahymena. J. Cell Biol.* **46**, 533–43. [215n]

Jerka-Dziadosz, M. & Frankel, J. (1970). The control of DNA synthesis in macronuclei and micronuclei of a hypotrich ciliate: a comparison of normal and regenerating cells. *J. exp. Zool.* **173**, 1–22. [69, 215]

Jockusch, B. M., Brown, D. F. & Rusch, H. F. (1970). Synthesis of nuclear protein in G2-phase. *Biochim. biophys. Res. Commun.* **38**, 279–83. [155n, 157]

John, B. & Lewis, K. R. (1969). The chromosome cycle. *Protoplasmatologia*, 6B, 1–125. [58]

Johnson, B. F. (1965). Autoradiographic analysis of regional wall growth of yeast, *Schizosaccharomyces pombe. Expl Cell Res.* **39**, 613–24. [190]

Johnson, B. F. & Gibson, E. J. (1966a). Autoradiographic analysis of regional cell wall growth of yeasts. II. *Pichia farinosa. Expl Cell Res.* **41**, 297–306. [190]

Johnson, B. F. & Gibson, E. J. (1966b). Autoradiographic analysis of regional wall growth of yeasts. III. *Saccharomyces cerevisiae. Expl Cell Res.* **41**, 580–91. [190]

Johnson, L. I., Chan, P., Lobue, J., Monette, F. C. & Gordon, A. S. (1967). Cell cycle analysis of rat lymphocytes cultured with phytohemagglutinin in diffusion chambers. *Expl Cell Res.* **47**, 210–18. [61]

Johnson, R. A. & Schmidt, R. R. (1963). Intracellular distribution of sulfur during the synchronous growth of *Chlorella pyrenoidosa. Biochim. biophys. Acta* **74**, 428–37. [199]

Johnson, R. A. & Schmidt, R. R. (1966). Enzymic control of nucleic acid synthesis during synchronous growth of *Chlorella pyrenoidosa.* I. Deoxythymidine monophosphate kinase. *Biochim. biophys. Acta* **129**, 140–4. [178]

Johnson, R. T. & Rao, P. N. (1971). Nucleo-cytoplasmic interactions in the achievement of nuclear synchrony in DNA synthesis and mitosis in multinucleate cells. *Biol. Rev.* **46**, 97–155. [251]

Johnson, T. C. & Holland, J. J. (1965). Ribonucleic acid and protein synthesis in mitotic HeLa cells. *J. Cell Biol.* **27**, 565–74. [133]

Jones, K. W. (1970). Chromosomal and nuclear location of mouse satellite DNA in individual cells. *Nature, Lond.* **225**, 912–19. [75]

Jones, R. F., Kates, J. R. & Keller, S. J. (1968). Protein turnover and macromolecular synthesis during growth and gametic differentiation in *Chlamydomonas reinhardtii. Biochim. biophys. Acta* **157**, 589–98. [88n]

Jung, C. & Rothstein, A. (1967). Cation metabolism in relation to cell size in synchronously grown tissue culture cells. *J. gen. Physiol.* **50**, 917–32. [199]

Kahane, I. & Razin, S. (1969). Synthesis and turnover of membrane protein and lipid in *Mycoplasma laidlawii. Biochim. biophys. Acta* **183**, 79–89. [187]

Kajiwara, K. & Mueller, G. C. (1964). Molecular events in the reproduction of animal cells. III. Fractional synthesis of DNA with 5-bromodeoxyuridine and its effect on cloning efficiency. *Biochim. biophys. Acta* **91**, 486–93. [76]

Kallenbach, N. R. & Ma, R-I. (1968). Initiation of deoxyribonucleic acid synthesis after thymine starvation of *Bacillus subilis. J. Bact.* **95**, 304–9. [108]

Kasten, F. H. & Strasser, F. F. (1966). Nucleic acid synthetic patterns in synchronized mammalian cells. *Nature, Lond.* **211**, 135–40. [81n]

Kasten, F. H., Strasser, F. F. & Turner, M. (1965). Nucleolar and cytoplasmic ribonucleic acid inhibition by excess thymidine. *Nature, Lond.* **207**, 161–4. [32]

Kates, J. R. & Jones, R. F. (1967). Periodic increases in enzyme activity in synchronized cultures of *Chlamydomonas reinhardtii. Biochim. biophys. Acta* **145**, 153–8. [164]

Kates, J. R., Chiang, K. S. & Jones, R. F. (1968). Studies on DNA replication during synchronized vegetative growth and gametic differentiation in *Chlamydomonas reinhardtii. Expl Cell Res.* **49**, 121–35. [45]

Kato, H. & Yosida, T. H. (1970). Nondisjunction of chromosomes in a synchronized cell population initiated by reversal of colcemid inhibition. *Expl Cell Res.* **60**, 459–64. [28]

Kauffman, S. L. (1968). Lengthening of the generation cycle during embryonic differentiation of the mouse neural tube. *Expl Cell Res.* **49**, 420–4. [63]

Kempner, E. S. & Marr, A. G. (1970). Growth in volume of *Euglena gracilis* during the division cycle. *J. Bact.* **101**, 561–7. [141]

Kessel, M. & Rosenberger, R. F. (1968). Regulation and timing of deoxyribonucleic acid synthesis in hyphae of *Aspergillus nidulans. J. Bact.* **95**, 2275–81. [70]

Kessler, D. (1967). Nucleic acid synthesis during and after mitosis in the slime mold, *Physarum polycephalum. Expl Cell Res.* **45**, 676–80. [122]

√ Killander, D. & Zetterberg, A. (1965a). Quantitative cytochemical studies on interphase growth. I. Determination of DNA, RNA and mass content of age determined mouse fibroblasts *in vitro* and of intercellular variation in generation time. *Expl Cell Res.* **38**, 272–84. [16, 17, 59, 88, 117, 130, 131]

Killander, D. & Zetterberg, A. (1965b). A quantitative cytochemical investigation of the relationship between cell mass and initiation of DNA synthesis in mouse fibroblasts *in vitro. Expl Cell Res.* **40**, 12–20. [88]

Killander, D., Ribbing, C., Ringertz, N. R. & Richards, B. M. (1962). The effects of X-radiation on nuclear synthesis of protein and DNA. *Expl Cell Res.* **27**, 63–9. [248]

Kim, J. H. & Perez, A. G. (1965). Ribonucleic acid synthesis in synchronously dividing populations of HeLa cells. *Nature, Lond.* **207**, 974–5. [119]

Kim, J. H. & Stambuk, B. K. (1966). Synchronization of HeLa cells by vinblastine sulfate. *Expl Cell Res.* **44**, 631–4. [29]

Kim, J. H., Gelbard, A. S. & Perez, A. G. (1968). Inhibition of DNA synthesis by actinomycin D and cycloheximide in synchronized HeLa cells. *Expl Cell Res.* **53**, 478–87. [86, 88n]

Kim, J. H., Kim, S. H. & Eidinoff, M. L. (1965). Cell viability and nucleic acid metabolism after exposure of HeLa cells to excess thymidine and deoxyadenosine. *Biochem. Pharmac.* **14**, 1821–9. [30]

Kimball, R. F. & Barka, T. (1959), Quantitative cytochemical studies on *Paramecium aurelia.* II. Feulgen microspectrophotometry of the macronucleus during exponential growth. *Expl Cell Res.* **17**, 173–82. [68, 81n]

Kimball, R. F. & Perdue, S. W. (1962). Quantitative cytochemical studies on *Paramecium.*

V. Autoradiographic studies on nucleic acid synthesis. *Expl Cell Res.* **27**, 405–15. [68, 123]

Kimball, R. F. & Prescott, D. M. (1962). Deoxyribonucleic acid synthesis and distribution during growth and mitosis of the macronucleus of *Euplotes*. *J. Protozool.* **9**, 88–92. [84]

Kimball, R. F., Caspersson, T. O., Svensson, G. & Carlson, L. (1959). Quantitative cytochemical studies of *Paramecium aurelia*. I. Growth in total dry weight measured by the scanning interference microscope and X-ray absorption methods. *Expl Cell Res.* **17**, 160–72. [17, 135, 141]

Kimball, R. F., Vogt-Köhne, L. & Caspersson, T. O. (1960). Quantitative cytochemical studies on *Paramecium aurelia*. III. Dry weight and ultraviolet absorbtion of isolated macronuclei during various stages of the interdivision interval. *Expl Cell Res.* **20**, 368–77. [68, 151]

King, D. W. & Barnhisel, M. L. (1967). Synthesis of RNA in mammalian cells during mitosis and interphase. *J. Cell Biol.* **33**, 265–72. [116n]

Kinsey, J. D. (1967). X-chromosome replication in early rabbit embryos. *Genetics, Princeton* **55**, 337–43. [76]

Kishimoto, S. & Lieberman, I. (1964). Synthesis of RNA and protein required for the mitosis of mammalian cells. *Expl Cell Res.* **36**, 92–101. [222]

Kjeldgaard, N. O. (1967). Regulation of nucleic acid and protein formation in bacteria. *Adv. microb. Physiol.* **1**, 39–95. [91]

Kjeldgaard, N. O., Maaløe, O. & Schaechter, M. (1958). The transition between different physiological states during balanced growth of *Salmonella typhimurium*. *J. gen. Microbiol.* **19**, 607–16. [106]

Klein, P. & Robbins, E. (1970). An ultra-sensitive assay for soluble sulfhydryl and its application to the study of glutathione levels during the HeLa life cycle. *J. Cell Biol.* **46**, 165–8. [199]

Klevecz, R. R. (1969*a*). Temporal order in mammalian cells. I. The periodic synthesis of lactate dehydrogenase in the cell cycle. *J. Cell Biol.* **43**, 207–19. [163, 168, 174, 179, 197]

Klevecz, R. R. (1969*b*). Temporal coordination of DNA replication with enzyme synthesis in diploid and heteroploid cells. *Science, N.Y.* **166**, 1536–8. [81n, 163, 179]

Klevecz, R. R. & Ruddle, F. H. (1968). Cyclic changes in enzyme activity in synchronized mammalian cell culture. *Science, N.Y.* **159**, 634–6. [179]

Klevecz, R. R. & Stubblefield, E. (1967). RNA synthesis in relation to DNA replication in synchronized Chinese hamster cell cultures. *J. exp. Zool.* **165**, 259–68. [119, 120]

Knaysi, G. (1940). A photomicrographic study of the rate of growth of some yeasts and bacteria. *J. Bact.* **40**, 247–53. [138]

Knaysi, G. (1941). A morphological study of *Streptococcus faecalis*. *J. Bact.* **42**, 575–86. [144]

Knutsen, G. (1965). Induction of nitrite reductase in synchronized cultures of *Chlorella pyrenoidosa*. *Biochim. biophys. Acta* **103**, 495–502. [180]

Knutsen, G. (1968). Repressed and derepressed synthesis of phosphatases during synchronous growth of *Chlorella pyrenoidosa*. *Biochim. biophys. Acta* **161**, 205–14. [178, 180]

Koch, A. L. (1966). Distribution of cell size in growing cultures of bacteria and the applicability of the Collins-Richmond principle. *J. gen. Microbiol.* **45**, 409–17. [142]

Koch, A. L. & Levy, H. R. (1955). Protein turnover in growing cultures of *Escherichia coli*. *J. Biol. Chem.* **217**, 947–57. [144]

Koch, J. & Stokstad, E. L. R. (1967). Incorporation of [^{3}H]thymidine into nuclear and mitochondrial DNA in synchronized mammalian cells. *Europ. J. Biochem.* **3**, 1–6. [39, 71]

Kogoma, T. & Nishi, A. (1965). Rhythmic variations in proteolytic activities during the cell cycle of *Escherichia coli*. *J. gen. appl. Microbiol. Tokyo* **11**, 321–9. [40n, 144, 176]

Kolodny, G. M. & Gross, P. R. (1969). Changes in patterns of protein synthesis during the mammalian cell cycle. *Expl Cell Res.* **56**, 117–21. [157]

Konrad, C. G. (1963). Protein synthesis and RNA synthesis during mitosis in animal cells. *J. Cell Biol.* **19**, 267–77. [116n, 132]

Kovacs, C. J. & Van't Hof, J. (1970). Synchronization of a proliferative population in a cultured plant tissue. *J. Cell Biol.* **47**, 536–9. [250]

Kraemer, P. D. (1966). Sialic acid of mammalian cell lines. *J. Cell Physiol.* **67**, 23–34. [31]

Kubitschek, H. E. (1968a). Linear cell growth in *Escherichia coli*. *Biophys. J.* **8**, 792–804. [50, 142]

Kubitschek, H. E. (1968b). Constancy of uptake during the cell cycle in *Escherichia coli*. *Biophys. J.* **8**, 1401–12. [50, 144]

Kubitschek, H. E. (1969a). Counting and sizing micro-organisms with the Coulter counter. In *Methods in Microbiology*. Vol. 1, pp. 592–610. Ed. by J. R. Norris and D. W. Ribbons. New York and London: Academic Press. [9]

Kubitschek, H. E. (1969b). Growth during the bacterial cell cycle: analysis of cell size distribution. *Biophys. J.* **9**, 792–809. [142]

Kubitschek, H. E. (1970). Evidence for the generality of linear cell growth. *J. theor. Biol.* **28**, 15–29. [146, 200]

Kubitschek, H. (1971). Control of cell growth in bacteria: experiments with thymine starvation. *J. Bact.* **105**, 472–6. [254]

Kubitschek, H. E., Bendigkeit, H. E. & Loken, M. R. (1967). Onset of DNA synthesis during the cell cycle in chemostat cultures. *Proc. natn. Acad. Sci. U.S.A.* **57**, 1611–17. [106]

Kuempel, P. L. (1970). Bacterial chromosome replication. *Adv. Cell Biol.* **1**, 3–56. [91]

Kuempel, P. L., Masters, M. & Pardee, A. B. (1965). Bursts of enzyme synthesis in the bacterial duplication cycle. *Biochem. biophys. Res. Commun.* **18**, 858–67. [167, 168, 169n, 176, 180]

Küenzi, M. T. & Fiechter, A. (1969). Changes in carbohydrate composition and trehalase-activity during the budding cycle of *Saccharomyces cerevisiae*. *Arch Mikrobiol.* **64**, 396–407. [177]

Kuhl, A. & Lorenzen, H. (1964). Handling and culturing of *Chlorella*. In *Methods in Cell Physiology*. Vol. 1, pp. 159–87. Ed. by D. M. Prescott. New York and London: Academic Press. [44]

Kusanagi, A. (1964). RNA synthetic activity in the mitotic nuclei. *Jap. J. Genet.* **39**, 254–8. [116n]

Lark, C. (1966). Regulation of deoxyribonucleic acid synthesis in *Escherichia coli*: dependence on growth rates. *Biochim. biophys. Acta* **119**, 517–25. [106]

Lark, C. & Lark, K. G. (1964). Evidence for two distinct aspects of the mechanism regulating chromosome replication in *Escherichia coli*. *J. molec. Biol.* **10**, 120–36. [110]

Lark, K. G. (1963). Cellular control of DNA biosynthesis. In *Molecular Genetics*. Vol 1, pp. 153–206. Ed. by J. H. Taylor. New York and London: Academic Press. [58]

Lark, K. G. (1966a). Chromosome replication in *Escherichia coli*. In *Cell Synchrony*, pp. 54–80. Ed. by I. L. Cameron and G. M. Padilla. New York and London: Academic Press. [91, 110]

Lark, K. G. (1966b). Regulation of chromosome replication and segregation in bacteria. *Bact. Rev.* **30**, 3–32. [91, 107]

Lark, K. G. (1969). Initiation and control of DNA synthesis. *A. Rev. Biochem.* **38**, 569–604. [91, 110]

Lark, K. G. & Lark, C. (1960). Changes during the division cycle in bacterial cell wall synthesis, volume and ability to concentrate free amino-acids. *Biochim. biophys. Acta* **43**, 520–30. [54]

Lark, K. G. & Lark, C. (1966). Regulation of chromosome replication in *Escherichia coli*: A comparison of the effect of phenethyl alcohol treatment with those of amino-acid starvation. *J. molec. Biol.* **20**, 9–20. [110]

Lark, K. G. & Maaløe, O. (1956). Nucleic acid synthesis and the division cycle of *Salmonella typhimurium*. *Biochim. biophys. Acta* **21**, 448–58. [40, 47, 102]

Lark, K. G. & Renger, H. (1969). Initiation of DNA replication in *Escherichia coli* 15T⁻: chronological dissection of the physiological processes required for initiation. *J. molec. Biol.* **42**, 221–35. [110]

Lark, K. G., Repko, T. & Hoffman, E. J. (1963). The effects of amino acid deprivation on subsequent deoxyribonucleic acid replication. *Biochim. biophys. Acta* **76**, 9–24. [92, 96]

Lawrence, P. A. (1968). Mitosis and the cell cycle in the metamorphic moult of the milk-weed bug, *Oncopeltus fasciatus*. A radioautographic study. *J. Cell Sci.* **3**, 391–404. [63]

Lea, D. E. (1955). *Action of Radiation on Living Cells*. Cambridge: Cambridge University Press. [234]

Lesser, B. & Brent, T. P. (1970). Cold storage as a method for accumulating mitotic HeLa cells without impairing subsequent synchronous growth. *Expl Cell Res.* **62**, 470–3. [49]

Levis, A. G., Danielli, G. A. & Piccini, E. (1965). Nucleic acid synthesis and the mitotic cycle in mammalian cells treated with nitrogen mustard in culture. *Nature, Lond.* **207**, 608–10. [249]

Lewis, M. R. & Lewis, W. H. (1915). Mitochondria (and other cytoplasmic structures) in tissue culture. *Am. J. Anat.* **17**, 339–401. [183]

Ley, K. D. & Tobey, R. A. (1970). Regulation of initiation of DNA synthesis in Chinese hamster cells. II. Induction of DNA synthesis and cell division by isoleucine and glutamine in G1-arrested cells in suspension culture. *J. Cell Biol.* **47**, 453–9. [34]

Liébecq-Hutter, S. (1965). Cultures synchrones de myoblastes et de fibroblastes d'embryons de poulet. *C. r. Séanc. Soc. Biol.* **159**, 768. [47, 63]

Lima-de-Faria, A. (1959). Differential uptake of tritiated thymidine into hetero- and euchromatin in *Melanoplus* and *Secale*. *J. biophys. biochem. Cytol.* **6**, 457–66. [73n, 75]

Lima-de-Faria, A. (1969). DNA replication and gene amplification in heterochromatin. In *Handbook of Molecular Cytology*, pp. 278–325. Ed. by A. Lima-de-Faria. Amsterdam and London: North-Holland Publishing Co. [73n, 75]

Lima-de-Faria, A., Bianchi, N. D. & Nowell, P. (1964). Replication in chromosomes from chronic granulocytic leukemia. *J. Cell Biol.* **23**, 54A. [73n]

Lin, L. & Wyss, O. (1965). Synchronization of *Azotobacter* cells. *Tex. Rep. Biol. Med.* **23**, 474–80. [40n]

Lindahl, P. E. & Sörenby, L. (1966). A new method for the continuous selection of cells in mitosis. *Expl Cell Res.* **43**, 424–34. [48].

Lindegren, C. C. & Haddad, S. A. (1954). Growth rate of individual yeast cells. *Genetica* **27**, 45–53. [141]

Linnartz-Niklas, A., Hempel, K. & Maurer, W. (1964). Autoradiographische Untersuchung über den Eiweiss- und RNAstoffwechsel tierischer Zellen während/der Mitose. *Z. Zellforsch. mikrosk. Anat.* **62**, 443–53. [116n]

Lison, L. & Pasteels, J. (1951). Études histophotométriques sur la teneur en acide désoxyribonucléique des noyaux au cours du développement embryonnaire chez l'oursin *Paracentrotus lividus*. *Arch. Biol, Liège* **62**, 2–43. [63]

Littlefield, J. W. (1962). DNA synthesis in partially synchronized L cells. *Expl Cell Res.* **26**, 318–26. [27, 34]

Littlefield, J. W. (1966). The periodic synthesis of thymidine kinase in mouse fibroblasts. *Biochim. biophys. Acta* **114**, 398–403. [178]

Littlefield, J. W. & Jacobs, P. S. (1965). The relation between DNA and protein synthesis in mouse fibroblasts. *Biochim. biophys. Acta* **108**, 652–8. [86n, 155n]

Loening, U. E. (1968). RNA structure and metabolism. *A. Rev. Pl. Physiol.* **19**, 37–70. [115]

Lorenzen, H. (1964). Synchronization of *Chlorella* with light-dark changes and periodical dilution of the populations to a standard cell number. In *Synchrony in Cell Division and Growth*, pp. 571–8. Ed. by E. Zeuthen. New York: Interscience. [44]

Louderback, A. L., Scherbaum, O. H. & Jahn, I. L. (1961). The effects of temperature shifts on the budding cycle of *Saccharomyces cerevisiae*. *Expl Cell Res.* **25**, 437–53. [46]

Lövlie, A. (1963). Growth in mass and respiration rate during the cell cycle of *Tetrahymena pyriformis*. *C. r. Trav. Lab. Carlsberg* **33**, 377–413. [135, 136, 141, 195, 196]

Lövlie, A. (1964). Genetic control of division rate and morphogenesis in *Ulva mutabilis* Föyn. *C. r. Trav. Lab. Carlsberg* **34**, 77–168. [45]

Løvtrup, S. & Iverson, R. M. (1969). Respiratory phases during early sea urchin development, measured with the automatic diver balance. *Expl Cell Res.* **55**, 25–32. [9]

Lowe-Jinde, L. & Zimmerman, A. M. (1971). The incorporation of phenylalanine and uridine in *Tetrahymena*: a pressure study. *J. Protozool.* **18**, 20–3. [214]

Lowy, B. A. & Leick, V. (1969). The synthesis of DNA in synchronized cultures of *Tetrahymena pyriformis* GL. *Expl Cell Res.* **57**, 277–88. [215]

Luck, D. J. L. (1963a). Formation of mitochondria in *Neurospora crassa*. *J. Cell Biol.* **16**, 483–99. [183]

Luck, D. J. L. (1963b). Genesis of mitochondria in *Neurospora crassa*. *Proc. natn. Acad. Sci. U.S.A.* **49**, 233–40. [183]

Luck, D. J. L. (1965). Formation of mitochondria in *Neurospora crassa*. *Am. Nat.* **99**, 241–53. [183]

Lyndon, R. F. (1967). The growth of the nucleus in dividing and non-dividing cells of the pea root. *Ann. Bot.* **31**, 133–46. [151]

Maaløe, O. & Hanawalt, P. C. (1961). Thymine deficiency and the normal DNA replication cycle. I. *J. molec. Biol.* **3**, 144–55. [95]

Maaløe, O. & Kjeldgaard, N. O. (1966). *Control of Macromolecular Synthesis*. New York: W. A. Benjamin. [91, 98, 107]

McBride, O. W. & Peterson, E. A. (1970). Separation of nuclei representing different phases of the growth cycle from unsynchronized mammalian cell cultures. *J. Cell Biol.* **47**, 132–9. [250]

McDonald, B. B. (1958). Quantitative aspects of deoxyribose nucleic acid (DNA) metabolism in an amicronucleate strain of *Tetrahymena*. *Biol. Bull. mar. biol. Lab., Woods Hole* **114**, 71–94. [67n]

McDonald, B. B. (1962). Synthesis of deoxyribonucleic acid by micro- and macronuclei of *Tetrahymena pyriformis*. *J. Cell Biol.* **13**, 193–203. [67n, 68]

MacDonald, H. R. & Miller, R. G. (1970). Synchronization of mouse L cells by a velocity sedimentation technique. *Biophys. J.* **10**, 834–42. [50]

McElroy, L. J., Wells, J. S. & Krieg, K. R. (1967). Mode of extension of cell surface during growth of *Spirillum volutans*. *J. Bact.* **93**, 499–501. [191]

Macgregor, H. C. & Callan, H. G. (1962). The actions of enzymes on lampbrush chromosomes. *Q. Jl. microsc. Sci.* **103**, 172–203. [79]

Mackenzie, T. B., Stone, G. E. & Prescott, D. M. (1966). The duration of G_1, S and G_2 at different temperatures in *Tetrahymena pyriformis* HSM. *J. Cell Biol.* **31**, 633–5. [67n]

Maclean, N. (1965). Ribosome numbers in a fission yeast. *Nature, Lond.* **207**, 322–3. [185]

McLeish, J. (1969). Changes in the amount of nuclear RNA during interphase in *Vicia faba*. *Chromosoma* **26**, 312–25. [121]

McQuade, H. A. & Friedkin, M. (1960). Radiation effect of thymidine-^{3}H and thymidine-^{14}C. *Expl Cell Res.* **21**, 118–25. [22]

Madreiter, H., Kaden, P. & Mittermayer, C. (1971). DNA polymerase, triphosphatase and deoxyribonuclease in a system of synchronized L cells. *Europ. J. Biochem.* **18**, 369–75. [254]

Maekawa, T. & Tsuchiya, J. (1968). A method for the direct estimation of the length of G_1, S, and G_2 phase. *Expl Cell Res.* **53**, 55–64. [61]

Mak, S. (1965). Mammalian cell cycle analysis using microspectrophotometry combined with autoradiography. *Expl Cell Res.* **39**, 286–9. [60]

Malamud, D. (1967). DNA synthesis and the mitotic cycle in frog kidney cells cultivated *in vitro*. *Expl Cell Res.* **45**, 277–80. [63]

Malinovsky, O. V. & Mitjushova, N. M. (1966). On synchronization of the division of yeast cultures by means of fractionation of the cells according to dimension (In Russian). *Tsitologiya* **8**, 563–6. [49]

Manor, H. & Haselkorn, R. (1967). Size fractionation of exponentially growing *Escherichia coli*. *Nature, Lond.* **214**, 983–6. [50, 125n]

Manton, I. (1959). Electron microscopical observations on a very small flagellate: the problem of *Chromulina pusilla* Butcher. *J. mar. biol. Ass. U.K.* **38**, 319–33. [183]

Manton, I. & Parke, M. (1960). Further observations on small green flagellates with special reference to possible relatives of *Chromulina pusilla* Butcher. *J. mar. biol. Ass. U.K.* **39**, 275–98. [183]

Marcus, P. I. & Robbins, E. (1963). Viral inhibition in the metaphase-arrest cell. *Proc. natn. Acad. Sci. U.S.A.* **50**, 1156–64. [29, 133]

Marshak, A. (1949). Recovery from ultra-violet light-induced delay in cleavage of *Arbacia* eggs by irradiation with visible light. *Biol. Bull. mar. biol. lab. Woods Hole* **97**, 315–22. [241]

Martin, P. G. (1966). The pattern of autosomal DNA replication in four tissues of the Chinese hamster. *Expl Cell Res.* **45**, 85–95. [73n, 75]

Martin, D. W. & Tomkins, G. M. (1970). The appearance and disappearance of the post-transcriptional repressor of tyrosine amino-transferase synthesis during the HTC cell cycle. *Proc. natn. Acad. Sci. U.S.A.* **65**, 1064–8. [173, 180]

Martin, D. W., Tomkins, G. M. & Bresler, M. A. (1969). Control of specific gene expression examined in synchronized mammalian cells. *Proc. natn. Acad. Sci. U.S.A.* **63**, 842–9. [173, 180]

Martin, D., Tomkins, G. M. & Granner, D. (1969). Synthesis and induction of tyrosine aminotransferase in synchronized hepatoma cells in culture. *Proc. natn. Acad. Sci. U.S.A.* **62**, 248–55. [119, 131, 132, 174, 179, 180]

Maruyama, Y. (1956). Biochemical aspects of cell growth of *Escherichia coli* as studied by the method of synchronous culture. *J. Bact.* **72**, 821–6. [54]

Maruyama, Y. & Lark, K. G. (1959). Periodic synthesis of RNA in synchronized cultures of *Alcaligenes faecalis*. *Expl Cell Res.* **18**, 389–91. [126]

Maruyama, Y. & Lark, K. G. (1962). Periodic nucleotide synthesis in synchronous cultures of bacteria. *Expl Cell Res.* **26**, 382–94. [54, 111]

Maruyama, Y. & Yanagita, T. (1956). Physical methods for obtaining synchronous culture of *Escherichia coli*. *J. Bact.* **71**, 542–6. [54]

Masters, M. (1970). Origin and direction of replication of the chromosome of *Escherichia coli* B/r. *Proc. natn. Acad. Sci. U.S.A.* **65**, 601–8. [98]

Masters, M. & Broda, P. (1971). Evidence for the bidirectional replication of the *Escherichia coli* chromosome. *Nature, New Biol. Lond.* In press. [252]

Masters, M. & Donachie, W. D. (1966). Repression and the control of cyclic enzyme synthesis in *Bacillus subtilis*. *Nature, Lond.* **209**, 476–9. [120, 168, 169n, 176]

Masters, M. & Pardee, A. B. (1965). Sequence of enzyme synthesis and gene replication during the cell cycle of *Bacillus subtilis*. *Proc. natn. Acad. Sci. U.S.A.* **54**, 64–70. [165, 169n, 172, 173, 176, 180]

Masters, M., Kuempel, P. L. & Pardee, A. B. (1964). Enzyme synthesis in synchronous cultures of bacteria. *Biochem. biophys. Res. Commun.* **15**, 38–42. [36]

Matney, T. S. & Suit, J. C. (1966). Synchronously dividing bacterial cultures. I. Synchrony following depletion and resupplementation of a required amino acid in *Escherichia coli*. *J. Bact.* **92**, 960–6. [34, 38]

Mauck, J. & Glaser, L. (1970). Turnover of the cell wall of *Bacillus subtilis* W-23 during logarithmic growth. *Biochem. biophys. Res. Commun.* **39**, 699–706. [192]

Mauro, F. & Madoc-Jones, H. (1969). Age response to X-radiation of murine lymphoma cells synchronized *in vivo*. *Proc. natn. Acad. Sci. U.S.A.* **63**, 686–91. [33]

May, J. W. (1962). Sites of cell-wall extension demonstrated by the use of fluorescent antibody. *Expl Cell Res.* **27**, 170–2. [190]

May, J. W. (1963). The distribution of cell wall label during growth and division of *Salmonella typhimurium*. *Expl Cell Res.* **31**, 217–220. [191]

Mayhew, E. (1966). Cellular electrophoretic mobility and the mitotic cycle. *J. gen. Physiol.* **49**, 717–25. [189]

Mayhew, E. H. & O'Grady, E. A. (1965). Electrophoretic mobilities of tissue culture cells in exponential and parasynchronous growth. *Nature, Lond.* **207**, 86–7. [189]

Mazia, D. (1961). Mitosis and the physiology of cell division. In *The Cell*. Vol. 3, pp. 77–412. Ed. by J. Brachet and A. E. Mirsky. New York and London: Academic Press. [201, 221]

Mazia, D. (1963). Synthetic activities leading to mitosis. *J. cell. comp. Physiol.* **62**, Suppl. 1. 123–40. [89, 230]

Mazia, D. & Zeuthen, E. (1966). Blockage and delay of cell division in synchronous populations of *Tetrahymena* by mercaptoethanol (monothioethylene glycol). *C. r. Trav. Lab. Carlsberg* **35**, 341–61. [209]

Mazia, D., Harris, P. J. & Bibring, T. (1960). The multiplicity of the mitotic centers and the time-course of their duplication and separation. *J. biophys. biochem. Cytol.* **7**, 1–20. [184]

Melnykovych, G., Bishop, C. F. & Swayze, M. A. B. (1967). Fluctuation of alkaline phosphatase activity in synchronized heteroploid cell cultures: effect of prednisolone. *J. Cell Physiol.* **70**, 231–6. [179]

Merriam, R. W. (1969). The intracellular distribution of the free amino acid pool in frog oocytes. *Expl Cell Res.* **56**, 259–64. [197]

Meselson, M. & Stahl, F. (1958). The replication of DNA in *Escherichia coli*. *Proc. natn. Acad. Sci. U.S.A.* **44**, 671–82. [92]

Meyenburg, H. K. von (1969). Energetics of the budding cycle of *Saccharomyces cerevisiae* during glucose limited aerobic growth. *Arch. Mikrobiol.* **66**, 289–303. [38]

Miller, A. O. A. (1967). Characterization of messenger ribonucleic acid in partially synchronized HeLa cells. *Archs. Biochem. Biophys.* **122**, 270–9. [121]

Miller, O. L., Stone, G. E. & Prescott, D. M. (1964). Autoradiography of water-soluble materials. In *Methods in Cell Physiology*. Vol. 1, pp. 371–9. Ed. by D. M. Prescott. New York and London: Academic Press. [10, 197]

Minutoli, F. & Hirshfield, H. I. (1968). DNA synthesis cycle in *Blepharisma*. *J. Protozool.* **15**, 532–5. [69]

Mitchison, J. M. (1957). The growth of single cells. I. *Schizosaccharomyces pombe*. *Expl Cell Res.* **13**, 244–62. [9, 16, 138, 190]

Mitchison, J. M. (1958). The growth of single cells. II. *Saccharomyces cerevisiae*. *Expl Cell Res.* **15**, 214–21. [9, 141, 250]

Mitchison, J. M. (1961). The growth of single cells. III. *Streptococcus faecalis. Expl Cell Res.* **22**, 208–25. [9, 144, 145]

Mitchison, J. M. (1963). Pattern of synthesis of RNA and other cell components during the cell cycle of *Schizosaccharomyces pombe. J. cell. comp. Physiol.* **62**, Suppl. 1, 1–13. [139]

Mitchison, J. M. (1969a). Enzyme synthesis in synchronous cultures. *Science, N.Y.* **165**, 657–63. [159, 160, 162, 166, 169n]

Mitchison, J. M. (1969b). Markers in the cell cycle. In *The Cell Cycle. Gene-Enzyme Interactions*, pp. 361–72. Ed. by G. M. Padilla, G. L. Whitson and I. L. Cameron. New York and London: Academic Press. [244]

Mitchison, J. M. (1970). Physiological and cytological methods for *Schizosaccharomyces pombe*. In *Methods in Cell Physiology*. Vol. 4, pp. 131–65. Ed. by D. M. Prescott. New York and London: Academic Press. [17, 50]

Mitchison, J. M. & Creanor, J. (1969). Linear synthesis of sucrase and phosphatases during the cell cycle of *Schizosaccharomyces pombe. J. Cell Sci.* **5**, 373–91. [89, 161, 167, 174, 177, 178, 180]

Mitchison, J. M. & Creanor, J. (1971a). Unpublished. [180]

Mitchison, J. M. & Creanor, J. (1971b). In preparation. [70n]

Mitchison, J. M. & Creanor, J. (1971c). Induction synchrony in the fission yeast *Schizosaccharomyces pombe. Expl Cell Res.* In press. [28, 31, 248, 250]

Mitchison, J. M. & Cummins, J. E. (1964). Changes in the acid-soluble pool during the cell cycle of *Schizosaccharomyces pombe. Expl Cell Res.* **35**, 394–401. [10, 138, 140]

Mitchison, J. M. & Gross, P. R. (1965). Selective synthesis of messenger RNA in a fission yeast during a step-down and its relation to the cell cycle. *Expl Cell Res.* **37**, 259–77. [124]

Mitchison, J. M. & Lark, K. G. (1962). Incorporation of ³H-adenine into RNA during the cell cycle of *Schizosaccharomyces pombe. Expl Cell Res.* **28**, 452–5. [124, 138]

Mitchison, J. M. & Swann, M. M. (1954). The mechanical properties of the cell surface. I. The cell elastimeter. *J. exp. Biol.* **31**, 443–60. [188–9]

Mitchison, J. M. & Swann, M. M. (1955). The mechanical properties of the cell surface. III. The sea urchin egg from fertilization to cleavage. *J. exp. Biol.* **32**, 734–50. [10]

Mitchison, J. M. & Vincent, W. S. (1965). Preparation of synchronous cell cultures by sedimentation. *Nature, Lond.* **205**, 987–9. [49]

Mitchison, J. M. & Walker, P. M. B. (1959). RNA synthesis during the cell life cycle of a fission yeast *Schizosaccharomyces pombe. Expl Cell Res.* **16**, 49–58. [124]

Mitchison, J. M. & Wilbur, K. M. (1962). The incorporation of protein and carbohydrate precursors during the cell cycle of fission yeast. *Expl Cell Res.* **26**, 144–57. [138, 141]

Mitchison, J. M., Cummins, J. E., Gross, P. R. & Creanor, J. (1969). The uptake of bases and their incorporation into RNA during the cell cycle of *Schizosaccharomyces pombe* in normal growth and after a step-down. *Expl Cell Res.* **57**, 411–22. [15, 16, 124, 200]

Mitchison, J. M., Kinghorn, M. L. & Hawkins, C. (1963). The growth of single cells. IV. *Schizosaccharomyces pombe* at different temperatures. *Expl Cell Res.* **30**, 521–7. [9, 140]

Mitchison, J. M., Passano, L. M. & Smith, F. H. (1956). An integration method for the interference microscope. *Q. Jl microsc. Sci.* **97**, 287–302. [9]

Mittermayer, C., Bosselmann, R. & Bremerskov, V. (1968). Initiation of DNA synthesis in a system of synchronized L-cells: rhythmicity of thymidine kinase activity. *Europ. J. Biochem.* **4**, 487–9. [178]

Mittermayer, C., Braun, R., Chayka, T. C., & Rusch, H. P. (1966). Polysome patterns and protein synthesis during the mitotic cycle of *Physarum polycephalum. Nature, Lond.* **210**, 1133–7. [122, 137, 141]

Mittermayer, C., Braun, R. & Rusch, H. P. (1964). RNA synthesis in the mitotic cycle of *Physarum polycephalum. Biochim. biophys. Acta* **91**, 399–405. [122]

Mittermayer, C., Braun, R. & Rusch, H. P. (1966). Ribonucleic acid synthesis *in vitro* in nuclei isolated from the synchronously dividing *Physarum polycephalum. Biochim. biophys. Acta* **114**, 536–46. [122]

Molloy, G. R. & Schmidt, R. R. (1970). Studies on the regulation of ribulose-1,5-diphosphate carboxylase synthesis during the cell cycle of the eucaryote *Chlorella. Biochem. biophys. Res. Commun.* **40**, 1125–33. [164, 169–70, 173, 174, 178]

Morales, M. & McKay, D. (1967). Biochemical oscillations in 'controlled' systems. *Biophys. J.* **7**, 621–5. [170]

Morimoto, H. & James, T. W. (1969). Sulfate-controlled synchrony in *Astasia longa. Expl Cell Res.* **58**, 195–200. [38]

Morishima, A., Grumbach, M. M. & Taylor, J. H. (1962). Asynchronous duplication of human chromosomes and the origin of sex chromatin. *Proc. natn. Acad. Sci. U.S.A.* **48**, 756–63. [73n]

Morris, N. R., Cramer, J. W. & Reno, D. (1967). A simple method for concentration of cells in the DNA synthetic period of the mitotic cycle. *Expl Cell Res.* **48**, 216–18. [50]

Morrison, D. C. & Morowitz, H. J. (1970). Studies on membrane biosynthesis in *Bacillus megaterium* KM. *J. molec. Biol.* **49**, 441–59. [188]

Mueller, G. C. (1969). Biochemical events in the animal cell cycle. *Fedn Proc. Fedn Am. Socs exp. Biol.* **28**, 1780–9. [29]

Mueller, G. C. & Kajiwara, K. (1966*a*). Early- and late-replicating deoxyribonucleic acid complexes in HeLa nuclei. *Biochim. biophys. Acta* **114**, 108–15. [28, 80]

Mueller, G. C. & Kajiwara, K. (1966*b*). Actinomycin D and *p*-fluorophenylalanine, inhibitors of nuclear replication in HeLa cells. *Biochim. biophys. Acta* **119**, 557–65. [88, 222]

Mueller, G. C., Kajiwara, K., Stubblefield, E. & Rueckert, R. R. (1962). Molecular events in the reproduction of animal cells. I. The effect of puromycin on the duplication of DNA. *Cancer. Res.* **22**, 1084–90. [86]

Müller, J. & Dawson, P. S. S. (1968). (1) The operational flexibility of the phased culture technique, as observed by changes in the cell cycle of *Candida utilis.* (2) The oxygen uptake of phased yeast cultures growing at different doubling times on nitrogen- and energy-limited media. *Can. J. Microbiol.* **14**, 1115–26, 1127–31. [38]

Murphree, S., Stubblefield, E. & Moore, E. C. (1969). Synchronized mammalian cell cultures. III. Variation of ribonucleotide reductase activity during the replication cycle of Chinese hamster fibroblasts. *Expl Cell Res.* **58**, 118–24. [179]

Murray, K. (1964). Histone nomenclature. In *The Nucleohistones*, pp. 15–20. Ed. by J. Bonner and P. Ts'o. San Francisco: Holden-Day, Inc. [153]

Murti, K. G. & Prescott, D. M. (1970). Micronuclear ribonucleic acid in *Tetrahymena pyriformis. J. Cell Biol.* **47**, 460–7. [253]

Nachtwey, D. S. (1965). Division of synchronized *Tetrahymena pyriformis* after emacronucleation. *C. r. Trav. Lab. Carlsberg* **35**, 25–35. [210]

Nachtwey, D. S. & Cameron, I. L. (1968). Cell cycle analysis. In *Methods in Cell Physiology.* Vol. 3, pp. 213–59. Ed. by D. M. Prescott. New York and London: Academic Press. [16, 19]

Nachtwey, D. S. & Dickinson, W. J. (1967). Actinomycin D: blockage of cell division of synchronized *Tetrahymena pyriformis. Expl Cell Res.* **47**, 581–95. [210]

Nachtwey, D. S. & Giese, A. C. (1968). Effects of ultraviolet light irradiation and heat shocks on cell division in synchronized *Tetrahymena. Expl Cell Res.* **50**, 167–76. [238, 239]

Nagata, T. & Meselson, M. (1968). Periodic replication of DNA in steadily growing *Escherichia coli*: the localised origin of replication. *Cold Spring Harb. Symp. quant. Biol.* **33**, 553–7. [92]

Nanney, D. L. (1967). Cortical slippage in *Tetrahymena. J. exp. Zool.* **166**, 163–70. [212]

Nass, M. M. K. (1969). Mitochondrial DNA: advances, problems and goals. *Science, N. Y.* **165**, 25–35. [71]

Natarajan, A. T. (1961). Chromosome breakage and mitotic inhibition induced by tritiated thymidine in root meristems of *Vicia faba. Expl Cell Res.* **22**, 275–81. [22]

Neifakh, A. A. & Rott, N. N. (1958). Synchronization of cell division in early embryos of the loach *Misgurnus fossilis*. (In Russian) *Dokl. Akad. Nauk SSR* **125**, 256–8. [39]

Nešković, B. A. (1968). Development phases in intermitosis and the preparation for mitosis of mammalian cells *in vitro. Int. Rev. Cytol.* **24**, 71–97. [181]

Newsome, J. (1966). The synthesis of ribonucleic acid in animal cells during mitosis. *Biochim. biophys. Acta* **114**, 36–43. [116, 121]

Newton, A. A. & Wildy, P. (1959). Parasynchronous division of HeLa cells. *Expl Cell Res.* **16**, 624–35. [39]

Newton, A., Dendy, P. P., Smith, C. L. & Wildy, P. (1962). A pool size problem associated with use of tritiated thymidine. *Nature, Lond.* **194**, 886–7. [15]

✓ Nias, A. H. W. & Fox, M. (1971). Synchronization of mammalian cells with reference to the mitotic cycle. *Cell Tissue Kinetics* **4**, 351–74. [25]

Niehaus, W. G. & Barnum, C. P. (1965). Incorporation of radioisotope, *in vivo*, into ribonucleic acid and histone of a fraction of nuclei preparing for mitosis. *Expl Cell Res.* **39**, 435–42. [155n]

Nilausen, K. & Green, H. (1965). Reversible arrest of growth in G_1 of an established fibroblast line (3T3). *Expl Cell Res.* **40**, 166–8. [34]

Nishi, A. & Hirose, S. (1966). Further observations on the rhythmic variation in peptidase activity during the cell cycle of various strains of *Escherichia coli. J. gen. appl. Microbiol., Tokyo* **12**, 293–7. [176]

Nishi, A. & Horiuchi, T. (1966). β-galactosidase formation controlled by episomal gene during the cell cycle of *Escherichia coli. J. Biochem., Tokyo* **60**, 338–40. [166, 180]

Nishi, A. & Kogoma, T. (1965). Protein turnover in the cell cycle of *Escherichia coli. J. Bact.* **90**, 884–90. [54, 144]

Nishi, A., Okamura, S. & Yanagita, T. (1967). Shift of cell-age distribution patterns in the later phases of *Escherichia coli* culture. *J. gen. appl. Microbiol., Tokyo* **13**, 103–19. [34, 36]

Nosoh, Y. & Takamiya, A. (1962). Synchronization of budding cycle in yeast cells, and effect of carbon monoxide and nitrogen-deficiency on the synchrony. *Pl. Cell Physiol., Tokyo* **3**, 53–66. [35]

Nygaard, O. F., Güttes, S. & Rusch, H. P. (1960). Nucleic acid metabolism in a slime mold with synchronous mitosis. *Biochim. biophys. Acta* **38**, 298–306. [70]

Odell, T. T., Jackson, C. W. & Reiter, R. S. (1968). Generation cycle of rat megakaryocytes. *Expl Cell Res.* **53**, 321–8. [81]

Oehlert, W., Seemayer, N. & Lauf, P. (1962). Autoradiographische Untersuchungen über den Generationszyklus der Zellen des Ehrlich-Ascitescarcinoma der weissen Maus, *Beitr. path. Anat.* **127**, 63–78. [82, 84]

Ogur, M., Minckler, S. & McClary, D. O. (1953). Desoxyribonucleic acid and the budding cycle in the yeast. *J. Bact.* **66**, 642–5. [35]

Ohara, H. & Terasima, T. (1969). Variations of cellular sulfhydryl content during cell cycle of HeLa cells and its correlation to cyclic change of X-ray sensitivity. *Expl Cell Res.* **58**, 182–5. [199]

Oishi, M., Yoshikawa, H. & Sueoka, N. (1964). Synchronous and dichotomous replications of the *Bacillus subtilis* chromosome during spore germination. *Nature, Lond.* **204**, 1069–73. [94]

Okada, S. (1967). A simple graphic method of computing the parameters of the life cycle of cultured mammalian cells in the exponential growth phase. *J. Cell Biol.* **34**, 915–16. [24]

Ord, M. J. (1968). The synthesis of DNA through the cell cycle of *Amoeba proteus*. *J. Cell Sci.* **3**, 483–91. [69, 81n, 83]

Ord, M. J. (1969). Control of DNA synthesis in *Amoeba proteus*. *Nature, Lond.* **221**, 964–6. [85]

Oster, G. & Pollister, A. W. (1956). Eds. *Physical techniques in biological research.* Vol. 3. Cells and tissues. New York and London: Academic Press. [10]

O'Sullivan, A. & Sueoka, N. (1967). Sequential replication of the *Bacillus subtilis* chromosome. IV. Genetic mapping by density transfer experiments. *J. molec. Biol.* **27**, 349–68. [99]

Osumi, M. & Sando, N. (1969). Division of yeast mitochondria in synchronous cultures. *J. Electron Microsc., Chiba Cy* **18**, 47–56. [184, 196]

Padilla, G. M. & Cameron, I. L. (1964). Synchronization of cell division in *Tetrahymena pyriformis* by a repetitive temperature cycle. *J. cell. comp. Physiol.* **64**, 303–8. [40]

Padilla, G. M. & Cook, J. R. (1964). The development of techniques for synchronizing flagellates. In *Synchrony in Cell Division and Growth*, pp. 521–35. Ed. by E. Zeuthen. New York: Interscience Publishers Inc. [45, 46]

Padilla, G. M. & James, T. W. (1960). Synchronization of cell division in *Astasia longa* on a chemically defined medium. *Expl Cell Res.* **20**, 401–15. [46]

Padilla, G. M., Cameron, I. L. & Elrod, L. H. (1966). The physiology of repetitively synchronized *Tetrahymena*. In *Cell Synchrony*, pp. 269–88. Ed. by I. L. Cameron and G. M. Padilla. New York and London: Academic Press. [40]

Padilla, G. M., Van Dreal, P. A. & Anderson, N. G. (1966). Studies in synchronized cells: radiation-induced division delay in the Flagellate *Astasia longa*. *Radiat. Res.* **28**, 157–65. [239, 240]

Pagoulatos, G. N. & Darnell, J. E. (1970). A comparison of heterogeneous nuclear RNA of HeLa cells in different periods of the cell growth cycle. *J. Cell Biol.* **44**, 476–83. [121]

Painter, P. R. & Marr, A. G. (1968). Mathematics of microbial populations. *A. Rev. Microbiol.* **22**, 519–48. [19]

Painter, R. B. & Schaefer, A. W. (1969). Rate of synthesis along replicons of different kinds of mammalian cells. *J. molec. Biol.* **45**, 467–79. [77, 79]

Painter, R. B., Drew, R. M. & Rasmussen, R. E. (1964). Limitations in the use of carbon-labelled and tritium-labelled thymidine in cell culture studies. *Radiat. Res.* **21**, 355–66. [32]

Painter, R. B., Jermany, D. A. & Rasmussen, R. E. (1966). A method to determine the number of DNA replicating units in cultured mammalian cells. *J. molec. Biol.* **17**, 47–56. [76]

Pardee, A. B. (1968). Control of cell division: models from microorganisms. *Cancer Res.* **28**, 1802–9. [91]

Pardue, M. L. & Gall, J. G. (1970). Chromosomal localization of mouse satellite DNA. *Science, N.Y.* **168**, 1356–8. [75]

Parsons, J. A. & Rustad, R. C. (1968). The distribution of DNA among dividing mitochondria of *Tetrahymena pyriformis*. *J. Cell Biol.* **37**, 683–93. [71, 183]

Pato, M. L. & Glaser, D. A. (1968). The origin and direction of replication of the chromosome of *Escherichia coli* B/r. *Proc. natn. Acad. Sci. U.S.A.* **60**, 1268–74. [99, 165, 166, 180]

Peacock, W. J. (1963). Chromosome duplication and structure as determined by autoradiography. *Proc. natn. Acad. Sci. U.S.A.* **49**, 793–801. [73n]

Pederson, T. & Robbins, E. (1970*a*). Absence of translational control of histone synthesis during the HeLa life cycle. *J. Cell Biol.* **45**, 509–13. [155]

Pederson, T. & Robbins, E. (1970*b*). RNA synthesis in HeLa cells. Pattern in hypertonic medium and its similarity to synthesis during G_2-prophase. *J. Cell Biol.* **47**, 734–44. [253]

Pederson, T. & Robbins, E. (1970*c*). Actinomycin-^{3}H binding during the HeLa cell life cycle. *J. Cell Biol.* **47**, 155*a*. [251]

Penman, S., Vesco, C. & Penman, M. (1968). Localization and kinetics of formation of nuclear heterodisperse RNA, cytoplasmic heterodisperse RNA and polyribosome-associated messenger RNA in HeLa cells. *J. molec. Biol.* **34**, 49–69. [87]

Petersen, A. J. (1964). DNA synthesis and chromosomal asynchrony. *J. Cell Biol.* **23**, 651–4. [73n]

Petersen, D. F. & Anderson, E. C. (1964). Quantity production of synchronized mammalian cells in suspension culture. *Nature, Lond.* **203**, 642–3. [26, 27]

Petersen, D. F., Anderson, E. C. & Tobey, R. A. (1968). Mitotic cells as a source of synchronized cultures. In *Methods in Cell Physiology.* Vol. 3, pp. 347–70. Ed. by D. M. Prescott. New York and London: Academic Press. [48, 49]

Petersen, D. F., Tobey, R. A. & Anderson, E. C. (1969). Essential biosynthetic activity in synchronized mammalian cells. In *The Cell Cycle. Gene-Enzyme Interactions,* pp. 341–59. Ed. by G. M. Padilla, G. L. Whitson, and I. L. Cameron. New York and London: Academic Press. [222, 223]

Pfeiffer, S. E. (1968). RNA synthesis in synchronously growing populations of HeLa S3 cells. II. Rate of synthesis of individual RNA fractions. *J. Cell Physiol.* **71**, 95–104. [119]

Pfeiffer, S. E. & Tolmach, L. J. (1967). Selecting synchronous populations of mammalian cells. *Nature, Lond.* **213**, 139–42. [29]

Pfeiffer, S. E. & Tolmach, L. J. (1968). RNA synthesis in synchronously growing populations of HeLa S3 cells. I. Rate of total RNA synthesis and its relationship to DNA synthesis. *J. Cell Physiol.* **71**, 77–94. [118, 120]

Pfleuger, O. H. & Yunis, J. J. (1966). Late replication patterns of chromosomal DNA in somatic tissues of the Chinese hamster. *Expl Cell Res.* **44**, 413–29. [62]

Pilgrim, Ch. (1964). Autoradiographische Bestimmung der Dauer der DNA-Verdopplung bei verschiedenen Zellarten von Maus und Ratte nach einer neuen Doppelmarkierungsmethode. *Anat. Anz.* **115**, 128–33. [23]

Pilgrim, Ch. & Maurer, W. (1965). Autoradiographische Untersuchung über die Konstanz der DNA-Verdopplungs-dauer bei Zellarten von Maus und Ratte durch Doppelmarkierung mit ^{3}H- und ^{14}C-thymidin. *Expl Cell Res.* **37**, 183–99. [62]

Pilgrim, Ch., Lang, W. & Maurer, W. (1966). Autoradiographische Untersuchungen der Dauer der S-phase und des Generationszyklus der Basal-Epithelium des Ohres der Maus. *Expl Cell Res.* **44**, 129–38. [62]

Pirson, A. & Lorenzen, H. (1966). Synchronized dividing Algae. *A. Rev. Pl. Physiol.* **17**, 439–58. [44, 45]

Pitelka, D. R. (1969). Centriole replication. In *Handbook of Molecular Cytology,* pp. 1199–218. Ed. by A. Lima-de-Faria. Amsterdam and London: North Holland Publishing Co. [184, 185]

Pittard, J. & Ramakrishnan, T. (1964). Gene transfer by F′ strains of *Escherichia coli.* IV. Effect of a chromosomal deletion on chromosome transfer. *J. Bact.* **88**, 367–73. [166]

Plaut, W. (1963). On the replication organisation of DNA in the polytene chromosome of *Drosophila melanogaster. J. molec. Biol.* **7**, 632–5. [73n]

Plaut, W., Nash, D. & Fanning, T. (1966). Ordered replication of DNA in polytene chromosomes of *Drosophila melanogaster. J. molec. Biol.* **16**, 85–93. [73n, 75]

Plesner, P. (1964). Nucleotide metabolism during synchronized cell division in *Tetrahymena pyriformis*. *C. r. Trav. Lab. Carlsberg* **34**, 1–76. [199]

Pontefract, R. D., Bergeron, G. & Thatcher, F. S. (1969). Mesosomes in *Escherichia coli*. *J. Bact.* **97**, 367–75. [100]

Popoff, M. (1908). Experimentelle Zellstudien. *Arch. exp. Zellforsch.* **1**, 245–379. [151]

Popoff, M. (1909). Experimentelle Zellstudien II. Ueber die Zellgrosse, ihre Fixierung und Vererbung. *Arch. exp. Zellforsch.* **3**, 124–80. [151]

Powell, W. F. (1962). The effect of ultraviolet irradiation and the inhibition of protein synthesis on the initiation of deoxyribonucleic acid synthesis in mammalian cells in culture. I. The overall process of deoxyribonucleic acid synthesis. *Biochim. biophys. Acta* **55**, 969–78. [86n]

Prasad, A. B. & Godward, M. B. E. (1965). Comparison of the developmental response of diploid and tetraploid *Phalaris* following irradiation of the dry seed. I. Determination of mitotic cycle time, mitotic time and phase time. *Radiat. Bot.* **5**, 465–74. [66, 83]

Prensky, W. & Smith, H. H. (1964). Incorporation of ^{3}H-arginine in chromosomes of *Vicia faba*. *Expl Cell Res.* **34**, 525–32. [156]

Prescott, D. M. (1955). Relations between cell growth and cell division. I. Reduced weight, cell volume, protein content, and nuclear volume of *Amoeba proteus* from division to division. *Expl Cell Res.* **9**, 328–37. [9, 16, 134, 135, 141, 151, 152]

Prescott, D. M. (1957). Change in the physiological state of a cell population as a function of culture growth and age (*Tetrahymena geleii*) *Expl Cell Res.* **12**, 126–34. [5]

Prescott, D. M. (1960). Relation between cell growth and cell division. IV. The synthesis of DNA, RNA and protein from division to division in *Tetrahymena*. *Expl Cell Res.* **19**, 228–38. [67n, 123, 235]

Prescott, D. M. (1961). The growth-duplication cycle of the cell. *Int. Rev. Cytol.* **11**, 255–82. [201]

Prescott, D. M. (1964a). The normal cell cycle. In *Synchrony in Cell Division and Growth*, pp. 71–108. Ed. by E. Zeuthen. New York: Interscience Publishers Inc. [123, 201]

Prescott, D. M. (1964b). Comments on the cell life cycle. *Natn. Cancer Inst. Monogr.* **14**, 57–72. [201]

Prescott, D. M. (1964c). Ed. *Methods in Cell Physiology*. Vol. 1. New York and London: Academic Press. [10]

Prescott, D. M. (1966a). Ed. *Methods in Cell Physiology*. Vol. 2. New York and London: Academic Press. [10]

Prescott, D. M. (1966b). The synthesis of total macronuclear protein, histone and DNA during the cell cycle in *Euplotes eurystomus*. *J. Cell Biol.* **31**, 1–9. [68, 81n, 154]

Prescott, D. M. (1968a). Ed. *Methods in Cell Physiology*. Vol. 3. New York and London: Academic Press. [10]

Prescott, D. M. (1968b). Regulation of cell reproduction. *Cancer Res.* **28**, 1815–20. [201]

Prescott, D. M. (1970). Structure and replication of eukaryotic chromosomes. *Adv. Cell Biol.* **1**, 57–117. [58]

Prescott, D. M. & Bender, M. A. (1962). Synthesis of RNA and protein during mitosis in mammalian tissue culture cells. *Expl Cell Res.* **26**, 260–8. [116, 121, 132]

Prescott, D. M. & Goldstein, L. (1967). Nuclear-cytoplasmic interaction in DNA synthesis. *Science, N.Y.* **155**, 469–70. [69, 85]

Prescott, D. M. & Kimball, R. F. (1961). Relation between RNA, DNA and protein syntheses in the replicating nucleus of *Euplotes*. *Proc. natn. Acad. Sci. U.S.A.* **47**, 686–93. [123, 153]

Prescott, D. M., Kimball, R. F. & Carrier, R. F. (1962). Comparison between the timing of micronuclear and macronuclear DNA synthesis in *Euplotes eurystomus*. *J. Cell Biol.* **13**, 175–6. [68]

Priest, J. H. & Callan, H. G. (1970). Relationship between DNA content, length of S phase and DNA fiber radioautograms. *J. Cell Biol.* **47**, 162*a*. [252]

Pritchard, R. H. & Lark, K. G. (1964). Induction of replication by thymine starvation at the chromosome origin in *Escherichia coli*. *J. molec. Biol.* **9**, 288–307. [95, 96, 108]

Pritchard, R. H., Barth, P. T. & Collins, J. (1969). Control of DNA synthesis in bacteria. *Symp. Soc. gen. Microbiol.* **19**, 263–97. [108]

Prudhomme, J. C., Gillot, S. & Daillie, J. (1967). Effets de l'actinomycine D sur la synthèse de l'ADN et de l'ARN dans la glande sérigène de *Bombyx mori*. *Expl Cell Res.* **48**, 186–9. [88n]

Puck, T. T. (1964). Phasing, mitotic delay, and chromosomal aberrations in mammalian cells. *Science, N.Y.* **144**, 565–6. [26]

Puck, T. T. & Marcus, P. I. (1956), Action of X-rays on mammalian cells. *J. exp. Med.* **103**, 653–66. [234n]

Puck, T. T. & Steffen, J. (1963). Life cycle analysis of mammalian cells. I. A method for localising metabolic events within the life cycle, and its application to the action of colcemide and sublethal doses of X-irradiation. *Biophys. J.* **3**, 379–97. [23]

Puck, T. T., Sanders, P. & Petersen, D. (1964). Life cycle analysis of mammalian cells. II. Cells from the Chinese hamster ovary grown in suspension culture. *Biophys. J.* **4**, 441–50. [223]

Quastler, H. & Sherman, F. G. (1959). Cell population kinetics in the intestinal epithelium of the mouse. *Expl Cell Res.* **17**, 420–38. [21]

Rajewsky, M. F. (1970). Synchronisation *in vivo*: kinetics of a malignant cell system following temporary inhibition of DNA synthesis with hydroxyurea. *Expl Cell Res.* **60**, 269–76. [33]

Ramareddy, G. & Reiter, H. (1970). Sequential loss of loci in thymine-starved *Bacillus subtilis* 168 cells. Evidence for a circular chromosome. *J. molec. Biol.* **50**, 525–32. [91]

Randall, J. & Disbrey, C. (1965). Evidence for the presence of DNA at basal body sites in *Tetrahymena pyriformis*. *Proc. R. Soc.* B **162**, 473–91. [185]

Rao, P. N. (1968). Mitotic synchrony in mammalian cells treated with nitrous oxide at high pressures. *Science, N.Y.* **160**, 774–6. [28]

Rao, P. N. & Engelberg, J. (1966). Effects of temperature on the mitotic cycle of normal and synchronized mammalian cells. In *Cell Synchrony*, pp. 332–52. Ed. by I. L. Cameron and G. M. Padilla. New York and London: Academic Press. [39, 61]

Rao, B. & Hinegardner, R. T. (1965). Analysis of DNA synthesis and X-ray-induced mitotic delay in sea urchin eggs. *Radiat. Res.* **26**, 534–7. [248]

Rao, M. V. N. & Prescott, D. M. (1967). Micronuclear RNA synthesis in *Paramecium caudatum*. *J. Cell Biol.* **33**, 281–5. [68, 123]

Rao, P. T. & Johnston, R. T. (1970). Mammalian cell fusion: studies on the regulation of DNA synthesis and mitosis. *Nature, Lond.* **225**, 159–64. [85]

Rapkine, L. (1931). Sur les processus chimiques au cours de la division cellulaire. *Annls. Physiol. Physicochim. biol.* **7**, 382–418. [199]

Rasch, E., Swift, H., & Klein, R. M. (1959). Nucleoprotein changes in plant tumor growth. *J. biophys. biochem. Cytol.* **6**, 11–34. [249]

Rasmussen, L. (1963). Delayed division in *Tetrahymena* as induced by short-time exposures to anaerobiosis. *C. r. Trav. Lab. Carlsberg* **33**, 53–71. [209, 210]

Rasmussen, L. (1967). Effects of metabolic inhibitors on *Paramecium aurelia* during the cell generation cycle. *Expl Cell Res.* **48**, 132–9. [225, 226]

Rasmussen, L. & Zeuthen, E. (1962). Cell division and protein synthesis in *Tetrahymena*, as studied with *p*-fluorophenylalanine. *C. r. Trav. Lab. Carlsberg* **32**, 333–58. [208, 209]

Ray, P. M. (1967). Radioautographic study of cell wall deposition in growing plant cells. *J. Cell Biol.* **35**, 659–74. [189]

Regan, J. D. & Chu, E. H. Y. (1966). A convenient method for assay of DNA synthesis in synchronized human cell cultures. *J. Cell Biol.* **28**, 139–43. [28]

Richards, B. M. & Bajer, A. (1961). Mitosis in endosperm. Changes in nuclear and chromosome mass during mitosis. *Expl Cell Res.* **22**, 503–8. [149]

Robbins, E. & Borun, T. W. (1967). The cytoplasmic synthesis of histones in HeLa cells and its temporal relationship to DNA replication. *Proc. natn. Acad. Sci. U.S.A.* **57**, 409–16. [155]

Robbins, E. & Gonatas, N. K. (1964). Histochemical and ultrastructural studies on HeLa cell cultures exposed to spindle inhibitors with special reference to the interphase cell. *J. Histochem. Cytochem.* **12**, 704–11. [29]

Robbins, E. & Morrill, G. A. (1969). Oxygen uptake during the HeLa cell life cycle and its correlation with macromolecular synthesis. *J. Cell Biol.* **43**, 629–33. [194, 195]

Robbins, E. & Scharff, M. (1966). Some macromolecular characteristics of synchronized HeLa cells. In *Cell Synchrony*, pp. 353–74. Ed. by I. L. Cameron and G. M. Padilla. New York and London: Academic Press. [47, 48, 181, 197, 200]

Robbins, E. & Scharff, M. D. (1967). The absence of a detectable G1 phase in a cultured strain of Chinese hamster lung cell. *J. Cell Biol.* **34**, 684–5. [60, 61, 130]

Robbins, E. & Shelanski, M. (1969). Synthesis of a colchicine-binding protein during the HeLa cell life cycle. *J. Cell Biol.* **43**, 371–3. [156]

Robbins, E., Jentzsch, G. & Micali, A. (1968). The centriole cycle in synchronized cells. *J. Cell Biol.* **36**, 329–39. [185]

Robinson, A. A. (1971). Unpublished. [168, 177]

Roelofsen, P. A. (1959). *The Plant Cell-wall.* Berlin: Borntraeger. [189]

Roelofsen, P. A. (1965). Ultrastructure of the wall in growing cells and its relation to the direction of growth. *Adv. Bot. Res.* **2**, 69–149. [189]

Rogers, A. W. (1967). *Techniques of Autoradiography.* Amsterdam and New York: Elsevier Publishing Co. [10]

Rogers, H. J. (1965). The outer layers of bacteria: the biosynthesis of structure. *Symp. Soc. gen. Microbiol.* **15**, 186–219. [192]

Romsdahl, M. M. (1968). Synchronization of human cell lines with colcemid. *Expl Cell Res.* **50**, 463–7. [28]

Ron, A. & Prescott, D. M. (1969). The timing of DNA synthesis in *Amoeba proteus. Expl Cell Res.* **56**, 430–4. [70]

Roodyn, D. B. & Wilkie, D. (1968). *The Biogenesis of Mitochondria.* London: Methuen. [71]

Rooney, D. W. & Eiler, J. J. (1967). Synchronization of *Tetrahymena* cell division by multiple hypoxic shocks. *Expl Cell Res.* **48**, 649–52. [210]

Rooney, D. W., Yen, B-C. & Mikita, D. J. (1971). Synchronization of *Chlamydomonas* division with intermittent hypothermia. *Expl Cell Res.* **65**, 94–8. [250]

Rosenberg, H. M. & Gregg, E. C. (1969). Kinetics of cell volume changes of murine lymphoma cells subjected to different agents. *Biophys. J.* **9**, 592–606. [30, 31, 32, 234n]

Ross, K. F. A. (1961). The immersion refractometry of living cells by phase contrast and interference microscopy. In *General Cytochemical Methods.* Vol. 2, pp. 1–59. Ed. by J. F. Danielli. New York and London: Academic Press. [10]

Ross, K. F. A. (1967). *Phase Contrast and Interference Microscopy for Cell Biologists.* London: Edward Arnold. [9]

Rudner, D., Rejman, E. & Chargaff, E. (1965). Genetic implications of periodic pulsations of the rate of synthesis and the composition of rapidly labelled bacterial RNA. *Proc. natn. Acad. Sci. U.S.A.* **54**, 904–11. [54, 125n, 126]

Rueckert, R. R. & Mueller, G. C. (1960). Studies on unbalanced growth in tissue cultures. I. Induction and consequences of thymidine deficiency. *Cancer Res.* **20**, 1584–91. [27, 28, 30]

Rusch, H. P. (1969). Some biochemical events in the growth cycle of *Physarum polycephalum. Fedn Proc. Fedn Am. Socs exp. Biol.* **28**, 1761–70. [56, 163, 178, 254]

Rusch, H. P. (1970). Some biochemical events in the life cycle of *Physarum polycephalum. Adv. Cell Biol.* **1**, 297–327. [227]

Rusch, H. P., Sachsenmaier, W., Behrens, K. & Gruter, V. (1966). Synchronization of mitosis by the fusion of the plasmodia of *Physarum polycephalum. J. Cell Biol.* **31**, 204–9. [248]

Rustad, R. C. (1960). Changes in the sensitivity to ultraviolet-induced mitotic delay during the cell division cycle of the sea urchin egg. *Expl Cell Res.* **21**, 596–602. [237]

Rustad, R. C. (1964). U.V.-induced mitotic delay in the sea urchin egg. *Photochem. & Photobiol.* **3**, 529–38. [234, 240]

Rustad, R. C. (1970). Variations in the sensitivity to X-ray-induced mitotic delay during the cell division cycle of the sea urchin egg. *Radiat. Res.* **42**, 498–512. [235, 236, 237]

Rustad, R. C. (1971). Radiation responses during the mitotic cycle of the sea urchin egg. In *Developmental Aspects of the Cell Cycle*, pp. 127–59. Ed. by I. L. Cameron, G. M. Padilla, and A. Zimmerman. New York and London: Academic Press. [237, 241]

Rustad, R. C. & Burchill, B. R. (1966). Radiation-induced mitotic delay in sea urchin eggs treated with puromycin and actinomycin D. *Radiat. Res.* **29**, 203–10. [243]

Ryter, A. (1968). Association of the nucleus and the membrane of bacteria: a morphological study. *Bact. Rev.* **32**, 39–54. [100]

Ryter, A., Hirota, Y. & Jacob, F. (1968). DNA-membrane complex and nuclear segregation in bacteria. *Cold Spring Harb. Symp. quant. Biol.* **33**, 669–76. [100]

Sachsenmaier, W. & Ives, D. H. (1965). Periodische Änderugen der Thymidinkinase Activität in Synchronen Mitosecyclus von *Physarum Polycephalum. Biochim. Z.* **343**, 399–406. [178]

Sachsenmaier, W., Bohnert, E., Clausnizer, B. & Nygaard, O. F. (1970). Cycle dependent variation of X-ray effects of synchronous mitosis and thymidine kinase induction in *Physarum polycephalum. Febs Lett. (Fed. eur. Biochem. Socs.)* **10**, 185–9. [238]

Sachsenmaier, W., Dönges, K. H., Rupff, H. & Czihak, G. (1970). Advanced initiation of synchronous mitosis in *Physarum polycephalum* following ultraviolet-irradiation. *Z. Naturf.* **25b**, 866–71. [239, 248]

Sachsenmaier, W., Fournier, D. V. & Gürtler, K. F. (1967). Periodic thymidine kinase production in synchronous plasmodia of *Physarum polycephalum*: inhibition by actinomycin and actidione. *Biochem. biophys. Res. Commun.* **27**, 655–60. [227]

Sachsenmaier, W., Scholz, R. & Grunet, J. (1968). Änderungen des Nucleotid-Spiegels im Verlauf des synchronen Mitosezyklus von *Physarum polycephalum. Hoppe–Seyler's Z. physiol. Chem.* **349**, 1257. [199]

Sadgopal, A. & Bonner, J. (1969). The relationship between histone and DNA synthesis in HeLa cells. *Biochim. biophys. Acta* **186**, 349–57. [155n, 156]

Sakai, H. (1960). Studies on sulfhydryl groups during cell division of sea urchin egg. III. —SH groups of KCl-soluble proteins and their change during cleavage. *J. biophys. biochem. Cytol.* **8**, 609–15. [213]

Sakai, H. & Dan, K. (1959). Studies on sulphydryl groups during cell division of sea urchin egg. 1. Glutathione. *Expl. Cell Res.* **16**, 24–41. [199, 213]

Salas, J. & Green, H. (1971). Proteins binding to DNA and their relation to growth in cultured mammalian cells. *Nature, New Biol. Lond.* **229**, 165–9. [253]

Salb, J. M. & Marcus, P. I. (1965). Translational inhibition in mitotic HeLa cells. *Proc. natn. Acad. Sci. U.S.A.* **54**, 1353–8. [133]

Samoshkina, N. A. (1968). A study of DNA synthesis in the period of ovicell division in mice (*in vitro* experiments). (In Russian). *Tsitologiya* **10**, 856–64. [66]

Sandberg, A. A., Sofuni, T., Takagi, N. & Moore, G. E. (1966). Chronology and patterns of human chromosome replication. IV. Autoradiographic studies of binucleate cells. *Proc. natn. Acad. Sci. U.S.A.* **56**, 105–10. [84]

✗ Sando, N. (1963). Biochemical studies on the synchronized culture of *Schizosaccharomyces pombe. J. gen. appl. Microbiol.*, *Tokyo* **9**, 233–41. [35, 36]

Satir, P. & Zeuthen, E. (1961). Cell cycle and the relationship of growth rate to reduced weight (RW) in the giant Amoeba *Chaos chaos* L. *C. r. Trav. Lab. Carlsberg* **32**, 241–64. [135]

Schaechter, M., Bentzon, M. W. & Maaløe, O. (1959). Synthesis of deoxyribonucleic acid during the division cycle of bacteria. *Nature, Lond.* **183**, 1207. [102]

Schaechter, M., Maaløe, O. & Kjeldgaard, N. O. (1958). Dependency on medium and temperature of cell size and chemical composition during balanced growth of *Salmonella typhimurium. J. gen. Microbiol.* **19**, 592–606. [108]

Schaechter, M., Williamson, J. P., Hood, J. R. & Koch, A. L. (1962). Growth, cell and nuclear division in some bacteria. *J. gen. Microbiol.* **29**, 421–34. [142, 143, 145]

Schaer, J. C., Ramseier, L. & Schindler, R. (1971). Studies on the division cycle of mammalian cells IV. Incorporation of labelled precursors into DNA of synchronously dividing cells in culture. *Expl Cell Res.* **65**, 17–22. [251]

Scharff, M. D. & Robbins, E. (1965). Synthesis of ribosomal RNA in synchronized HeLa cells. *Nature, Lond.* **208**, 464–6. [119, 130, 132]

Scharff, M. D. & Robbins, E. (1966). Polyribosome disaggregation during metaphase. *Science, N.Y.* **151**, 992–5. [28, 133]

Scherbaum, O. H. (1960). Synchronous division of micro-organisms. *A. Rev. Microbiol.* **14**, 283–310. [201n]

Scherbaum, O. H. (1963a). Comparison of synchronous and synchronized cell division. *Expl Cell Res.* **33**, 89–98. [56]

Scherbaum, O. H. (1963b). Chemical pre-requisites for cell division. In *The Cell in Mitosis*, pp. 125–37. Ed. by L. Levine. New York and London: Academic Press. [201n]

Scherbaum, O. H. (1964). Biochemical studies on synchronized *Tetrahymena*. In *Synchrony in Cell Division and Growth*, pp. 177–95. Ed. by E. Zeuthen. New York: Interscience Publishers Inc. [201n]

Scherbaum, O. H. & Loefer, J. B. (1964). Environmentally induced growth oscillations in Protozoa. In *Biochemistry and Physiology of Protozoa*. Vol. 3, pp. 9–59. Ed. by S. H. Hutner. New York and London: Academic Press. [201n]

Scherbaum, O. & Rasch, G. (1957). Cell size distribution and single cell growth in *Tetrahymena pyriformis* GL. *Acta path. microbiol. Scand.* **41**, 161–82. [21]

Scherbaum, O. & Zeuthen, E. (1954). Induction of synchronous cell division in mass cultures of *Tetrahymena pyriformis*. *Expl Cell Res.* **6**, 221–7. [40, 201]

Schiff, S. O. (1965). Ribonucleic acid synthesis in neuroblasts of *Chortophaga viridifasciata* (de Geer) as determined by observations of individual cells in the mitotic cycle. *Expl Cell Res.* **40**, 264–76. [116n]

Schildkraut, C. L. & Maio, J. J. (1968). Studies on the intranuclear distribution and properties of mouse satellite DNA. *Biochim. biophys. Acta* **161**, 76–93. [76]

Schindler, R., Ramseier, L., Schaer, J. C. & Grieder, A. (1970). Studies on the division cycle of mammalian cells. III. Preparation of synchronously dividing cell populations by isotonic sucrose gradient centrifugation. *Expl Cell Res.* **59**, 90–6. [50]

Schmid, P. (1967). Temperature adaptation of the growth and division process of *Tetrahymena pyriformis*. I. Adaptation. II. Relationship between cell growth and cell replication. *Expl Cell Res.* **45**, 460–70; 471–86. [135]

Schmid, W. (1963). DNA replication pattern of human chromosomes. *Cytogenetics.* **2**, 175–93. [73n]

Schmidt, R. R. (1966). Intracellular control of enzyme synthesis and activity during synchronous growth of *Chlorella.* In *Cell Synchrony*, pp. 189–235. Ed. by I. L. Cameron and G. M. Padilla. New York and London: Academic Press. [44]

Schmidt, R. R. (1969). Control of enzyme synthesis during the cell cycle of *Chlorella.* In *The Cell Cycle. Gene-Enzyme Interactions*, pp. 159–77. Ed. by G. M. Padilla, G. L. Whitson and I. L. Cameron. New York and London: Academic Press. [44, 141]

Schneider, D. O. & Johns, R. M. (1966). Enhancement of radiation-induced mitotic inhibition by BUdR incorporation in L cells. *Radiat. Res.* **28**, 657–67. [241]

Schneider, L. K. & Rieke, W. O. (1967). DNA replication patterns and chromosomal protein synthesis in opossum lymphocytes *in vitro. J. Cell Biol.* **33**, 497–510. [73n, 154]

Scholander, P. F., Claff, C. L., Sveinsson, S. L. & Scholander, S. I. (1952). Respiratory studies of single cells. III. Oxygen consumption during cell division. *Biol. Bull. mar. biol. Lab., Woods Hole* **102**, 185–99. [193]

Scholander, P. F., Leivestad, H. & Sundnes, G. (1958). Cycling in the oxygen consumption of cleaving eggs. *Expl Cell Res.* **15**, 505–11. [193]

Schwarzacher, H. G. (1963). Sex chromatin in living human cells *in vitro. Cytogenetics* **2**, 117–28. [181]

Schwarzacher, H. G. & Schnedl, W. (1965). Der Zellzyklus in Fibroblastenkulturen von Menschen. *Z. Zellforsch. mikrosk. Anat.* **67**, 165–73. [81n]

Scopes, A. W. & Williamson, D. H. (1964). The growth and oxygen uptake of synchronously dividing cultures of *Saccharomyces cerevisiae. Expl Cell Res.* **35**, 361–71. [141, 195, 196]

Scott, D. B. M. & Chu, E. (1958). Synchronized division of growing cultures of *Escherichia coli. Expl Cell Res.* **14**, 166–74. [38, 40n]

Seed, J. (1962). The synthesis of deoxyribonucleic acid and nuclear protein in normal and tumour strain cells. *Proc. R. Soc.* B **156**, 41–56. [151]

Seed, J. (1964). The relation between RNA, DNA and protein in normal embryonic cell nuclei and spontaneous tumour cell nuclei. *J. Cell Biol.* **20**, 17–23. [151]

Seed, J. (1966). The synthesis of DNA, RNA and nuclear protein in normal and tumour strain cells. I. Fresh embryo human cells. II. Fresh embryo mouse cells. III. Mouse ascites tumour cells. IV. HeLa tumour strain cells. *J. Cell Biol.* **28**, I. 233–48, II. 249–56, III. 257–62, IV. 263–76. [151]

Selman, G. G. & Waddington, C. H. (1955). The mechanism of cell division in the cleavage of the newt's egg. *J. exp. Biol.* **32**, 700–33. [10, 188]

Senger, H. & Bishop, N. I. (1969). Light-dependent formation of nucleic acids and its relation to the induction of synchronous cell division in *Chlorella.* In *The Cell Cycle. Gene-Enzyme Interactions*, pp. 179–202. Ed. by G. M. Padilla, G. L. Whitson and I. L. Cameron. New York and London: Academic Press. [44]

Shall, S. & McClelland, A. J. (1971). Synchronization of mouse fibroblast LS cells grown in suspension culture. *Nature, New Biol. Lond.* **229**, 59–61. [250]

Shapiro, I. M. & Levina, L. Y. (1967). Autoradiographical study on the time of nuclear protein synthesis in human leucocyte blood culture. *Expl Cell Res.* **47**, 75–87. [154]

Shen, S. R-C. & Schmidt, R. R. (1966). Enzymic control of nucleic acid synthesis during synchronous growth of *Chlorella pyrenoidosa.* II. Deoxycytidine monophosphate deaminase. *Archs Biochem. Biophys.* **115**, 13–20. [168]

Shepherd, G. R., Noland, B. J. & Hardin, J. M. (1971). Histone acetylation in synchronized mammalian cell cultures. *Biochim. biophys. Acta* **228**, 544–9. [250]

Showacre, J. L. (1968). Staging of the cell cycle with time-lapse photography. In *Methods in Cell Physiology*. Vol. 3, pp. 147–59. Ed. by D. M. Prescott. New York and London: Academic Press. [16]

Sinclair, R. & Bishop, D. H. L. (1965). Synchronous culture of strain-L mouse cells. *Nature, Lond.* **205**, 1272–3. [49]

Sinclair, W. K. (1967). Hydroxyurea: effects on Chinese hamster cells grown in culture. *Cancer Res.* **27**, 297–308. [33]

Sinclair, W. K. (1968). Cyclic X-ray response in mammalian cells *in vitro*. *Radiat. Res.* **33**, 620–43. [234, 235, 236, 242]

Sinclair, W. K. & Morton, R. A. (1963). Variation in X-ray response during the division cycle of partially synchronized Chinese hamster cells in culture. *Nature, Lond.* **199**, 1158–60. [47]

Sinclair, W. K. & Ross, D. W. (1969). Modes of growth in mammalian cells. *Biophys. J.* **9**, 1056–70. [131]

Sisken, J. E. (1964). Methods for measuring the length of the mitotic cycle and the timing of DNA synthesis for mammalian cells in culture. In *Methods in Cell Physiology*. Vol. 1, pp. 387–401. Ed. by D. M. Prescott. New York and London: Academic Press. [21]

Sisken, J. E. & Iwasaki, T. (1969). The effects of some amino acid analogs on mitosis and the cell cycle. *Expl Cell Res.* **55**, 161–7. [225]

Sisken, J. E. & Kinosita, R. (1961). Timing of DNA synthesis in the mitotic cycle *in vitro*. *J. biophys. biochem. Cytol.* **9**, 509–18. [61]

Sisken, J. E. & Morasca, L. (1965). Intrapopulation kinetics of the mitotic cycle. *J. Cell Biol.* **25**, 179–90. [88]

Sisken, J. E. & Wilkes, E. (1967). The times of synthesis and the conservation of mitosis-related proteins in cultured human amnion cells. *J. Cell Biol.* **34**, 97–110. [225]

Sisken, J. E., Morasca, L. & Kibby, S. (1965). Effects of temperature on the kinetics of the mitotic cycle of mammalian cells in cultures. *Expl Cell Res.* **39**, 103–16. [61]

Sitz, T. O., Hopkins, H. A. & Schmidt, R. R. (1970). Synchronous culture production by density selection. *Science, N.Y.* **170**, 97. [51]

Sitz, T. O., Kent, A. B., Hopkins, H. A. & Schmidt, R. R. (1970). Equilibrium density-gradient procedure for selection of synchronous cells from asynchronous cultures. *Science, N.Y.* **168**, 1231–2. [51]

Smets, L. (1969). Discrepancies between precursor uptake and DNA synthesis in mammalian cells. *J. Cell Physiol.* **74**, 63–6. [15]

Smith, B. J. (1970). Light satellite-band DNA in mouse cells infected with polyoma virus. *J. molec. Biol.* **47**, 101–6. [76]

Smith, C. L. & Dendy, P. P. (1962). Relation between mitotic index, duration of mitosis, generation time and fraction of dividing cells in a cell population. *Nature, Lond.* **193**, 555–6. [19]

Smith, D., Tauro, P., Schweizer, E. & Halvorson, H. O. (1968). The replication of mitochondrial DNA during the cell cycle in *Saccharomyces lactis*. *Proc. natn. Acad. Sci. U.S.A.* **60**, 936–42. [71]

Smith, D. W. & Hanawalt, P. C. (1967). Properties of the growing point region in the bacterial chromosome. *Biochim. biophys. Acta* **149**, 519–31. [100]

Smith, H. H., Fussell, C. P. & Kugelman, B. H. (1963). Partial synchronisation of nuclear division in root meristems with 5-amino-uracil. *Science, N.Y.* **142**, 595. [28]

Smith, H. S. & Pardee, A. B. (1970). Accumulation of a protein required for division during the cell cycle of *Escherichia coli*. *J. Bact.* **101**, 901–9. [40n, 108, 111, 228, 229]

Sober, H. A. (1968). Ed. *Handbook of Biochemistry*. Cleveland, Ohio: Chemical Rubber Co. [98]

Soeder, C. J. (1965). Elektronenmikroskopische Untersuchung der Protoplastenteilung bei *Chlorella fusca* Shihira et Kraus. *Arch. Mikrobiol.* **50**, 368–77. [183]

Solter, D., Škreb, N. & Damjanov, I. (1971). Cell cycle analysis in the mouse egg-cylinder. *Expl Cell Res.* **64**, 331–4. [251]

Sonenshein, G. E., Shaw, C. A. & Holt, C. E. (1970). Synthesis of genes for rRNA during the G_2 period in *Physarum. J. Cell Biol.* **47**, 198a [251]

Sorokin, C. (1963). On the variability in the activity of the photosynthetic mechanism. In *Photosynthesis Mechanisms in Green Plants.* Publ. 1145. by Natn. Acad. Sci. Natn. Res. Council U.S.A. [44]

Sorokin, C. (1964). Aging at the cellular level. *Experientia* **20**, 353–63. [44]

Spalding, J., Kajiwara, K. & Mueller, C. G. (1966). The metabolism of basic proteins in HeLa cell nuclei. *Proc. natn. Acad. Sci. U.S.A.* **56**, 1535–42. [155n, 156]

Stambrook, P. J. & Sisken, J. E. (1970). Induced alterations in rates of total RNA synthesis during G1 and S of synchronized Chinese hamster cells. *J. Cell Biol.* **47**, 200a. [252]

Stanier, R. Y. (1961). La place des bactéries dans le monde vivant. *Annls Inst. Pasteur, Paris* **101**, 297–312. [6n]

Stanners, C. P. & Till, J. E. (1960). DNA synthesis in individual L-strain mouse cells. *Biochim. biophys. Acta* **37**, 406–19. [19, 60, 81n]

Stebbing, N. (1969). The pool in relation to protein synthesis, with special reference to *Schizosaccharomyces pombe.* Ph.D. thesis, Univ. of Edinburgh. [140, 197, 254, 255]

Stebbing, N. (1971). Changes in pool and macromolecular components of the fission yeast, *Schizosaccharomyces pombe,* during the cell cycle. *J. Cell Sci.* Submitted. [254]

Steed, P. & Murray, R. G. E. (1966). The cell wall and cell division of gram-negative bacteria. *Can. J. Microbiol.* **12**, 263–70. [111]

Stein, G. & Baserga, R. (1970). Continued synthesis of non-histone chromosomal proteins during mitosis. *Biochem. biophys. Res. Commun.* **41**, 715–22. [253]

Steinberg, W. & Halvorson, H. O. (1968). Timing of enzyme synthesis during outgrowth of spores of *Bacillus cereus.* I. Ordered enzyme synthesis. II. Relationship between ordered enzyme synthesis and deoxyribonucleic acid replication. *J. Bact.* **95**, 469–78, 479–89. [111, 116]

Steinert, M. & Van Assel, S. (1967). Replications coordonées des acides désoxyribonucléiques nucléaire et mitochondrial chez *Crithidia luciliae. Archs int. Physiol. Biochem.* **75**, 370–1. [183]

Stern, H. (1966). The regulation of cell division. *A. Rev. Pl. Physiol.* **17**, 345–78. [201]

Stern, H. & Kirk, P. L. (1948). The oxygen consumption of the microspores of *Trillium* in relation to the mitotic cycle. *J. gen. Physiol.* **31**, 243–8. [194]

Stone, G. E. (1968). Synchronized cell division in *Tetrahymena pyriformis* following inhibition with vinblastine. *J. Cell Biol.* **39**, 556–63. [29]

Stone, G. E. & Prescott, D. M. (1964). Cell division and DNA synthesis in *Tetrahymena pyriformis* deprived of essential amino-acids. *J. Cell Biol.* **21**, 275–81. [67n, 87]

Stone, G. E., Miller, O. L. & Prescott, D. M. (1965). ^{3}H-thymidine derivative pools in relation to macromolecular DNA synthesis in *Tetrahymena pyriformis. J. Cell Biol.* **25**, 171–7. [197]

Stonehill, E. H. & Hutchison, D. J. (1966). Chromosomal mapping by means of mutational induction in synchronous populations of *Streptococcus faecalis. J. Bact.* **92**, 136–43. [38, 111]

Stubblefield, E. (1964). DNA synthesis and chromosomal morphology of Chinese hamster cells cultured in media containing *N*-deacetyl-*N*-methyl colchicine (Colcemid). In *Cytogenetics of Cells in Culture,* pp. 223–48. Ed. by R. J. C. Harris. New York and London: Academic Press. [33]

Stubblefield, E. (1965). Quantitative tritium autoradiography of mammalian chromosomes. I. The basic method. *J. Cell Biol.* **25** (3), 137–47. [73n]

Stubblefield, E. (1968). Synchronization methods for mammalian cell culture. In *Methods in Cell Physiology*. Vol. 3, pp. 25–43. Ed. by D. M. Prescott. New York and London: Academic Press. [28]

Stubblefield, E. & Brinkley, B. R. (1967). Architecture and function of the mammalian centriole. *Symp. int. Soc. Cell Biol.* **6**, 175–218. [184]

Stubblefield, E. & Mueller, G. C. (1962). Molecular events in the reproduction of animal cells. II. The focalized synthesis of DNA in the chromosomes of HeLa cells. *Cancer Res.* **22**, 1091–9. [73n]

Stubblefield, E. & Murphree, S. (1967). Synchronized mammalian cell cultures. II. Thymidine kinase activity in colcemid synchronized fibroblasts. *Expl Cell Res.* **48**, 652–6. [179]

Stubblefield, E., Klevecz, R. & Deaven, L. (1967). Synchronized mammalian cell cultures. I. Cell replication cycle and macromolecular synthesis following brief colcemid arrest of mitosis. *J. Cell Physiol.* **69**, 345–54. [131]

Studzinski, G. P. & Lambert, W. C. (1969). Thymidine as a synchronizing agent. I. Nucleic acid and protein formation in synchronous HeLa cultures treated with excess thymidine. *J. Cell Physiol.* **73**, 109–18. [30]

Sueoka, N. (1966). Synchronous replication of the chromosome in *Bacillus subtilis*. In *Cell Synchrony*, pp. 38–53. Ed. by I. L. Cameron and G. M. Padilla. New York and London: Academic Press. [91, 94]

Sueoka, N. & Quinn, W. C. (1968). Membrane attachment of the chromosome replication origin in *Bacillus subtilis*. *Cold Spring Harb. Symp. quant. Biol.* **33**, 695–705. [91, 100, 101]

Summers, L. G. (1963). Variation of cell and nuclear volume of *Tetrahymena pyriformis* with three parameters of growth: age of culture, age of cell and generation time. *J. Protozool.* **10**, 288–93. [135, 151]

Sussman, A. S. & Halvorson, H. O. (1966). *Spores, their Dormancy and Germination*. New York: Harper and Row. [56]

Swann, M. M. (1951). Protoplasmic structures and mitosis. I. The birefringence of the metaphase spindle and asters of the living sea-urchin egg. *J. exp. Biol.* **28**, 417–33. [10]

Swann, M. M. (1953). The mechanism of cell division. A study with carbon monoxide on the sea-urchin egg. *Q. Jl microsc. Sci.* **94**, 369–79. [198, 230]

Swann, M. M. (1954). The mechanism of cell division: experiments with ether on the sea-urchin egg. *Expl Cell Res.* **7**, 505–17. [230]

Swann, M. M. (1955). The mechanism of cell division: the action of 2,4-dinitrophenol and certain glycolytic inhibitors on the sea-urchin egg. *Proc. R. phys. Soc. Edinb.* **24**, 5–7. [231]

Swann, M. M. (1957). The control of cell division: a review. I. General mechanisms. *Cancer Res.* **17**, 727–58. [198, 201, 230]

Swann, M. M. (1958). The control of cell division: a review. II. Special mechanisms. *Cancer Res.* **18**, 1118–60. [201]

Swann, M. M. & Mitchison, J. M. (1953). Cleavage of sea-urchin eggs in colchicine. *J. exp. Biol.* **30**, 506–14. [33]

Swift, H. (1950). The constancy of desoxyribosenucleic acid in plant nuclei. *Proc. natn. Acad. Sci. U.S.A.* **36**, 643–53. [58]

Swift, H. (1953). Quantitative aspects of nuclear nucleoprotein. *Int. Rev. Cytol.* **2**, 1–76. [58]

Swift, H. (1964). The histones of polytene chromosomes. In *The Nucleohistones*, pp. 169–83. Ed. by J. Bonner and P. Ts'o. San Francisco: Holden-Day Inc. [155]

Sylvén, B., Tobias, C. A., Malmgren, H., Ottoson, R. & Thorell, B. (1959). Cyclic variations in the peptidase and catheptic activities of yeast cultures synchronized with respect to cell multiplication. *Expl Cell Res.* **16**, 75–87. [35, 177]

Takahashi, M., Yagi, Y., Moore, G. E. & Pressman, D. (1969). Immunoglobulin production in synchronized cultures of human hematopoietic cell lines. I. Variation of cellular immunoglobulin with the generation cycle. *J. Immun.* **103**, 834–43. [157]

Tamiya, H. (1963*a*). Control of cell division in microalgae. *J. cell. comp. Physiol.* **62**, Suppl. **1**, 157–74. [41, 43]

Tamiya, H. (1963*b*). Cell differentiation in *Chlorella*. *Symp. Soc. exp. Biol.* **17**, 188–214. [41, 42, 43]

Tamiya, H. (1964). Growth and cell division of *Chlorella*. In *Synchrony in Cell Division and Growth*, pp. 247–306. Ed. by E. Zeuthen. New York: Interscience Publishers Inc. [41]

Tamiya, H. (1966). Synchronous cultures of algae. *A. Rev. Pl. Physiol.* **17**, 1–26. [45]

Tamiya, H. & Morimura, Y. (1964). Synchronous cultures of *Chlorella* starting from homogeneous populations. In *Synchrony in Cell Division and Growth*, pp. 565–9. New York: Interscience Publishers Inc. [42, 51]

Tamiya, H., Iwamura, T., Shibata, K., Hase, E. & Nihei, T. (1953). Correlation between photosynthesis and light-independent metabolism in growth of *Chlorella*. *Biochim. biophys. Acta.* **12**, 23–40. [40]

Tamiya, H., Morimura, Y., Yokota, M. & Kunieda, R. (1961). Mode of nuclear division in synchronous cultures of *Chlorella*: comparison of various methods of synchronization. *Pl. Cell Physiol., Tokyo* **2**, 383–403. [43]

Tandler, B., Erlandson, R. A., Smith, A. L. & Wynder, E. L. (1969). Riboflavin and mouse hepatic cell structure and function. II. Division of mitochondria during recovery from simple deficiency. *J. Cell Biol.* **41**, 477–93. [183]

Tauro, P. & Halvorson, H. O. (1966). Effect of gene position on the timing of enzyme synthesis in synchronous cultures of yeast. *J. Bact.* **92**, 652–61. [36, 49, 160, 170n, 171, 177]

Tauro, P., Halvorson, H. L. & Epstein, R. L. (1968). Time of gene expression in relation to centromere distance during the cell cycle of *Saccharomyces cerevisiae*. *Proc. natn. Acad. Sci. U.S.A.* **59**, 277–84. [164, 170n, 171, 173, 176]

Tauro, P., Schweizer, E., Epstein, R. & Halvorson, H. O. (1969). Synthesis of macromolecules during the cell cycle in yeast. In *The Cell Cycle. Gene-Enzyme Interactions*, pp. 101–18. Ed. by G. M. Padilla, G. L. Whitson and I. L. Cameron. New York and London: Academic Press. [124]

Taylor, E. W. (1963). Relation of protein synthesis to the division cycle in mammalian cell cultures. *J. Cell Biol.* **19**, 1–18. [222]

Taylor, E. W. (1965). Control of DNA synthesis in mammalian cells in cultures. *Expl Cell Res.* **40**, 316–32. [86, 88]

Taylor, J. H. (1960*a*). Nucleic acid synthesis in relation to the cell division cycle. *Ann. N.Y. Acad. Sci.* **90**, 409–21. [116n, 132]

Taylor, J. H. (1960*b*). Asynchronous duplication of chromosomes in culture cells of Chinese hamster. *J. biophys. biochim. Cytol.* **7**, 455–64. [73n]

Taylor, J. H. (1963). The replication and organisation of DNA in chromosomes. In *Molecular Genetics*. Pt. 1, pp. 65–111. Ed. by J. H. Taylor. New York and London: Academic Press. [58, 73n]

Taylor, J. H. (1968). Rates of chain growth and units of replication in DNA of mammalian chromosomes. *J. molec. Biol.* **31**, 579–94. [77]

Terasima, T. & Tolmach, L. J. (1961). Changes in X-ray sensitivity of HeLa cells during the division cycle. *Nature, Lond.* **190**, 1210–11. [81, 119]

Terasima, T. & Tolmach, L. J. (1963*a*). Growth and nucleic acid synthesis in synchronously dividing populations of HeLa cells. *Expl Cell Res.* **30**, 344–62. [116n, 131–2, 235]

Terasima, T. & Tolmach, L. J. (1963*b*). Variation in several responses of HeLa cells to X-irradiation during the division cycle. *Biophys. J.* **3**, 11–33. [47]

Terasima, T. & Yasukawa, M. (1966). Synthesis of G1 protein preceding DNA synthesis in cultured mammalian cells. *Expl Cell Res.* **44**, 669–71.

Terry, O. W. & Edmunds, L. N. (1970). Phasing of cell division by temperature cycles in *Euglena* cultured autotrophically under continuous illumination. *Planta* **93**, 106–27. [46]

Thompson, L. H. & Humphrey, R. M. (1970). Proliferation kinetics of mouse L-P59 cells irradiated with ultraviolet light; a time lapse photographic study. *Radiat. Res.* **41**, 183–201. [240]

Thompson, L. R. & McCarthy, B. J. (1968). Stimulation of nuclear DNA and RNA synthesis by cytoplasmic extracts *in vitro*. *Biochim. biophys. Res. Commun.* **30**, 166–72. [84]

Thormar, H. (1959). Delayed division in *Tetrahymena pyriformis* induced by temperature changes. *C. r. Trav. Lab. Carlsberg* **31**, 207–25. [204, 205]

Thrasher, J. D. (1966). Analysis of renewing epithelial cell populations. In *Methods in Cell Physiology*. Vol. 2, pp. 324–57. Ed. by D. M. Prescott. New York and London: Academic Press. [21]

Till, J. E., Whitmore, G. F. & Gulyas, S. (1963). Deoxyribonucleic acid synthesis in individual L-strain mouse cells. II. Effects of thymidine starvation. *Biochim. biophys. Acta* **72**, 277–89. [27]

Tobey, R. A. & Ley, K. D. (1970). Regulation of initiation of DNA synthesis in Chinese hamster cells. I. Production of a stable, reversible G$_1$-arrested population in suspension culture. *J. Cell Biol.* **46**, 151–7. [34]

Tobey, R. A., Anderson, E. C. & Petersen, D. F. (1966). RNA stability and protein synthesis in relation to the division of mammalian cells. *Proc. natn. Acad. Sci. U.S.A.* **56**, 1520–7. [222, 223, 234]

Tobey, R. A., Anderson, E. C. & Petersen, D. F. (1967*a*). The effect of thymidine on the duration of G1 in Chinese hamster cells. *J. Cell Biol.* **35**, 53–9. [26, 61]

Tobey, R. A., Anderson, E. C. & Petersen, D. F. (1967*b*). Differential response of two derivatives of the BHK21 hamster cell to thymidine. *J. Cell Physiol.* **69**, 341–4. [27]

Tobey, R. A., Anderson, E. C. & Petersen, D. F. (1967*c*). Properties of mitotic cells prepared by mechanically shaking monolayer cultures of Chinese hamster cells. *J. Cell Physiol.* **70**, 63–8. [47]

Tobey, R. A., Petersen, D. F., Anderson, E. C. & Puck, T. T. (1966). Life cycle analysis of mammalian cells. III. The inhibition of division in Chinese hamster cells by puromycin and actinomycin. *Biophys. J.* **6**, 567–80. [222]

Tobia, A., Schildkraut, C. J. & Maio, J. J. (1970). DNA replication in synchronized cultured mammalian cells. I. Time of synthesis of molecules of different average guanine and cytosine content. *J. molec. Biol.* **54**, 499–515 [76, 251].

Tolmach, L. J. & Marcus, P. I. (1960). Development of X-ray induced giant HeLa cells. *Expl Cell Res.* **20**, 350–60. [234]

Troy, M. R. & Wimber, D. E. (1968). Evidence for a constancy of the DNA synthetic period between diploid-polyploid groups in plants. *Expl Cell Res.* **53**, 145–54. [82]

Turner, M. K., Abram, R. & Lieberman, I. (1968). Levels of ribonucleotide reductase activity during the division cycle of the L cell. *J. biol. Chem.* **243**, 3725–8. [178]

Utakoji, T. & Hsu, T. C. (1965). DNA replication pattern in somatic and germ-line cells of the male Chinese hamster. *Cytogenetics*, **4**, 295–315. [76]

Van Assel, S. & Steinert, M. (1971). Nuclear and kinetoplastic DNA replication cycles in normal and synchronously dividing *Crithidia luciliae*. *Expl Cell Res.* **65**, 353–8. [252]

Van't Hof, J. (1963). DNA, RNA and protein synthesis in the mitotic cycle of pea root meristem cells. *Cytologia* **28**, 30–5. [66, 116n]

Van't Hof, J. (1965a). Cell population kinetics of excised roots of *Pisum sativum*. *J. Cell Biol.* **27**, 179–89. [66]

Van't Hof, J. (1965b). Relationships between mitotic cycle duration, S period duration and the average rate of DNA synthesis in the root meristem cells of several plants. *Expl Cell Res.* **39**, 48–58. [21, 66–7, 82]

Van't Hof, J. (1967). Studies on the relationships between cell populations and growth kinetics of root meristems. *Expl Cell Res.* **46**, 335–47. [66]

Van't Hof, J. (1968). Experimental procedures for measuring cell population kinetics parameters in plant root meristems. In *Methods in Cell Physiology*. Vol. 3, 95–117. Ed. by D. M. Prescott. New York and London: Academic Press. [24]

Van't Hof, J. & Kovacs, C. J. (1970). Mitotic delay in two biochemically different G1 cell populations in cultured roots of pea (*Pisum sativum*). *Radiat. Res.* **44**, 700–12. [255]

Verbin, R. S. & Farber, E. (1967). Effect of cycloheximide on the cell cycle of the crypts of the small intestine of the rat. *J. Cell Biol.* **35**, 649–58. [224]

Vielmetter, W., Messer, W. & Schütte, A. (1968). Growth direction and segregation of the *E. coli* chromosome. *Cold Spring Harb. Symp. quant. Biol.* **33**, 585–98. [99]

Villadsen, I. S. & Zeuthen, E. (1970). Synchronization of DNA in *Tetrahymena* populations by temporary limitation of access to thymine compounds. *Expl Cell Res.* **61**, 302–10. [215]

Villiger, M., Czihak, G., Tardent, P. & Baltzer, F. (1970). Feulgen microspectrophotometry of spermatozoa and blastula nuclei of different sea-urchin species. *Expl Cell Res.* **60**, 119–26. [65]

Volpe, P. (1969). Derepression of ornithine-δ-transaminase synchronized with the life cycle of HeLa cells cultivated in suspension. *Biochem. biophys. Res. Commun.* **34**, 190–5. [179]

Wade, J. & Satir, P. (1968). The effect of mercaptoethanol on flagellar morphogenesis in the amoeboflagellate *Naegleria gruberi* (Schardinger). *Expl Cell Res.* **50**, 81–92. [227]

Wagenaar, E. B. (1966). High mitotic synchronization induced by 5-aminouracil in root cells of *Allium cepa* L. *Expl Cell Res.* **43**, 184–90. [28]

Wain, W. H. (1971). Synthesis of soluble protein during the cell cycle of the fission yeast *Schizosaccharomyces pombe*. *Expl Cell Res.* In press. [254]

Walker, P. M. B. (1954). The mitotic index and the interphase processes. *J. exp. Biol.* **31**, 8–15. [20, 21]

Walker, P. M. B. (1971). 'Repetitive' DNA in higher organisms. *Prog. Biophys. molec. Biol.* In press. [75]

Walker, P. M. B. & Mitchison, J. M. (1957). DNA synthesis in two ciliates. *Expl Cell Res.* **13**, 167–70. [67n, 68]

Walker, P. M. B. & Richards, B. M. (1959). Quantitative microscopical techniques for single cells. In *The Cell*. Vol. 1, pp. 91–138. Ed. by J. Brachet and A. E. Mirsky. New York and London: Academic Press. [10]

Walker, P. M. B. & Yates, H. B. (1952). Nuclear components of dividing cells. *Proc. R. Soc.* B **140**, 274–99. [16, 58]

Walter, C. (1969). Oscillations in controlled biochemical systems. *Biophys. J.* **9**, 863–72. [170]

Walters, R. A. & Petersen, D. F. (1968a). Radiosensitivity of mammalian cells. I. Timing and dose dependence of radiation-induced delay. *Biophys. J.* **8**, 1475–86. [234, 236]

Walters, R. A. & Petersen, D. F. (1968*b*). Radiosensitivity of mammalian cells. II. Radiation effects on macromolecular synthesis. *Biophys. J.* **8**, 1487–1504. [243]

Walther, W. G. & Edmunds, L. N. (1970). Periodic increase in deoxyribonuclease activity during the cell cycle in synchronized *Euglena. J. Cell Biol.* **46**, 613–17. [164]

Wanka, F. & Moors, J. (1970). Selective inhibition by cycloheximide of nuclear DNA synthesis in synchronous cultures of *Chlorella. Biochem. biophys. Res. Commun.* **41**, 85–90. [86n]

Wanka, F., Joosten, H. F. P. & De Grip, W. J. (1970). Composition and synthesis of DNA in synchronously growing cells of *Chlorella pyrenoidosa. Arch. Mikrobiol.* **75**, 25–36. [251]

Warmsley, A. M. H. & Pasternak, C. A. (1970). The use of conventional and zonal centrifugation to study the life cycle of mammalian cells. *Biochem. J.* **119**, 493–9. [50, 119, 131]

Warmsley, A. M. H., Phillips, B. & Pasternak, C. A. (1970). The use of zonal centrifugation to study membrane formation during the life cycle of mammalian cells. Synthesis of 'marker' enzymes and other components of cellular organelles. *Biochem. J.* **120**, 683–8. [157, 163, 178, 184]

Warren, L. & Glick, M. C. (1968). Membranes of animal cells. II. The metabolism and turnover of membranes. *J. Cell Biol.* **37**, 729–46. [187]

Watanabe, I. & Okada, S. (1967). Effects of temperature on growth rate of cultured mammalian cells (L5178Y). *J. Cell Biol.* **32**, 309–23. [16, 19, 61]

Watanabe, Y. (1963). Some factors necessary to produce division conditions in *Tetrahymena pyriformis. Jap. J. med. Sci. Biol.* **16**, 107–24. [204]

Watanabe, Y. & Ikeda, M. (1965*a*). Isolation and characterization of the division protein in *Tetrahymena pyriformis. Expl Cell Res.* **39**, 443–52. [214]

Watanabe, Y. & Ikeda, M. (1965*b*). Further confirmation of 'division protein' fraction in *Tetrahymena pyriformis. Expl Cell Res.* **39**, 464–9. [214]

Watson, W. E. (1965). An autoradiographic study of the incorporation of nucleic-acid precursors by neurones and glia during nerve regeneration. *J. Physiol.* **180**, 741–53. [11]

Wegener, K., Hollweg, S. & Maurer, W. (1964). Autoradiographische Bestimmung der DNA-Verdopplungszeit und anderer Teil-phasen des Zell-zyklus bei fetalen Zellarten der Ratte. *Z. Zellforsch. mikrosk. Anat.* **63**, 309–26. [62]

Weiss, L. (1966). Studies on cell deformability. II. Effects of some proteolytic enzymes. *J. Cell. Biol.* **30**, 39–43. [189]

Westerveld, A. & Freeke, M. A. (1971). Cell cycle of multinucleate cells after cell fusion. *Expl Cell Res.* **65**, 140–4. [251]

Wheeler, G. P., Bowdon, B. J., Wilkoff, L. J. & Dulmadge, E. A. (1967). The cell cycle of leukemia L1210 cells *in vivo* and *in vitro. Proc. Soc. exp. Biol. Med.* **126**, 903–6. [61]

Whitfield, J. F. & Rixon, R. H. (1959). Effects of X-irradiation on multiplication and nucleic acid synthesis in cultures of L-strain mouse cells. *Expl Cell Res.* **18**, 126–37. [248]

Whitfield, J. F. & Youdale, T. (1965). Synchronization of cell division in suspension cultures of L-strain mouse cells. *Expl Cell Res.* **38**, 208–10. [34]

Whitmore, G. F. & Gulyas, S. (1966). Synchronization of mammalian cells with tritiated thymidine. *Science, N.Y.* **151**, 691–4. [33]

Whitmore, G. F., Stanners, C. P., Till, J. E. & Gulyas, S. (1961). Nucleic acid synthesis and the division cycle in X-irradiated L-strain mouse cells. *Biochim. biophys. Acta* **47**, 66–77. [33, 234n, 242, 248]

Whitmore, G. F., Till, J. E., Gwatkin, R. B. L., Siminovitch, L. & Graham, A. F. (1958). Increase of cellular constituents in X-irradiated mammalian cells. *Biochim. biophys. Acta* **30**, 583–90. [248]

Whitson, G. L., Padilla, G. M. & Fisher, W. D. (1966a). Cyclic changes in polyribosomes of synchronized *Tetrahymena pyriformis*. *Expl Cell Res.* **42**, 438–46. [40]

Whitson, G. L., Padilla, G. M. & Fisher, W. D. (1966b). Morphogenetic and macro-molecular aspects of synchronized *Tetrahymena*. In *Cell Synchrony*, pp. 289–306. Ed. by I. L. Cameron and G. M. Padilla. New York and London: Academic Press. [40]

Williamson, D. H. (1966). Nuclear events in synchronously dividing yeast cultures. In *Cell Synchrony*, pp. 81–101. Ed. by I. L. Cameron and G. M. Padilla. New York and London: Academic Press. [70]

Williamson, D. H. & Moustacchi, E. (1971). The synthesis of mitochondrial DNA during the cell cycle in the yeast *Saccharomyces cerevisiae*. *Biochem. biophys. Res. Commun.* **42**, 195–201. [71]

Williamson, D. H. & Scopes, A. W. (1960). The behaviour of nucleic acids in synchronously dividing cultures of *Saccharomyces cerevisiae*. *Expl Cell Res.* **20**, 338–49. [35, 51, 124]

Williamson, D. H. & Scopes, A. W. (1961a). Synchronization of division in cultures of *Saccharomyces cerevisiae* by control of the environment. *Symp. Soc. gen. Microbiol.* **11**, 217–42. [35]

Williamson, D. H. & Scopes, A. W. (1961b). Protein synthesis and nitrogen uptake in synchronously dividing cultures of *Saccharomyces cerevisiae*. *J. Inst. Brew.* **67**, 39–42. [141]

Wimber, D. E. (1960). Duration of the nuclear cycle in *Tradescantia paludosa* root tips as measured with ^{3}H-thymidine. *Am. J. Bot.* **47**, 828–34. [66]

Wimber, D. E. (1961). Asynchronous replication of deoxyribonucleic acid in root tip chromosomes of *Tradescantia paludosa*. *Expl Cell Res.* **23**, 402–7. [73n, 81]

Wimber, D. E. (1966). Duration of the nuclear cycle in *Tradescantia* root tips at three temperatures as measured with ^{3}H-thymidine. *Am. J. Bot.* **53**, 21–4. [66, 67]

Wimber, D. E. & Quastler, H. (1963). A ^{14}C- and ^{3}H-thymidine double labeling technique in the study of cell proliferation in *Tradescantia* root tips. *Expl Cell Res.* **30**, 8–22. [23, 66]

Wolf, B., Newman, A. & Glaser, D. A. (1968a). On the origin and direction of replication of the *Escherichia coli* K12 chromosome. *J. molec. Biol.* **32**, 611–29. [97]

Wolf, B., Pato, M. L., Ward, C. B. & Glaser, D. A. (1968b). On the origin and direction replication of the *Escherichia coli* chromosome. *Cold Spring Harb. Symp. quant. Biol.* **33**, 575–84. [97, 99]

Wolff, S. (1969a). Strandedness of chromosomes. *Int. Rev. Cytol.* **25**, 279–95. [89, 243]

Wolff, S. (1969b). The transition from interphase to prophase induced by trypsinization of isolated fixed *Vicia faba* nuclei. *Expl Cell Res.* **57**, 457–62. [89]

Wolfsberg, M. F. (1964). Cell population kinetics in the epithelium of the forestomach of the mouse. *Expl Cell Res.* **35**, 119–32. [62]

Wolpert, L. & O'Neill, C. H. (1962). Dynamics of the membrane of *Amoeba proteus* studied with labelled specific antibody. *Nature, Lond.* **196**, 1261–6. [187]

Woodard, J., Gelber, B. & Swift, H. (1961). Nucleoprotein changes during the mitotic cycle in *Paramecium aurelia*. *Expl Cell Res.* **23**, 258–64. [68, 81n, 123, 135]

Woodard, J., Rasch, E. & Swift, H. (1961). Nucleic acid and protein metabolism during the mitotic cycle in *Vicia faba*. *J. biophys. biochem. Cytol.* **9**, 445–62. [17, 151, 153]

Wunderlich, F. & Peyk, D. (1969). Antimitotic agents and macronuclear division of ciliates. II. Endogeneous recovery from colchicine and colcemid – a new method of synchronization in *Tetrahymena pyriformis* GL. *Expl Cell Res.* **57**, 142–4. [28]

Xeros, N. (1962). Deoxyriboside control and synchronization of mitosis. *Nature, Lond.* **194**, 682–3. [26]

Yagil, G. & Feldman, M. (1969). The stability of some enzymes in cultured cells. *Expl Cell Res.* **54**, 29–36. [174]

Yanagita, T. & Kaneko, K. (1961). Synchronization of bacterial cultures by preincubation of cells at high densities. *Pl. Cell Physiol., Tokyo* **2**, 443–9. [36]

Yang, S-J., Hahn, G. M. & Bagshaw, M. A. (1966). Chromosome aberrations induced by thymidine. *Expl Cell Res.* **42**, 130–5. [26]

Yasmineh, W. G. & Yunis, J. J. (1970). Localisation of mouse satellite DNA in constitutive heterochromatin. *Expl Cell Res.* **59**, 69–75. [76]

Yeoman, M. M. (1970). Early development in callus cultures. *Int. Rev. Cytol.* **29**, 383–409. [250]

Yeoman, M. M. & Evans, P. K. (1967). Growth and differentiation of plant tissue cultures. II. Synchronous cell division in developing callus cultures. *Ann. Bot.* **31**, 323–32. [34]

Yoshikawa, H. (1965). DNA synthesis during germination of *Bacillus subtilis* spores. *Proc. natn. Acad. Sci. U.S.A.* **53**, 1476–83. [107]

Yoshikawa, H. (1970). Temperature-sensitive mutants of *Bacillus subtilis*. I. Multiforked replication and sequential transfer of DNA by a temperature-sensitive mutant. *Proc. natn. Acad. Sci. U.S.A.* **65**, 206–13. [91]

Yoshikawa, H., O'Sullivan, A. & Sueoka, N. (1964). Sequential replication of the *Bacillus subtilis* chromosome. II. Regulation of initiation. *Proc. natn. Acad. Sci. U.S.A.* **52**, 973–80. [94]

Yoshikura, H., Hirokawa, Y. & Yamada, M. (1967). Synchronized cell division induced by medium change. *Expl Cell Res.* **48**, 226–38. [34]

Young, C. W. (1966). Inhibitory effects of acetoxycycloheximide, puromycin and pactamycin upon synthesis of protein and DNA in asynchronous populations of HeLa cells. *Molec. Pharmacol.* **2**, 50–5. [86n]

Young, C. W., Hendler, F. J. & Karnofsky, D. A. (1969). Synthesis of protein for DNA replication and cleavage events in the sand dollar embryo. *Expl Cell Res.* **58**, 15–26. [64, 86n]

Young, I. E. & Fitz-James, P. C. (1959). Pattern of synthesis of deoxyribonucleic acid in *Bacillus cereus* growing synchronously out of spores. *Nature, Lond.* **183**, 372–3. [111]

Yu, C. K. & Sinclair, W. K. (1967). Mitotic delay and chromosomal aberrations induced by X-rays in synchronized Chinese hamster cell *in vitro*. *J. Natn. Cancer Inst.* **39**, 619–31. [235]

Zalik, S. E. & Yamada, T. (1967). The cell cycle during lens regeneration. *J. exp. Zool.* **165**, 385–94. [63]

Zech, L. (1966). Dry weight and DNA content in sisters of *Bursaria truncatella* during the interdivision interval. *Expl Cell Res.* **44**, 599–605. [136–7]

Zeitz, L., Ferguson, R. & Garfinkel, E. (1968). Analysis of DNA synthesis and X-ray induced mitotic delay in sea-urchin eggs. *Radiat. Res.* **34**, 200–8. [248]

Zellweger, A. & Braun, R. (1971). RNA of *Physarum*. II. Template replication and transcription in the mitotic cycle. *Expl Cell Res.* **65**, 424–32. [252]

Zetterberg, A. (1966a). Synthesis and accumulation of nuclear and cytoplasmic proteins during interphase in mouse fibroblasts *in vitro*. *Expl Cell Res.* **42**, 500–11. [148, 149]

Zetterberg, A. (1966b). Nuclear and cytoplasmic nucleic acid content and cytoplasmic protein synthesis during interphase in mouse fibroblasts *in vitro*. *Expl Cell Res.* **43**, 517–25. [121, 148]

Zetterberg, A. (1966c). Protein migration between cytoplasm and cell nucleus during interphase in mouse fibroblasts *in vitro*. *Expl Cell Res.* **43**, 526–36. [148, 150]

Zetterberg, A. & Killander, D. (1965a). Quantitative cytochemical studies on interphase growth. II. Derivation of synthesis curves from the distribution of DNA, RNA and mass value of individual mouse fibroblasts *in vitro*. *Expl Cell Res.* **39**, 22–32. [21, 59–60, 81n, 117, 130, 131]

Zetterberg, A. & Killander, D. (1965b). Quantitative cytophotometric and autoradiographic studies on the rate of protein synthesis during interphase in mouse fibroblasts *in vitro. Expl Cell Res.* **40**, 1–11. [130]

Zeuthen, E. (1946). Oxygen uptake during mitosis. Experiments on the eggs of the frog *Rana platyrrhina. C. r. Trav. Lab. Carlsberg* **25**, 191–228. [192, 193]

Zeuthen, E. (1952). Segmentation, nuclear growth and cytoplasmic storage in eggs of Echinoderms and Amphibia. *Pubbl. Staz. zool. Napoli* **23**, Suppl. 47–69. [33]

Zeuthen, E. (1953). Growth as related to the cell cycle in single-cell cultures of *Tetrahymena pyriformis. J. Embryol. exp. Morph.* **1**, 239–49. [9, 135, 195, 196]

Zeuthen, E. (1955). Mitotic respiratory rhythms in single eggs of *Psammechinus miliaris* and of *Ciona intestinalis. Biol. Bull. mar. Biol. Lab., Woods Hole* **108**, 366–85. [9, 192]

Zeuthen, E. (1958). Artificial and induced periodicity in living cells. *Adv. biol. med. Physics.* **6**, 37–73. [201n]

Zeuthen, E. (1960). Cycling in oxygen consumption in cleaving eggs. *Expl Cell Res.* **19**, 1–6. [9, 193]

Zeuthen, E. (1961). The cartesian diver balance. In *General Cytochemical Methods.* Vol. II, pp. 61–91. Ed. by J. F. Daneilli. New York and London: Academic Press. [9, 214]

Zeuthen, E. (1964). The temperature-induced division synchrony in *Tetrahymena.* In *Synchrony in Division and Growth,* pp. 99–158. Ed. by E. Zeuthen. New York: Interscience Publishers Inc. [39, 40, 201, 204, 205, 206, 210]

Zeuthen, E. (1970). Independent synchronization of DNA synthesis and of cell division in same culture of *Tetrahymena* cells. *Expl Cell Res.* **61**, 311–25. [215]

Zeuthen, E. (1971). Recent developments in the studies of the *Tetrahymena* cell cycle. *Adv. Cell Biol.* **2**. In press. [201n]

Zeuthen, E. & Rasmussen, L. (1971). Synchronized cell division in Protozoa. *In Research in Protozoology.* Vol. 4. In press. Ed. by T. T. Chen. Oxford: Pergamon Press. [181, 201, 210]

Zeuthen, E. & Scherbaum, O. H. (1954). Synchronous division in mass cultures of the ciliate protozoon *Tetrahymena pyriformis,* as induced by temperature changes. *Colston Pap.* **7**, 141–55. [202]

Zeuthen, E. & Williams, N. E. (1969). Division-limiting morphogenetic processes in *Tetrahymena.* In *Nucleic Acid Metabolism, Cell Differentiation* and *Cancer Growth,* pp. 203–16. Ed. by E. V. Cowdry and S. Seno. Oxford: Pergamon Press. [207, 213]

Zimmerman, A. M. (1969). Effects of high pressure on macromolecular synthesis in synchronized *Tetrahymena.* In *The Cell Cycle, Gene-Enzyme Interactions,* pp. 203–25. Ed. by G. M. Padilla, G. L. Whitson and I. L. Cameron. New York and London: Academic Press. [209]

Zusman, D. & Rosenberg, E. (1970). DNA cycle of *Myxococcus xanthus. J. molec. Biol.* **49**, 609–19. [111]

Index

acid phosphatase, 168, 178, 179, 180; in *Schizosaccharomyces*, 161, 167, 178

actinomycin, inhibitor of RNA synthesis, 122, 133; binding of, by chromatin, 252; and DNA synthesis, 87-8; transition points after, 210, 222, 223, 225-7; and volume increase, 32

adenine, pool of, 13, 199, 200

age distribution in culture, and mitotic index, 18-19

agglutinin: binding site for, only exposed in mitosis, 255

alanine, in *Chlorella* cell cycle, 198

Alcaligenes faecalis: DNA synthesis in, 111; synchronized by filtration, 55

alcohol dehydrogenase, 168, 177, 179

alkaline phosphatase, 168, 169, 176, 178, 179, 180; in *Schizosaccharomyces*, 161, 167, 174, 177

amethopterin, folic acid antagonist: inhibits DNA synthesis, 28

amino-acids: pool of, 13, 140, 197, 200, 254, 255; pulse-labelling of growing cells with, 130

α-aminoadipic acid reductase, 177

aminolevulinic acid dehydrase, 176

aminolevulinic acid synthetase, 176

aminopterin, synchronisation by, 81

5-aminouracil, inhibitor of DNA synthesis: partial synchronisation by, 28

Amoeba: DNA cycle in, 69-70, 73, 85; growth curves for, 134-5, 147; histones of, 156; nucleus of, 150, 151, 152; protein synthesis in, 141-2; return of RNA from cytoplasm to nucleus in, 121, 253; synchronised by temperature cycles, 46; turnover of membrane material in, 187

amphibian embryos: cell membrane in, 188; DNA synthesis in, 63, 72, 82, 83, 249, 252; *see also* frog cells

anaerobiosis, synchronisation by, 209, 210

Anthocnidaris, sulphydryl in egg proteins of, 213

arginine: in *Chlorella* cell cycle, 198; and synthesis of ornithine transcarbamylase, 255; tritiated, in histones, 22, 153 (may enter DNA, 11, 155)

arginosuccinase, 177

artichoke tubers, synchronised cells of, 34

Ascites cells, 88, 151, 248; synchrony in multinucleate, 84

aspartate transcarbamylase, 168, 172, 176, 177, 178, 180

aspartokinase, 177

Aspergillus, multinucleate cells of, 70, 84

Astasia: division delay after irradiation of, 239, 240; methods of synchronising, 38, 46; oxygen uptake at mitosis of, 194

ATP: pool of, 198-9; synthesis of, inhibited by carbon monoxide, 230

autoradiography, 5, 10; double-layered, 18; of individual DNA molecules, 77, 93

azide, metabolic inhibitor: delay of division by, 209, 232

Azotobacter: growth curves of, 142; synchronised by cold shock, 40n

Bacillus cereus: cell-wall growth in, 191; growth curves of, 142; synchrony after spore germination in, 111

Bacillus megaterium: cell membrane in, 186, 188; cell-wall growth in, 191, 192; growth curves of, 142; synchronised by cold shock, 40n

Bacillus subtilis: chromosome of, 91, 100; DNA synthesis in, 94, 99, 100, 101, 107, 111; enzyme potential in, 180; enzyme synthesis in, 120, 166, 168, 176, 247

bacteria: cell-wall growth in, 190-2; chromosone of, 91, 93-4, 112, 114; lack histones, 153; synchronisation of, 51-6; *see also individual species*

balanced growth, concept of, 4; inhibitor blocks and, 29-33, 49, 253; in synchrony by starvation, 38

beans; binucleate cells in roots of, 84;

mammalian cells: cell membranes in, 186, 254–5; division delay after irradiation of, 234n, 235, 243, 248; DNA cycle in, 58–62, 71, 72, 77, 81, 88n, 249, 251; enzyme synthesis in, 162, 175; growth of, 130–3, 146; growth of, in absence of DNA synthesis and division, 31, 247; histone synthesis in, 153, 155n; methods of synchronising, 34, 47, 49, 250; RNA synthesis in, 116n; thymidine pool in, 16; transition points in, 222; *see also individual species*

mammary gland, ovarian hormones and DNA cycle in cells of, 62

markers, for stages of cell cycle, 232, 244–5

membrane elution, synchronisation by, 51–4

mercaptoethanol, inhibitor of initial duplication of centriole, 184; and development of flagella in *Naegleria*, 227; and 'division protein', 209, 210

meristems of plants: cell cycle in, 24; DNA cycle in, 67, 72

mesosomes, prokaryotic chromosomes attached to cell membrane by, 100

metaphase, arrest of cells in, 28–9, 32–3

methionine, labels 18S component of rRNA, 119

micronuclei of Ciliates: DNA cycle in, 68–9, 72, 215, no RNA synthesis during mitosis in, 123, 253

microscopy: interference, 9, 128; polarised-light, 10

microsomes, synthesis of proteins of, 155, 157, 184

microtubules: colchicine-binding protein as constituent of, 156; in growth of nucleus? in oral apparatus of *Tetrahymena*, 212–13

mitochondria: division of, 182–3; DNA of, 71, 72, 73, 86, 252; RNA of, 116–17; synthesis of proteins of, 157, 184

mitomycin C, inhibitor of DNA synthesis: and cell division in prokaryotes, 110

mitosis, 58–9; abnormalities of, caused by excess thymidine, 22, 26; cell membranes at, 254–5; determining phases of DNA cycle by labelling at, 21–2; inducers of, 85, 251; interaction of nucleus and cytoplasm in initiation of, 251; morphogenetic abnormalities caused by pulses or shocks during, 232; no RNA synthesis during, 116, 119, 121–2, 126 (except of tRNA, 253); oxygen uptake during, 192; protein synthesis during, 132, 137, 141, 253; time spent in, in mammalian cells, 60, and in plants, 66; viruses in cells arrested in, 133

mitotic index, 18–19, 22, 26, 27; interpretation of peaks in, 57, 224; in synchronised culture of *Schizosaccharomyces*, 50

mouse cells: binucleate, synchrony of, 84; cell membrane of, 255; division delay after irradiation of, 240; DNA cycle in, 59–60, 62, 66, 81, 84, 251; enzyme synthesis in, 163, 178; growth rates of, 130, 131; growth rates of nucleus and cytoplasm of, 148–50, 151; histones in, 156; nuclear changes in, 182; pools in, 197, 199; protein synthesis in, 157, 253; RNA synthesis in, 119, 121; synchronised by colchicine, 32, 33; transition points in, 234; turnover of membrane material in, 187

multinucleate cells: of *Aspergillus* hyphae, DNA cycle in, 70; percentage of, in synchronous growth of *Chlorella*, 45; synchrony of nuclei in, 84, 251

Mycoplasma hominis, chromosome of, 95

Myxococcus xanthus, periodic DNA synthesis in, 111

myxomycetes, *see Physarum*

Naegleria: development of flagella in, 227; synchronised by heat shock, 39

nalidixic acid, inhibitor of DNA synthesis: and cell division in prokaryotes, 110

Navicula, synchronised by silicon depletion, 36

Neurospora: enzyme synthesis in, 166; lacks thymidine kinase, 11; mitochondria of, 183, 184

Newcastle disease virus, unable to multiply in cells arrested in mitosis, 133

newt cells, DNA cycle in, 63

Nitella, chloroplast division in, 183

nitrite reductase, 180

nitrogen, total: and total protein, 141

nitrogen mustard, cell division blocked by, 249

nitrosoguanidine, mutagen: for locating origin of chromosome in *E. coli*, 98–9

nitrous oxide, under pressure: blocks cells in metaphase, 28

nuclear bodies, in *E. coli* grown at different rates, 105

nuclear division: cytoplasmic stimulus to, 83–6; and DNA cycle, 59, 63; not clearly defined in prokaryotes, 106; positions of cell division and, in life cycle of *Chlorella*, 43, 44, 45, and of *Saccharomyces*, 70

nuclear membrane: initiation of DNA replication near, in eukaryotes, 112; lacking in prokaryotes, 91

nucleic acids, synthesis of, in *Astasia*